Governance and Environment in Western Europe
Politics, Policy and Administration

Kenneth Hanf
and
Alf-Inge Jansen (eds)

LONGMAN

Addison Wesley Longman Limited
Edinburgh Gate
Harlow
Essex CM20 2JE
United Kingdom
and Associated Companies throughout the world

*Published in the United States of America
by Addison Wesley Longman, New York*

© Addison Wesley Longman Limited 1998

The right of Kenneth Hanf and Alf-Inge Jansen to
be identified as editors of this work has been
asserted by them in accordance with the Copyright,
Designs and Patents Act 1998.

All rights reserved; no part of this publication may be reproduced,
stored in a retrieval system, or transmitted in any form or by
any means, electronic, mechanical, photocopying, recording,
or otherwise without either the prior written permission of the
Publishers or a licence permitting restricted copying in the United
Kingdom issued by the Copyright Licensing Agency Ltd.,
90 Tottenham Court Road, London W1P 9HE.

First published 1998

ISBN 0 582 36820 0

British Library Cataloguing-in-Publication Data

A catalogue record for this book is available from the British Library

Library of Congress Cataloging-in-Publication Data

Governance and environmental quality : environmental politics, policy
  and administration in Western Europe / Kenneth Hanf and Alf-Inge
  Jansen (eds.).
      p.   cm.
  Includes bibliographical references and index.
  ISBN 0-582-36820-0 (pbk.)
  1. Environmental policy--Europe, Western--Case studies.   I. Hanf,
Kenneth. II. Jansen, Alf-Inge.
HC240.9.E5G68 1998
363.7'0094--dc21   98-21367
                                                                              CIP

Set by Aina M. Kristensen in 11pt Times
Produced by Addison Wesley Longman Singapore (Pte) Ltd.,
Printed in Singapore

Contents

Foreword	vii
Preface	ix
1. Environmental policy – the outcome of strategic action and institutional characteristics *Kenneth Hanf and Alf-Inge Jansen*	1
2. Britain: Coming to terms with sustainable development? *Neil Carter and Philip Lowe*	17
3. Denmark: Consensus seeking and decentralisation *Mikael Skou Andersen, Peter Munk Christiansen and Søren Winter*	40
4. France: Fragmented policy and consensual implementation *Corinne Larrue and Lucien Chabason*	60
5. Germany: The engine in European environmental policy? *Heinrich Pehle and Alf-Inge Jansen*	82
6. Greece: Administrative symbols and policy realities *Calliope Spanou*	110
7. Italy: Environmental policy in a fragmented state *Rudolf Lewanski*	131
8. The Netherlands: Joint regulation and sustainable development *Kenneth Hanf and Egbert van de Gronden*	152
9. Norway: Balancing environmental quality and interest in oil *Alf-Inge Jansen and Per Kristen Mydske*	181
10. Spain: Environmental policy and public administration. A marriage of convenience officiated by the EU? *Nuria Font and Francesc Morata*	208
11. Sweden: From environmental restoration to ecological modernisation *Lennart J. Lundqvist*	230

12. The European Union: Environmental policy and the prospects for sustainable development
Alan Butt Philip 253

**13. Environmental challenges and institutional changes.
An interpretation of the development of environmental policy in Western Europe.**
Alf-Inge Jansen, Oddgeir Osland and Kenneth Hanf 277

List of contributors 327

Index 329

Foreword

As chairman of one of the research programmes of the Norwegian Science Council on government and administration I am pleased that the Council has sponsored this book. Sustainable development is the overriding goal in environmental policy and administration and this book contributes to the understanding of the conditions for achieving that goal.

The range of countries covered in this book – representing as they do different stages of development in environmental policy, different state and institutional traditions – provides an interesting opportunity to understand how different types of countries confronting roughly similar problems of environmental management have responded politically and (re)organised their administrative system for carrying out these policy decisions. The readers are well-advised to prepare themselves for a *tour de force* of the European development of the role of state in environmental policy and, therefore, in public policy in general.

Each of the ten country chapters enables the reader to appreciate the experience of each country separately while, at the same time, the book provides a basis for a comparison of the different national institutional responses to the range of environmental problems confronting such countries. While the Norwegian chapter stands on its own merits, I find it particularly interesting to see how Norway's and the other nine countries' responses to the environmental problems are put into a comparative perspective and discussed within such a perspective.

Students of environmental policy and administration as well as of state – society relations, be they occupied with these challenges on a theoretical basis or in practice, at national or local levels, will profit from this book. It will increase their understanding of governing our environment.

Morten Fjeldstad
Oslo
Winter 1997

Preface

This book is an outcome of a project emanating from a workshop, organised by the editors, on comparative research on environmental policy and administration. Two or three participants from ten Western European countries were invited to the workshop, that took place at Drøbak, a charming town on the Oslofjord.

In pursuing this project we have incurred several debts which we gladly acknowledge. We are grateful to the Norwegian Science Council, which sponsored the workshop that first brought most of the contributors together, and later helped to co-finance the project. We also wish to thank the Faculty of Social Sciences and the Department of Administration and Organisation Theory of the University of Bergen for their financial support. A special thank is due to the head of the department, Torodd Strand, for his encouragement.

The following persons have read two or more chapters of the book: Graham Cox, Eilev S. Jansen, Nina Kleiv and Peter Mills. They all gave us valuable advice and comments. Rajesh Parashar and Russel Shuler have been helpful in teaching some of us subtleties of the English language. We are grateful to Aina Kristensen who took on the thankless task of putting the manuscript into camera-ready form. Finally, we wish to acknowledge our gratitude to the staff of Addison Wesley Longman for their ability to combine efficiency with good will.

Bergen/Rotterdam, April 1998

Alf-Inge Jansen					Kenneth Hanf

Chapter 1

Environmental policy – the outcome of strategic action and institutional characteristics

Kenneth Hanf and Alf-Inge Jansen

Protecting the environment – a challenge for governance

This book is intended to provide an overview of developments in the area of environmental politics, policy and administration in a selection of ten Western European countries: Denmark, France, Germany, Greece, Italy, the Netherlands, Norway, Spain, Sweden and the United Kingdom. In the light of the growing importance of the European Union for what both its members and other countries do in this field, a chapter on the EU has also been included. The range of countries covered, representing as they do different stages of development in environmental policy as well as different state and institutional traditions, provides an interesting opportunity to examine how different countries, which confront roughly similar types of environmental problems, have responded politically and (re)organised their administrative systems for carrying out these policies. These developments will be illuminated in a comparative perspective, and various country-specific as well as more general characteristics will be discussed.

Around 1970 the environment had been established as a political category in most Western European societies. Thereby various issues (e.g. nature protection, pollution, cultural heritage) came to be categorised and redefined as issues pertaining to the environment. A new policy field, that of environmental policy, was established.[1]

The time of birth and establishment as well as the extension of the environmental policy field has varied between countries in Western Europe, but from around 1970 there has in general been a continuous public debate on how to create institutional arrangements for making and implementing policies that reflect and go along with, not against, the holism of nature and the wholeness of the environment.

A significant impetus to this ongoing debate was given by the World Commission on Environment and Development. The Commission emphasised that if we are to deal effectively with the environmental challenge, we will need to follow a policy strategy that co-ordinates and integrates the efforts to combat environmental degradation, and notably a policy strategy that integrates environmental policy and economic policy.

Authorities in Western European countries have publicly seconded this way of seeing the challenge and defining the overriding problem. Notions of sustainable development have at various levels of government been taken on board as the official framework, or strategic objective of the efforts of both economic development and environmental protection. Likewise, with its Fifth Action Programme, the European Union has officially declared that its environmental policy is on 'the way towards sustainability'.

In a number of countries, the government, with reference to various terms of 'sustainability', has argued that new policy instruments and administrative procedures should be applied. Moreover, it is argued that new divisions of labour and authority between levels of government and among different agencies as well as between these governmental actors and societal actors (e.g. firms and interest organisations, media organisations) should be put into operation. By means of such changes, policies are said to be less sectoral and more integrative as well as able to be guided by a multi-media strategy.

However, words differ from deeds. Observations from many countries indicate that in practice there is considerable variation in the manner, area and extent to which such changes take place, and that this variation seems to be dependent on environmental and institutional features specific to a given country. In this connection it should be pointed out that environmental policy is obviously just one of many policy fields in which government is called upon to act. Although commitment to environmental quality has been officially stated by a country's government and even appears firmly anchored in the social value system of the country, shifting economic conditions and balance of political support and opposition can turn it into an empty gesture when it comes to shaping policies that are implemented. A general commitment to these objectives is 'costless' compared to the difficult choices to be faced in operationalizing them in the form of concrete decisions through which the balance between economic development and environmental quality is to be struck. It is through the political process that the general commitment is tested, e.g. in concrete cases where the alternatives are either to choose short-term or immediate pain caused by environmentally sound but economically painful decisions, or to sacrifice long-term environmental benefits for more quickly realised and visible economic pay-off.

The need to confront increasingly serious national and transboundary problems of environmental deterioration occurs at a time when societies and their political leaders are being pressed to make difficult and painful decisions to adjust their post-war welfare states to what appear to be new realities. A substantive and organisational re-calibration of the welfare state has been placed on the agenda, and a re-examination of the relation between the state and society, and the role that government can be expected to play (or should play) in providing solutions to societal problems, has followed. Taken together these developments have, for instance, in many countries changed the relationship between government regulators and business, and are reshaping the way in which governments go about defining and implementing environmental quality objectives. Furthermore, the relationships between central and sub-national governmental actors, for instance municipalities, have also been the object of substantial change.

As if these changes were not enough, in one arena after another it has become apparent that the satisfaction of domestically generated demands is impossible without co-operation with other national governments in a variety of international arenas. For a number of countries the impact of the environmental policy of the European Union has been most immediately relevant for many aspects of the governance of environmental quality.

Throughout Western Europe the debate about, and increasingly the practice of, this general reorganising and restructuring of the familiar institutional arrangements and processes of the welfare state can be characterised by such terms as: decentralisation, deregulation, privatisation, rehabilitation of the market etc. Consequently, institutional changes and the shift in paradigm to a more preventive policy in the service of environmental policy, while primarily generated by developments specific to the environmental policy field itself, have occurred in the context of and, indeed, have reflected the changes underway in the broader institutional and political system.

In line with the above we view environmental policy as embedded in a system of governance characteristic of the country in question. Therefore, we believe that environmental policy, concerned either with specifically combating pollution, or more generally with promoting good environmental quality, must contend with demands made on limited societal and political resources by other problems and policy fields. We also assume that the processes of change that we have described will be filtered through and shaped by the particular problem agendas, political forces and institutional arrangements of a given country. Indeed, how these 'new imperatives' are perceived, translated into policy strategies and converted into policies that are implemented will vary according to features of governance of the country.

The core of governance has to do with determining what ends and values should be chosen and the means by which those ends and values should be pursued, i.e. the direction of the social unit, e.g. society, community or organisation. Governance includes activities such as efforts to influence the social construction of shared beliefs about reality; the creation of identities and institutions (Therborn 1991); the allocation and regulation of rights and obligations among interested parties; and the distribution of economic means and welfare services. Governance, in other words, is the shaping and sustaining of the arrangements of authority and power within which actors make decisions and frame policies that are binding on individual and collective actors within different territorial bounds, such as those of the state, county and municipality. In this book we shall focus on how these ten Western

European countries have responded to the environmental imperatives in terms of governance.

Institutions and the infusion of values into societal and state activities

There are a number of ways of conceptualising relationships of authority and power. These different conceptions and their related principles of organisation permit recourse to different forms of governance in constituting human societies (Ostrom 1986:111). In most Western European societies the state-centred mode of governance has been seen as representing the dominant tradition of governance. However, in spite of their common features, states differ, and, as we understand it, they differ primarily because their institutional characteristics differ. We shall in the following, therefore, state how we view institutions and their significance for social activities generally and for the development of the state and state–society relations specifically.

Institutions are here conceived as both patterns of human activity and symbolic systems, cognitive constructions and normative rules through which actors categorise that activity and infuse it with meaning and value (Friedland and Alford 1991:232). Central institutions in current Western European societies are the nuclear family, Christian religion, representative democracy and the market economy. These institutions all have their distinctive symbolic systems, categories, beliefs, norms and values, which actors take for granted and draw upon when conducting certain activities. For example, when participating in elections in democracies, individuals must possess concepts of citizenship, elections, the party system, etc., and when acting as citizens they tend to vote according to what they think is right for the community – according to the social values they believe should be the community's or the nation's. In the same manner, when buying in a market, an individual has to have concepts of buyers and sellers, of price and contract, and as a buyer, or consumer, he or she tends to calculate and act according to his or her own preferences. In these two contexts, individuals tend to act in accordance with two different logics, the logic of representative democracy and that of the market, and the logic inherent in these sets of relationships is institution specific. The rules that constitute what is meaningful and what modes of social conduct are to be sanctioned differ according to institutions. Therefore, behaviour appropriate in one institutional setting would be inappropriate in another.

These examples illustrate a basic point: as actors draw on different institutions when conducting various types of activities, the various activities are infused with the values of and regulated by different institutional logics. Characteristically, therefore, some of the most important struggles between interest groups and organisations are over the appropriate relationships between institutions, over the institutional logic by which the various social activities should be regulated and over categories of persons to which they should apply (Friedland and Alford 1991:256)

History has taught us that institutions often are transformed incrementally and peacefully, but when important institutional changes take place, they often do so rapidly and are usually contested. It took centuries before institutions like the nuclear family, Christian religion, the state and the market economy were established as central institutions. In many parts of Europe this institutionalisation took place through conflicts, even wars. The nation state emerged as the solution of the crisis of the medieval order and the product of the confessional wars. The characteristic of the state was sovereignty, i.e. the exclusive right of the ruler to give laws to everybody in general and in particular without being bound by any positive law himself (Grimm 1986:92). The state even had right to use 'swords' to enforce these laws. This sovereignty appeared as a property of the monarch and was administered by personal servants. The post-medieval community was divided into state (the monarch and his men) and society (comprising everybody else in the role as private members and subject to the sovereign power of the state).

State–society relations were crucially transformed during the eighteenth, nineteenth and twentieth centuries. In some countries rights concerning individual freedom which were associated with civil society (e.g. freedom of speech, the right to own property and to conclude valid contracts and right to justice) were established during the eighteenth and the first decades of the nineteenth century. The liberal state of the nineteenth century, which succeeded the absolutist monarchies, reduced the functions and the rank of the state. The liberal assumption was that social justice and economic welfare were best produced by individual freedom. Mercantilistic and other measures of regulating production and trade of private goods were banished and were, like social regulation in general, to be left to the market mechanism (Grimm 1986:102). In German territories where the bourgeoisie did not win any significant influence on the domain of the state, compensation was sought by a particular emphasis on rule by law (the *Rechtstaat*). Institutional characteristics of the *Rechtstaat* emerged in a majority of the ten countries focused upon in this book, and administrative apparatuses were further developed towards the standard Weberian model of bureaucracy predominantly staffed by lawyers. The establishment of *civil rights* represented the emergence of the status of citizenship, which was instrumental in providing the legal and moral bulwark for the emergence of contract as the primary basis of economic organisation, i.e. allowing market forces to govern the movement of labour, land, goods and capital. While this first distinct step in the emergence of citizenship represented an institutional differentiation within the state, it simultaneously was complementary to, resultant from and instrumental in the further institutionalisation of the market economy (Marshall 1964:87ff.; King and Waldron 1988:418–421). The development of the institutional characteristics of the state were related to the development of those of market economy.

A new phase of increased state activity occurred during the last part of the nineteenth century, when the state in most of these ten countries became increasingly involved in the technical and economical modernisation of society. The state recruited officials from new professions (e.g. engineers, agronomists, medical doctors) and set up administrative units (central agencies) that were organised to apply symbolic systems and practices more tied to the logics of substantive consequence than to that of law and rule following (Eckhoff and Jacobsen 1960; Jacobsen 1970; Fischer and Lundgren 1975; Burrage 1992 and March and Olsen 1995:154–155). This meant that from now on the state administrative apparatus included distinctly different types of organisations that were infused with values from and constrained in their action according to different and partly contradictory institutions.

During the same period, crucial steps in the transformation of the institution of citizenship were also taken. From primarily being *state subjects* with limited *civil rights,* citizens gradually became participants in the governing of the state. The core element of this process was the establishing of political citizenship. That meant that, in addition to established civil rights, *political rights* (right to vote and be eligible as representatives in legislative bodies) and structures of competitive mass politics in representative democracies (e.g. electoral laws, party systems, mass media) were created. From this time onwards it was not enough that state activity was conducted according to the letter of the law and was technically effective, it had also to be accepted by a majority of the citizens, or at least by a majority of their elected representatives in the parliament. The institution of representative democracy was incorporated into the institutional order of the state.

A crucial next stage in the institutional development of the state and also in state – society relations took place as the experience of economic *laissez-faire* and the consequences of industrial capitalism provoked the return of state intervention in society. The typical pattern of development in a majority of these countries was that the state in the last half of the nineteenth century adopted legislative and controlling measures to restrict *laissez-faire* and to protect the weaker members of society, e.g. in the field of labour relations, social security and consumer and competitor protection (Grimm 1986:103). This transfer of competence from the market system to the state was speeded up to dampen the economic and social crisis between the two World Wars and was institutionalised in the post-World War II era as the welfare state. New social rights (e.g. sickness and unemployment benefit, retirement pensions, equal rights to health care and education) may be seen as instruments aimed at rectifying social degradation and insecurity inherent in the capitalist market economy, and a transfer of responsibilities from the private sphere to the public sphere (Eriksen 1996:61). The adoption of the welfare state implied that new relations between citizens and between citizens and state were established; i.e. new *social rights* as well as new social obligations were institutionalised.[2] By providing minimum standards in these areas, the state offsets the vagaries of market processes and corrects the gross inequalities of distribution arising from the market (King and Waldron 1988:419). The institutionalisation of this extension of citizenship made the role of the citizen irrevocably distinct from the role of the consumer.

In this sketch we have noted how various types of social activities are infused with different values and are regulated by different institutional logics. In fact, society can be conceived as constituted through multiple institutional logics (Friedland and Alford 1991:243), and we see the institutional structure of societies, like those of the ten countries examined in this book, primarily as the historical result of the strategies pursued and power exerted by various interest groups and organisations.[3] The interrelated development of, particularly, two institutional orders, that of the state and that of the market economy, has been pointed out. More specifically, we have taken note of how different institutions with their distinctive logics have historically been incorporated into the institutional order of the state. The current state appears to be constituted through a number of institutional logics. Some of these, such as those of military and law and order enforcing institutions of the early states, socio-economic modernisation institutions and various types of rights constituting the core of values and practices of citizenship, mass democracy and the welfare state, are potentially contradictory. A crucial governmental task is to balance and co-ordinate the organisations that in terms of public policies give expression to and reflect these different institutions in ways which make the activities of these organisations more compatible with and conducive to the requirements of collective well-being (Offe and Preuss 1991:145).

All ten of the states on which we focus have incorporated these different institutional logics, but the developmental pattern of this incorporation differs. The time at which and the extent to which the liberal state was established and the capitalistic market economy grew, the way and degree to which the state became involved in the modernisation of society and the time at which and manner in which critical steps in the development and structuring of competitive mass politics were taken (Rokkan 1970:72–144), as well as the degree to which social rights were established, have varied.

Also other types of country-specific institutional features have developed in these countries. For instance, there are unitary states and federal states (among the former ones, France developed as more centralised than the Netherlands and Norway), and there are significant differences between these states as to the extent of parliamentarism and other types of representative and responsive government as well as to the degree to which a democracy is characterised by corporatively structured bargaining democracy (Rokkan 1976). Similarly there is a strong emphasis on constitutionalism in the conduct of government and a concomitant formal and legalistically oriented policy process in a country like Germany (Katzenstein 1987:382–384), while in Britain policy traditionally has been conceived as a series of problems and thus constituting cases that should be judged on their own merits, i.e. the particular has preference over the general, the commonsensical over the principled (Weale 1992:81). Moreover, while there are widely established consensus-making routines both at the national level and at the sectoral level in Germany, the Netherlands and the Nordic countries, such institutions have been weak or almost non-existent in Spain and Italy.

We expect to find that variation in how these ten Western European societies have responded to the environmental challenge, in terms of policy content (environmental goals and values) and of instruments and measures as well as in terms of organisation for making and implementing environmental policy, significantly reflects institutional characteristics as well as strategic action by individuals, interest groups and organisations specific to the different countries. (style + structure)

Institutions, organisations and strategic action

In the ten countries examined in this book environmental policy has developed as an identifiable policy sector with its own characteristic processes and structures of governance. Our focus is primarily on the organisation of practices for environmental policy making and administration and the arrangements of power and authority as well as values and perceptions of problems and solutions represented by this organisation. Since organisations take centre stage in our analysis, we will state how we view organisations and how they here will be seen as related to institutions and strategic action.

One of our basic assumptions is that organisations are mobilisers of resources for the attainment of goals and the pursuit of values. We further assume that it is equally characteristic of organisations that they promote only some goals and some interests, and not all. By organising, a division of labour and also of rights and obligations is established, along with corresponding expectations as to what are correct and reasonable decisions and policies as well as proper procedures and rules of the game. Concurrently some actors and some points of view are defined in, while others are defined out. Some lines of conflict are defined as important, while others are regarded as peripheral or irrelevant.

This characteristic of organisation is generally referred to as an organisation's mobilisation of bias.[4] In many cases the mobilisation of bias can be seen as predominantly the result of calculation and planning, i.e. the organisation functions as an instrument designed to achieve specified goals. However, the development of tasks and division of labour, the distribution of rights and obligations, authority and responsibility, stable expectations and patterns of proper procedures of action and rules of the game as well as patterns of interest and conflict are developed by informal as well as formal organisation, and they are established over time. As stated by Philip Selznick almost 50 years ago, goals and procedures therefore tend to achieve an established, value-impregnated status (Selznick 1949:257). In general, organisations tend to be infused with values beyond the instrumental or technical requirements of the task at hand (Selznick 1957:16–17). This characteristic constitutes the core aspect of the institutional character of an organisation. We take it that the organisation of the environmental policy sector as a whole or the organisations that are parts of this policy sector, as the result of design and institutionalisation, represent and promote specific goals and values as well as perceptions of problems and (of their) solutions, and that they in this manner represent institutional logics that are more or less compatible.

It is useful to distinguish between institutions and organisations. While we conceive of institutions as both symbolic systems and patterns of human activity, we here focus on organisations as actors, such as individuals and groups, that draw on institutions to categorise and order reality as well as infuse their action with meaning and values. In this way institutions shape social action. On the other hand, when organisations, groups and individuals as actors are drawing on institutions, they confirm and sustain these very institutions. By reproducing the institutions in this way, actors also reproduce the conditions that make such action possible. In other words: institutions are both medium and outcome of the action of organisations, groups and individuals (Giddens 1984:16–25).

This particular focus on what we have pointed out as characteristics of institutions follows from our assumption that these characteristics are crucial because they shape action by limiting what is perceived by actors as available future options (Krasner 1988:71). We take it that state institutions, the values, understandings and ways of action taken for granted, which these institutions represent, will be the principal influences on how the state and its component parts function in terms of environmental politics, policy and administration. However, individual, organisational and group action, institutionally shaped as it may be, is the engine that drives political life. The relation of authority and power is, therefore, not pre-determined; people can always 'act otherwise' (Lukes 1977:6). Actors are not determined; actors can make a difference (Giddens 1982:30). Consequently, we find it necessary to focus also on characteristics of environmental politics, i.e. the activities through which environmental problems are defined and accepted as tasks to be dealt with by the authorities and also through which public resources (financial, organisational, authority) are assigned to solve these problems. In sum: we assume that the practice of a country's policy system is the outcome of the interaction between the strategic activities of the various actors and the institutional structure within which they act.

A framework for comparative analysis of environmental policy

What is presented in the following ten chapters is not simply a description of formal institutions and processes found in each country. We were interested in the extent to which and the ways in which these institutional arrangements, and the environmental policies they 'produce', are similar or different. The following chapters map the distinctiveness of each of the ten countries as well as provide data to be used in the concluding chapter.

Each chapter begins with a short overview of some of the salient socio-economic and geographical features of the country as a lead-in to a description of the type of environmental problems with which the country has been confronted. This brief examination of some features of the country; its geographical location, exposure to imported pollution, peculiar ecological features influencing its vulnerability to environmental damage, its socio-economic structure (e.g. degree of industrialisation/farming/tourism, urbanisation/concentration of population), gives a general idea of the type of environmental stresses or pressure to which the country (or particular parts of it) is exposed and to which its policy process would have to respond. Such a description also suggests the kinds of social and economic interests that are likely to be affected by these problems and the environmental policy measures that are applied. This in turn suggests something of the socio-economic configuration of support or opposition to environmental measures.

It is assumed that both the relevance of environmental quality issues and the agenda of specific problems facing the country, will be, in some way, related to characteristics like those mentioned above. However, problems are not phenomena 'out there'. Problems are perceived and constructed by somebody as a discrepancy between what is (or would be) the desirable and the existing state of affairs. We assume that whether or not environmental problems are translated into demands for public action, and ultimately into a supply of appropriate policy measures for meeting these demands, will be an outcome of the interaction between the strategies of the various actors and the institutional structure of the politico-administrative system of the country. In this sense it is clear that there is no necessary relation between environmental conditions and environmental policy. Still it is important to note that the geographical location and features of the country and its socio-economic structure, combined with the peculiar set of relevant public policies, will have important consequences for the condition of the environment and, as filtered through the political process, for the agenda and activities of environmental policy.

In this sense, therefore, 'environmental problems' and 'environmental policy' are themselves clearly products of the political process. Countries will vary among themselves and over time as to what is considered to be an 'environmental problem' as well as with respect to what should be done, and by whom, about it. For this reason each country chapter traces the emergence and changing dimensions of the 'environmental problem' as an object of social concern and political action. Consequently, the description of the overall situation in a country with regard to factors relevant for the determination of the quality of its environment is followed by a brief historical sketch of the development of public awareness of and action with regard to environmental problems, including an examination of the societal actors who have contributed to the shaping of environmental politics in the country. This overview addresses such questions as: What kinds of issues emerged or were promoted by which type of societal (or governmental) actors or groups? What were the initial responses to these concerns or demands for action? How have things been changing (especially with the advent of 'modern environmentalism' in the sixties, or whenever that happened in a given country)? In order to answer these questions regarding the dynamics of institutional and policy development, each author was asked to describe, in broad lines, the interweaving of institutions, problem perceptions and definitions, and policy measures over time.

Another set of questions has to do with the 'politics' of environmental policy and administration more narrowly defined. Changes in and operations of institutional arrangements do not take place in a disembodied manner without support or stimulation (as well as opposition) from particular (coalitions of) actors seeking to promote their particular interests. For this reason we were interested in learning: Who have been the carriers or generators of institutional and policy changes? How have they attempted to influence the direction of policy change? How does (or has) the constellation of political forces differ(ed) from one level of government to another or with respect to different aspects of environmental policy? Also, in this connection, where and in what way has the balance been struck between economic considerations (development) and environmental quality? Has there been a move towards a policy strategy aiming at pollution prevention and integrated pollution control? What are the relationships between environmental policy (and its institutional champions) and other policy fields and their supporting actors?

Public policy is in this book viewed as the course of action taken by government over time to deal with a particular societal problem. As such it can be described along three dimensions: intentions, commitments of resources and operational activities.

The intentions of public authorities in attempting to deal with environmental problems can be discerned, in the first instance, in the general policy strategy of the national government. Such an environmental policy strategy would contain definitions of the problems, identification of causes of these problems, selection of the appropriate approach for dealing with them and the choice of instruments most likely to bring about the quality objectives pursued. Such a policy strategy would also contain assumptions regarding the relation between environmental protection and economic development. Of particular interest in this connection would be to know to what extent a given country is committed to a policy of sustainable development and in what way this commitment has been translated into measures designed to achieve this goal.

An analysis of policy along the two dimensions of resources committed and daily operations would reveal how this national policy strategy in a given country has been put into actual practice. In this regard, the authors were asked to examine the organisational arrangements that have been developed, and the procedures that are followed, as well as to consider types of resources committed to the realisation of policy objectives. Such an examination of the structural and processual characteristics of the environmental policy system helps us understand why policy has developed the way it has in a particular country. A look at the procedures used for applying the different policy instruments suggests the kinds of opportunities available for the interested parties to enter the decision process as ally or opponent of the different policy objectives. In this way an indication can be given for the constellation of political forces surrounding the actors responsible for performing the various policy functions.

Against the backdrop of the national environmental situation and its policies for handling environmental problems, each chapter proceeds by looking at the main features of the country's environmental administration. This section of the national chapters provides a summary of the formal institutional arrangements for formulating and carrying out environmental policy. These summaries of the overall division of labour and functions between different governmental actors begin with a description of the jurisdiction and powers of the administrative actors primarily responsible for environmental policy at the different governmental levels. Here, too, an effort is made to give some indication of the relative 'political weight' these environmental actors have in the overall policy scheme of things.

Of course, decisions regarding how the environmental administration of a country is to be organised and the way in which its decisions are to be made are not made in a vacuum. The already existing organisation of the political system, with its mobilisation of prevailing bias, will provide a (limiting) framework within which institutional arrangements for a 'new' policy sector are shaped. Therefore, the initial development of an environmental policy system is placed in the context of the more general politico-administrative system; traditional ways of organising and doing things, out of which these new policy structures emerge.

Although the chapters thus give an overview of the institutional arrangements on national, provincial and municipal levels, the authors were requested to pay particular attention to the linkages between these levels of action. Interesting in this regard is the ways in which the pattern of inter-relationships between centre and periphery (national and sub-national actors) works to constrain or define the set of action possibilities within which individual actors operate and work together with other actors.

As noted above, national governments are increasingly forced to operate in pursuit of national policy objectives in a broader context of international co-operation and the obligations they have taken upon themselves as signatories to various international environmental agreements. Such international commitments can function both as a set of constraints upon national policy and as a set of opportunities for achieving environmental quality objectives that would otherwise be beyond the capacity of any given country acting alone. Most obviously, international environmental agreements require adjustments in the substance of policy measures formulated and implemented by a country.

The effects of decisions and policies by a supranational organisation like the European Union are without precedence in Western European history.

The environmental directives of the European Union must be translated into appropriate legal measures by each of the member states. The importance of European policy for the distribution of costs and benefits within a member state affects the nature of political conflict surrounding environmental issues, i.e. decision-making with regard to such European Union policy can mobilise groups and redirect old participants, leading to new coalitions and shifting patterns of political activity at both the national and European levels. Furthermore, European Union policies can lead to pressures for change in the institutional arrangements and policy activities of particular countries. Therefore, in each of the country studies, some attention is paid to the impact of the EU's environmental policy upon domestic developments. Moreover, as has been mentioned, we have included a chapter (chapter 12) which examines the development of EU policy and the institutional arrangements through which this is shaped. In structure this chapter is similar to the country chapters.

While such a general overview is indispensable to our understanding of how environmental policy is administered, the story would be grossly incomplete without paying particular attention to how this system works in practice. By looking at how the different parts of the system (described above) are 'put together' in the daily operations of carrying out general policy and its programmes in concrete cases, it is possible to follow the way in which policy 'moves' from one level of decision-making to another (a process which can, of course, move in both directions) and how the different actors interact with one another.

Of particular interest in this connection is the role of local authorities in implementing national policies, and the network of inter-relations joining them with national policy makers. While we assume that the key 'bottom line' pay-offs are produced at the local level, this does not mean that we expect local officials (or local representatives of any level of government) to determine these results all by themselves. Rather it is assumed that these outputs are joint products of the interactions between different governmental actors, and between public officials and private actors. Nevertheless, in some of the ten countries under investigation the local level of government has been acquiring an increasing importance. In each chapter the emerging role of local communities in the system of environmental policy and administration is described and discussed.

An overview

This rest of the book is divided into 12 chapters. There are ten country chapters, a chapter on the European Union, and a concluding chapter in which an attempt is made to compare and discuss the developments and experiences of the ten countries in terms of some of the questions raised earlier.

The material presented in the following country studies will, it is hoped, enable the reader to follow the general line of development in each country with regard to environmental politics, policy and administration. At the same time the presentation has been organised in a way so as to lay the basis for a comparison of the different national institutional responses to the range of environmental problems confronting countries at a roughly similar level of social and economic development.

The analytical framework developed in the preceding section represents a 'maximum programme' for the effort that started with an international meeting at Drøbak in Norway. Since then the initial analytical framework has been developed further to guide the other authors in their revisions of their chapters. Although we were interested in comparing the experiences in different countries, we did not try to impose particular concepts and developments on the national self-understandings. Our approach was not intended to meet the requirements, and therefore we could not utilise the advantages, of the cross-national aggregate analysis mode. A basic methodological concern has been to let each country 'speak for itself' in order to see if, and in what way, a number of issues, that are either analytically interesting or indeed visible in certain countries, are in fact taking place in a particular country. The approach we have used in terms of guidelines and stimuli given to authors of the country chapters was aimed at ensuring that each chapter contained country-specific information that was presented in terms of a common set of analytical categories. The result is a set of case studies where, while each tells its own story, there is also substantial commonality and consistency in presentation to allow suggestive comparisons among the countries.[5] Needless to say, the case chapters do not provide sufficient information to shed light on all the relevant developments that have occurred in environmental policy. Nor can we in any systematic way answer all the questions we have raised in our framework.

The concluding chapter is a synthesis of developments of environmental politics and policy in Western Europe in which we differentiate between cross-country developments and country-specific characteristics. As to the latter, we try to show how environmental politics and governmental response in terms of policy and implementation measures as well as of environmental policy organisation in each of these countries have been shaped by the interaction of institutional features and actors' interests and strategies. Moreover, we examine how the environmentalist conceptualisation of the environmental *problematique* was reconstructed and transformed into the policy strategy of ecological modernisation. We give an overview of how this policy strategy has been put into practice, and we discuss how the development of environmental politics and policy is related to developments in other policy fields and to the effort to restructure the relation between the different institutional orders of these societies.

Notes

1. We use the concept of environmental policy area to refer to a subclass of the environmental policy field, e.g. issues pertaining to air pollution constitute one such environmental policy area, issues pertaining to nature protection would be another. By the concept of environmental policy sector we refer to the actors that try to affect public policy on environmental issues, the pattern of interaction among these actors as well as their substantive environmental position (e.g. the environmental values and policy positions they promote and represent).

2. With the establishing of social rights we can, in the words of T.H. Marshall, speak of the historical development of three layers of citizenship rights; the first layer comprised civil rights, the second layer political rights and from the pre- and post-World War II era there is a third layer of social rights, 'the whole range from the right to a modicum of economic welfare and security to the right to share the full in the social heritage and to live the life of a civilised being according to the standards prevailing in the society.' (Marshall 1964:72).

3. It follows from the institutional perspective that this structure in turn differentially affects the power, political consciousness and strategies of these social groups (Alford & Friedland 1974:326)

4. Nowhere is this more succinctly expressed than in Schattschneider's epigrammatic formulation: Organisation is itself a mobilisation of bias in preparation for action (Schattschneider 1960:30).

5. Our approach was partly inspired by arguments made by F.G. Castles (1989:1–14).

References

Alford, R.R. and Friedland, R. (1974) 'Nations, Parties, and Participation: A Critique of Political Sociology', *Theory and Society* 1: 307–28.

Burrage, M. (1992) 'States as Users of Knowledge: A comparison of Lawyers and Engineers in France and Britain', in R. Torstendahl (ed.) *State Theory and State History*. London: Sage Publications Ltd., pp. 168–205.

Castles, F.G. (1989) 'Introduction. Puzzles of Political Economy', in F.G. Castles (ed.) *The Comparative History of Public Policy*. Cambridge: Polity Press.

Eckhoff, T. and Jacobsen, K.D. (1960) *Rationality and Responsibility in Administrative and Judicial Decision-Making*. Copenhagen: Munksgaard.

Eriksen, E.O. (1996) 'Justification of Needs in the Welfare State', in E.O. Eriksen and J. Loftager (eds) *The Rationality of the Welfare State*. Oslo: Scandinavian University Press, pp. 55–75.

Fischer, W. And Lundgren, P. (1975) 'The Recruitment and Training of Administrative and Technical Personnel', in C. Tilly (ed.) *The Formation of National States in Western Europe*. Princeton, New Jersey: Princeton University Press, pp. 456–561.

Friedland, R. and Alford, R.A. (1991) 'Bringing Society Back In: Symbols, Practices, and Institutional Contradictions', in W.W. Powell and P.J. DiMaggio (eds) *Institutionalism in Organizational Analysis*. London: The University of Chicago Press, pp. 232–63.

Giddens, A. (1982) 'Power, the Dialectic Control and Class Structuration', in A. Giddens and G. Mackenzie (eds) *Social Class and the Division of Labour*. Cambridge: Cambridge University Press.

Giddens, A. (1984) *The Constitution of Society*. Cambridge: Polity Press.

Grimm, D. (1986) 'The Modern State: Continental Traditions', in F.X. Kaufmann, G. Majone, V. Ostrom (eds) *Guidance, Control and Evaluation in the Public Sector*. Berlin: Walter de Gruyter, pp. 89–109.

Jacobsen, K.D. (1970) 'Institusjonelle betingelser for planlegging', in A.J. Stokke (ed) *Beslutningsprosesser i norsk offentlig administrasjon*. Oslo: Universitetsforlaget, pp. 15–28.

Katzenstein, P.J. (1987) *Policy and Politics in West Germany. The Growth of a Semisovereign State*. Philadelphia: Temple University Press.

King, D. and Waldron, J. (1988) 'Citizenship, Social Citizenship and the Defence of Welfare Provision', *British Journal of Political Science*, 18: 415–42.

Krasner, S.D. (1988) 'Sovereignity – An Institutional Perspective', *Comparative Political Studies*, 21: 66–94.

Lukes, S. (1977) 'Power and Structure', in S. Lukes *Essays in Social Theory*. London: The Macmillan Press Ltd.

March, J.G. and Olsen, J.P. (1995) *Democratic Governance*. New York: The FreePress.

Marshall, T.H. (1964) *Class, Citizenship and Social Development*. Garden City, New York: Doubleday & Company Inc.

Offe, C. and Preuss, U.K. (1991) 'Democratic Institutions and Moral Resources', in D. Held (ed.) *Political Theory Today*. Stanford, Cal.: Stanford University Press, pp. 143–71.

Ostrom, V. (1986) 'Constitutional Considerations with Particular Reference to Federal Systems', in F.-X. Kaufmann, G. Majone, V. Ostrom (eds) *Guidance, Control and Evaluation in the Public Sector*. Berlin: Walter de Gruyter, pp. 111–25.

Rokkan, S. (1970) 'Nation-Building, Cleavage Formation and the Structuring of Mass Politics', in S. Rokkan *Citizens Elections Parties. Approaches to the Comparative Study of the Process of Development*. Oslo: Universitetsforlaget, pp. 72–144.

Rokkan, S. (1976) '"Votes Count, Resources Decide": Refleksjoner over territorialitet vs. funksjonalitet i norsk og europeisk politikk', in *Makt og motiv. Et festskrift til Jens Arup Seip*. Oslo: Gyldendal, pp. 216–24.

Schattschneider, E.E. (1960) *The Semi-Sovereign People*. New York: Holt, Rinehart and Winston.

Selznick, P. (1949) *TVA and the Grass Roots*. Berkeley: University of California Press.

Selznick, P. (1957) *Leadership in Administration*. Evanston, Ill.: Row, Peterson.

Therborn, G. (1991) 'Cultural Belonging, Structural Location, and Human Action', in M. Bertilsson og A. Molander (ed.) *Handling, norm och rationalitet. Om förhållandet mellom samhällsvetenskap och praktisk filosofi*. Bergen: Ariadne Forlag, pp. 93–115.

Weale, A. (1992) *The New Politics of Pollution*. Manchester: Manchester University Press.

Chapter 2

Britain: Coming to terms with sustainable development?

Neil Carter and Philip Lowe

In Britain, the central features of environmental policy and the policy-making process have, until recently, been characterised by continuity rather than change. Complacent governments were content to allow environmental policy to evolve gradually and reactively in a largely apolitical culture that favoured administrative informality and technical considerations over legislative rules and standards. Yet, since the late 1980s, a degree of instability has permeated British environmental policy. As the abusive tag of Britain as the 'Dirty Man of Europe' acquired wider usage, if not always validity, both at home and abroad, the politicisation of the environment as an issue encouraged widespread questioning of many of the long-established features of the policy arena. In response to a variety of pressures, the government introduced institutional reforms, implemented several policy initiatives and created a cross-sectoral Environment Agency as a means of rationalising the prevailing structures of environmental regulation.

The extent to which these actions reflect a change in the substance of British environmental policy and regulation is a matter of some debate (Gray 1995). At the very least, the interaction between a burgeoning environmental politics and a long established administrative structure is forcing the Government to re-examine its entire approach to environmental policy. In order to examine these issues this chapter is divided into three main sections. The opening section briefly describes the salient socio-economic and geographical features which provide the context for understanding the type of environmental problems confronting British policy makers. The second section sets out the peculiar administrative and legislative traditions characterising the evolution of policy relating to environmental protection. The third section analyses the various pressures for reform of environmental policy and regulation that have emerged over the past decade and assesses the resulting changes in policy.

The context and the nature of the problem

Through history and geology, the United Kingdom presents a diversity of rural and urban environments. South East England is one of the most densely populated regions in Europe, whereas the Highlands of Scotland include some of Europe's last remaining wilderness. As the first nation to industrialise and urbanise Britain also has a certain legacy of environmental problems and attitudes. A long-urbanised nation has developed strong cultural attachments to the countryside. Rural areas were viewed by the Victorian middle classes, who formed the first amenity and conservation groups, as places of retreat from the grimy, polluted, insanitary and unruly industrial towns. Their wildlife, landscapes, historic buildings and settlements needed to be protected from the onslaught of urbanisation. Rural conservation, historic preservation and landscape protection thus figure prominently in the British environmental movement. At the same time, as the nineteenth-century state and civic leaders sought to grapple with some of the appalling health and amenity problems of urban-industrial Britain, various professional and regulatory fields were established, such as public health, pollution control, sanitary engineering and town planning. These were set up largely on a devolved basis in keeping with strong traditions of local administration, which has structured the response to subsequent environmental problems.

Since the mid-1970s the established panoply of environmental protection has encountered new issues thrown up by the restructuring of the national economy and the move away from many of the post-war nostrums of economic governance. The drastic industrial restructuring of the 1980s diminished some environmental pressures; large-scale closure of plant, especially in traditional heavy and extractive industries, eliminated long-standing sources of air and water pollution but left behind extensive areas of dereliction. Periods of high energy costs and interest rates encouraged increasingly cost-conscious manufacturers to look for ways to reduce their energy and raw material costs which, in turn, alleviated some problems of pollution and waste disposal. One of the few traditional industries to grow steadily throughout was agriculture, reflecting its highly protected status within the European Community; not surprisingly, therefore, farm pollution and the impact of agricultural intensification on wildlife, the landscape and the welfare of farm animals emerged as major issues in the 1980s. The recession of the early 1980s brought a slowdown in public and private investment in pollution control and environmental improvements. The low level of industrial investment meant slower replacement of inefficient or obsolete equipment, which is often a prime source of pollution. Public expenditure cuts meant that much needed capital investment in waste treatment facilities was either shelved or indefinitely postponed. Successive governments felt unable to apply significant measures introduced in the *Control of Pollution Act* 1974, and ten years after it was passed much of this major piece of environmental legislation had still not been implemented.

The shortcomings were exposed when the economy recovered. The reversal, in the mid-1980s, of the progressive improvement over many years of river quality and a halt in the downward trend of sulphur emissions, suggested that the scope for living off past investments in pollution abatement had been exhausted, and that continued progress in curbing pollution would demand major new investment in control measures and treatment facilities. The institutional context in which such decisions had to be made, however, had been transformed by the extensive programme of privatisation of the 1980s. Many of the utilities and infrastructure providers were now private companies, and new and often complicated regulatory regimes were demanded to protect the public interest.

Three further threats to the environment arise from the changes in the economic structure and its regional distribution. First, the growing technology-based industries, such as microelectronics and biotechnology, while holding out the promise of more efficient and cleaner industrial processes, have created novel pollution and regulatory problems of their own. Second, the consumer-oriented booms of the 1980s and 1990s alongside continuing tight constraints on public expenditure have emphasised a different set of problems related to modern lifestyles and consumption pressures. These problems of 'private affluence amidst private squalor' include congestion and pollution arising from ever-expanding car ownership and use, the dispersal of a growing volume of household and packaging waste, the loss of rural land to house building, and noise intrusion from high-tech leisure pursuits in urban neighbourhoods and the countryside. Third, there is a more polarised geography of amenity, reflecting the sharply contrasting economic fortunes of different localities. Traditional industrial towns face a legacy of extensive dereliction and land contamination; inner cities are characterised by urban decay and declining or overstretched infrastructure; smaller settlements and the countryside experience diffuse development pressures as well as threats to landscape and wildlife from agricultural intensification and afforestation.

These divergent pressures, mediated through local political and state structures, yield great variations in environmental politics. For example, in most industrial cities, a strongly pro-development outlook, articulated by local business communities, trade unionists and politicians, tends to predominate, and is often linked to a concern not to make environmental controls too stringent for established or incoming firms. In contrast, throughout much of southern and central England, now largely dependent on a post-industrial service economy, anti-development preservationism predominates, expressed mainly by middle-class residents and oriented towards the protection of amenity. In the remoter rural areas, especially of Scotland and Wales, conflict is often found between agrarian, forestry, mineral and tourism interests, on the one hand, and, on the other, national pressure groups and statutory agencies promoting conservation.

Administrative and legislative traditions

The administrative context provides the backcloth, the prevailing norms and procedures, within which policy and institutional change takes place. An appreciation of this backcloth is vital for an adequate understanding of recent developments and current tensions in British environment policy; for the administrative system shapes and is in turn shaped by its continual engagement with policy problems and wider political forces.

The British administrative structure can be broadly divided into central and sub-central government. The executive organisations of central government are the central departments and ministries of state, including their regional outstations and the territorial ministries of Scotland, Wales and Northern Ireland, headed by ministers of the Crown. The main institutions of sub-central government include local authorities, the National Health Service, the few remaining public corporations and various non-departmental organisations; adjudicatory bodies, advisory bodies and 'quangos', which are executive agencies separated by statute from ministries.

The two components of sub-central government particularly relevant to a discussion of environmental policy are local government and the non-departmental organisations. Local government has responsibilities for such matters as transport and strategic planning, development control, leisure services and housing. Local government has traditionally enjoyed a considerable degree of autonomy in policy development and implementation which, in certain fields such as land-use planning and environmental health, it still retains despite concerted efforts by successive Conservative governments to curb the powers and spheres of action of local government between 1979-97. The various non-departmental organisations, particularly quangos, play a key role in administration at both national and local levels and, as we shall see, include a number of important environmental bodies.

Britain has the oldest system of environmental protection of any industrialised nation and probably one of the most elaborate. A diverse structure of agencies and statutes has evolved in an ad hoc and pragmatic manner since the mid-Victorian period. By the early twentieth century these included instruments to safeguard public health, contain industrial pollution, preserve historic monuments and protect wild birds. A detailed land-use planning system and protective measures and agencies for important landscapes and habitats had been added by mid-century. Legislation in the 1950s and 1960s rationalised and extended controls over air and water pollution and bird protection, promoted rural and urban conservation, and addressed the growing problems of noise and toxic wastes. Thus Britain, with this long tradition of environmental concern, was a pathfinder in many areas of legislation and regulation.

As the environment gained political importance in the late 1960s and early 1970s, the environmental lobby was instrumental in pressing for an accelerated pace of reform. Existing environmental agencies were strengthened. Major pieces of legislation were introduced, including comprehensive measures for pollution control and wildlife protection. The planning system was overhauled and provision made for public participation in preparing development plans and determining planning permissions. A new inquiry procedure, involving Planning Inquiry Commissions, was introduced to be used when development raised considerations of national importance or presented novel technical features. The Department of Environment was set up and a number of advisory bodies were created, the most significant being the Royal Commission on Environmental Pollution, which established itself as an authoritative forum for providing scientific opinion and advice to Government on environmental matters.

In retrospect, the above period represents a peak of reforming activity in relation to environmental protection, unsurpassed until the late 1980s (see below). The environmental record of the Thatcher Government before 1988 was largely negative. Thatcher came to office in 1979 preaching the neo-liberal virtues of deregulation. This outlook was not favourable to the extension, or even concerted application, of environmental controls. Actual deregulation was most consistently pursued in the field of urban planning, but attempts to relax planning controls in green belts and rural areas met with strong opposition, notably from Conservative MPs and local councils (Marsden *et al* 1993). The one major item of legislation was the *Wildlife and Countryside Act* 1981, introduced partly in response to the EC Birds Directive but also for the first time addressing the threat posed by modern agriculture to rural landscapes and semi-natural habitats, although the safeguards it introduced were both weak and cumbersome (Lowe *et al* 1986).

The policy process and the administrative structure

There is considerable debate within the environmental politics literature on the relationship between policy styles and issues. The debate has tended to polarise between those who believe that environmental policy conforms to a national policy style and those who claim that there is a distinctive environmental style which differs from other policy fields. The latter position is most closely associated with the work of Lowi (1964) who argued that it is the nature of the policy itself that shapes the way a problem is processed. In other words, distinct types of politics flow from specific types of policy issue, presumably with parallels in other countries facing similar problems. In contrast, the former approach attempts to identify distinctly national processes on the assumption that a national policy system approaches environmental problems much as it would do any other problem. Britain has been subjected to a number of studies from this perspective (Richardson and Watts 1985; Vogel 1986).

In particular, J Richardson *et al* (1982) argued that policy makers have a favoured behavioural policy style, a routinised way of dealing with issues. Their typology of policy styles leans heavily on two factors: whether a government has an anticipatory/active or a reactive stance towards problems; and whether it favours reaching a consensus or imposing decisions. A comparative study of six West European countries suggested that their policy-making approaches could be encompassed within this typology, with distinct national variations, although countries did in fact have more than one policy style.

So how might we characterise the environmental policy process in Britain? There have been a number of significant changes since the late 1980s (see below) but to assess these and to appreciate the considerable degree of continuity, it is important to identify the characteristic features of Britain's traditional approach to environmental policy. First, *government structures and legislation relating to environmental protection have been (and largely remain) an accretion of common law, statutes, agencies, procedures and policies*. There has been no overall environmental policy other than the sum of these individual elements, most of which have been pragmatic and incremental responses to specific problems and to the evolution of scientific knowledge. A 'tactical rather than a strategic approach' (G Richardson *et al* 1982:33) has, for instance, dominated the history of pollution control in Britain. The preparation of the environment White Paper, *This Common Inheritance* (1990), represented something of a departure, bringing together as it did all the facets of policy in a single document (although it was criticised for being largely a compilation of existing measures rather than a strategy, see below).

Environmental problems are intrinsically cross-sectoral and this, as well as the traditions of British public administration, also accounts for the fragmented nature of an environmental policy that has emerged piecemeal. There are numerous examples of administrative divisions and inconsistencies explicable only by reference to their provenance, rather than on grounds of principle. England, for example, seems to be the only country to make an administrative distinction between landscape conservation (the responsibility of the Countryside Commission) and nature conservation (that of English Nature) (Baldock 1987). As late as 1969 there were as many as ten different ministries concerned with environmental protection. Critics of the British Government's environmental record point accusingly at this absence of a single focus of interest and expertise within the machinery of government. To some extent, this is a problem common to environmental policy in all countries because environmental concerns overlap with almost every other policy field, notably trade, agriculture, energy, economy, health and transport.

The formation of the Department of Environment (DoE) in 1970 only partly resolved this problem. It emerged from a reorganisation of the machinery of government which brought together many of the functions that had previously been carried out by several departments and agencies, but provided the new department with few additional powers. The result was a wide-ranging 'super-ministry' combining environmental responsibilities with those more associated in continental states with a number of very different ministries: thus the DoE addresses policies for planning and regional development, local government, new towns, housing, construction, inner cities, environmental protection, water, the countryside, conservation, and the management of government-owned land and property. The image created by this unitary environmental administration, and its incorporation within an administrative structure that also includes local government, gives a false impression that the British conception of the environment is one of an all-pervading, integrated and genuinely transversal policy sector. In reality, the approach remains fragmented. The DoE is a Department *of* the Environment not *for* the Environment (in other words, not a Ministry for Environmental Protection). Within its wide range of functions are several, including housing and mineral planning, that may prove detrimental to the environment. It has been estimated that only about 10 per cent of the DoE's staff actually deal with environmental issues and several of its key environmental responsibilities, e.g. for industrial pollution, water quality, and nature and landscape conservation, are discharged by separate quangos operating at arm's length even if under its auspices. In short, environmental issues often struggle to reach the top of even the DoE's political agenda. In the late 1980s, for example, public discontent with the poll tax (a local government tax that proved a key factor in the downfall of Mrs Thatcher) meant that frequently the attention of DoE ministers was directed more at its local government responsibilities rather than in responding to the burgeoning public concern about the environment.

Although the DoE is the most important department dealing with environmental issues, several other departments have significant responsibilities in this area. The Department of Transport looked after marine pollution control, vehicle emissions and transport noise, although, since 1997, the DoE and the Department of Transport have been combined in a very broad ranging Department of the Environment, Transport and the Regions (DTER). The Ministry of Agriculture, Fisheries and Food (MAFF) deals with agri-environmental matters, countryside stewardship schemes, farm waste management, pesticide control and animal welfare. The Department of Trade and Industry has responsibility for many aspects of energy policy. The Health and Safety Executive, a non-departmental body, controls nuclear installations regulation, pesticide safety and industrial hazards. Not surprisingly, this fragmentation has produced glaring procedural inconsistencies, such as the differing rules governing public access to information (DoE 1986). Richardson and Watts (1985:8) suggest that a fragmented approach to environmental issues arises from 'a very reactive approach to problem solving', a characteristic of British policy making generally which, they argue, makes for a simpler and more flexible style than that found, say, in Germany.

In part this flexibility stems from a second feature of environmental policy: *regulation has been highly devolved and decentralised.* In most cases, responsibility for taking action has fallen to the various tiers of local government or to an administrative agency. The roles allocated to these bodies normally combine policy development, research and advice with regulation and implementation, but central government usually retains responsibility for promoting general policies, exercises financial control and often has reserve powers.

Indeed, in the past the range of administrative agencies was wide and, despite recent rationalisations, remains confusingly so. They play a prominent part in environmental management in the UK and, in combining policy advice and implementation, they concentrate considerable expertise in their respective fields. The Department of the Environment is responsible for some of them, including English Nature, the Countryside Commission and, most important, the Environment Agency – set up in 1996 through an amalgamation of separate agencies responsible for industrial pollution control, water and fisheries protection, and waste regulation. The Department is the lead department for UK environmental policy. However, the geographical remit of these agencies is either England and Wales, or just England. Scotland has its own Scottish Environment Agency. Likewise, there is Scottish Natural Heritage and the Countryside Council for Wales, each of which combines responsibilities for landscape protection and nature conservation which are handled separately in England. These regional agencies are the responsibilities respectively of the Scottish and Welsh Offices. There are other environmental agencies outside of the ambit of the DoE. For example, English Heritage, which is responsible for historic preservation and heritage conservation, is within the purview of the Department for Culture, Media and Sport, but its Scottish and Welsh equivalents are the responsibilities of the relevant territorial ministries. The Forestry Commission, which promotes forestry and has to balance the interests of afforestation and conservation, has a Great Britain remit and reports to the Secretaries of State for Wales and for Scotland, and the Minister of Agriculture in England.

Britain is unusual in the extent to which functions that are carried out primarily by government agencies in most countries are performed by private bodies in Britain (Lowe and Goyder 1983). These include the National Trust which buys amenity land and historic buildings to retain them for public use; and the Royal Society for Nature Conservation (RSNC) and the Royal Society for the Protection of Birds (RSPB) which purchase land to maintain as nature reserves. The Tidy Britain group, although a voluntary organisation, is the official body for government litter prevention and abatement policy. The RSNC helps English Nature administer sites of special scientific interest (SSSIs) which are areas of protected wildlife habitat. By statute, bodies such as the Society for the Protection of Ancient Buildings and the Victorian Society must be consulted when buildings of historical significance are threatened by development. Many of these organisations receive significant proportions of their funding from the government in exchange for carrying out these duties.

Clearly, the highly devolved and decentralised regulatory structure reflects and, in turn, perpetuates the fragmented and piecemeal character of environmental policy.

Third, *environmental control, like many other aspects of state regulation in Britain, has been pervaded by administrative rather than judicial procedures.* The approach pursued is informal, accommodative and technocratic rather than formal, confrontational and legalistic (Hawkins 1984; Ashby and Anderson 1981). Legislation tends to be broad and discretionary. Regulatory agencies are usually given wide scope to determine environmental objectives and considerable latitude in their enforcement. In particular, and in contrast with many other industrialised countries, there has been an avoidance of, indeed distaste for, legislatively prescribed standards and quality objectives:

> It has long been traditional to rely upon, where practicable, the characteristics of the local natural environment as a sensible disposal and dispersal route for potential pollutants. This underlying approach in theory requires that agencies should be given complete independence and discretion to determine, in the light of local circumstances, the degree of seriousness of a potential pollutant and the appropriate control measures. (Macrory 1986:8)

Regulatory officials wish to protect the considerable discretion allowed to them by their relative insulation from the scrutiny of both parliament and the judiciary. Consequently, when laws are broken, in the vast majority of cases, officials prefer not to prosecute.

Fourthly, *the approach traditionally adopted in dealing with private concerns has sought to foster co-operation and striven to achieve the objectives of environmental policy through negotiation and persuasion.* Again the British approach emerges as distinctive. In a comparison of British and American environmental practice, Vogel (1986) has noted that, despite key similarities in political and cultural traditions, common environmental conflicts and even shared organisational responses, there are sharp differences in the administration of environmental controls in the United Kingdom and the United States:

> in many important ways, each nation controls industrial emissions in much the same manner that it regulates everything else. (...) Americans rely heavily on formal rules, often enforced in the face of strong opposition from the institutions affected by them, while the British continue to rely on flexible standards and voluntary compliance – including, in many cases, self-regulation. (p.77)

Thus, where other countries have made extensive use of uniform standards of emissions and environmental quality backed up by law, British governments have preferred implementation through consent by means of industry self-regulation and informal agreement. For 'the British are reluctant to adopt rules and regulations with which they cannot guarantee compliance' (ibid, p.77). Instead, regulations are made in such a way that officials can negotiate arrangements with firms that will not be disallowed by their superiors or the courts. Consequently, government officials seek to 'persuade' industrial and farming interests of the need to modify their behaviour. This predilection for co-operation rather than enforcement has often meant that the first instinct of policy makers has been to encourage voluntary measures (as in the removal of detergent effluent from inland rivers), or self-regulation (e.g. the Pesticides Safety Precaution Scheme, as it existed until 1985), or the promulgation of codes of practice (such as the National Farmers' Union straw burning code which eventually was superseded by a statutory ban). Notions of 'best practicable means' of controlling pollution or requirements such as the one directing conservation and planning authorities to take into account the needs of agriculture and forestry (under Section 37 of the *Countryside Act* 1968) ensure that regulatory authorities are sensitive to the economic and practical constraints that private organisations face. Indeed, the latest notion of pollution control; Best Available Technique not Entailing Excessive Costs (BATNEEC), enshrines just this principle.

According to Vogel (1986), the continued dominance of this 'mid-Victorian style of regulation' (p.26), is built on three features: a highly respected civil service, a business community that is prepared to defer to public authority, and a public that is generally unsuspicious of the motives or power of industry. The main advantage that defenders claim for the voluntaristic approach is the low level of conflict existing between regulator and regulated, a crucial factor in enabling policies to be implemented effectively. Vogel could not find a single corporate executive in a British-based firm who 'could cite an occasion when his firm had been required to do anything it regarded as unreasonable'. Not surprisingly, he concluded that the business community is 'among the most consistent defenders of its nation's system of environmental controls' against EC efforts to harmonise Britain's regulatory policies with those of other members (pp.21–2).

Fifth, *environmental policy making, where it impinges on major economic interests, has traditionally taken place in relatively closed policy communities* (Ward and Samways 1992). For Vogel (1986) this represents a policy style that is less 'pro-industry' than it is biased in favour of the status quo: 'as a general rule, the people most closely consulted about a particular policy are those recognised by the government as having a disproportionate stake in its outcome' (p.272). In most areas of environmental policy these will be representatives of producer interests: industrialists, trade associations, farmers and landowners. Consequently, the negative view of the voluntaristic tradition is the belief, widely held among environmentalists, that government officials have been 'captured' by industrial managers and agricultural interests. The absence of tough regulatory controls is seen by many commentators as evidence that the environmental lobby has had very little influence on policy compared to the economic interests who are supposedly being regulated (McCormick 1991). The common explanation for this situation is that most environmental policies are decided between civil servants and industry representatives in private arenas that are largely invisible not just to the public, but also to Parliament and even to the relevant departmental minister.

Thus the control of pollution has traditionally been based on voluntary agreements negotiated between regulatory officials and representatives of industry. The application of the long-established principle of 'best practicable means' has minimised prosecutions. Furthermore, in the past, the fragmented structure of the various regulatory agencies responsible for pollution control meant that there was no one authority responsible for monitoring 'pollution in the round' (Weale *et al* 1991). Not only did this weaken government officials in their dealings with industry, but the veil of secrecy surrounding these regulatory processes minimised public involvement. This secrecy persisted because it was argued both by government officials and by industrialists that voluntary compliance could only work if details of agreements were confidential. Consequently, it was hardly surprising that the public remained so uncritical of industry. When Vogel argued in 1986 that 'there is in Britain today no significant domestic pressure to change the way British pollution-control policy is either made or enforced' (quoted in McCormick 1991:93), the public did not (and largely still does not) know what damage industrial production was causing to the environment.

Similarly, in agricultural policy there has been a well-established policy community since the 1940s. Policy has been made within a closed group consisting of MAFF officials and leaders of the National Farmers Union (NFU) and, to a lesser extent, the Country Landowners Association (CLA) who all accepted the consensus that agriculture needed to be supported and production increased (Cox *et al* 1986; Smith 1989). Intrinsic to that consensus was a preference on the part of MAFF officials for dealing with the NFU rather than with environmental (or consumer) groups. Elsewhere, the close ties between the Department of Transport and the road lobby have, until recently, resulted in the transport policy of successive governments favouring public investment in roads rather than railways (Hamer 1987). Energy policy, likewise, was for long dominated by the large utilities.

Producer interests do not hold sway in all fields of environmental policy. There are some established fields to do with positive environmental planning and management where moderate environmental groups also enjoy insider status, including close ties with relevant quangos. These fields include nature conservation, landscape protection, historic preservation and countryside recreation. The most prominent of the insider groups, notably the RSPB and the National Trust, emphasise their own direct practical contributions to conservation rather than a lobbying role. By and large, these traditional fields of conservation policy have impinged little on producer interests.

Only in land-use planning, where there is greater institutionalised opportunity for participation by environmental and community pressure groups, have the interests of producers been more obviously compromised. In consequence, the machinery of land-use planning, particularly the system of public inquiries, has become the focus for direct conflict and the airing of fundamental contradictions between the interests of environmental protection and economic development. Notable public inquiries include those concerning the Windscale and Sizewell B nuclear power stations and the Twyford Down motorway extension.

Developments since the late 1980s

By the late 1980s, pressures from a number of different directions had built up to stimulate a major reappraisal of policy and structures. Four significant developments can be identified: the politicisation of environmental issues; a growing openness in the policy process; the gathering momentum of European political integration; and the gradual emergence of a governmental sustainable development strategy. Each has contributed to a move away from the ad hoc, piecemeal and reactive approach to traditional policy, towards a more strategic, integrated and anticipatory stance.

In Britain, the emergence of the contemporary environmental movement dates back to the late 1960s, and in that sense, this is when *the politicisation of the environment* began. Yet there is a paradox surrounding the environmental lobby in Britain. On the one hand, as McCormick (1991) asserts, 'Britain has the oldest, strongest, best-organised and most widely supported environmental lobby in the world' (p.34). Membership grew rapidly during the 1980s from a figure of around 2.5–3 million in 1982 to between 4–5 million in the early 1990s, roughly one in ten adult Britons (Rawcliffe 1995). Yet, at least up to the early 1980s, the movement had failed to 'translate its numerical power into appreciable and consistent direct political influence of the kind enjoyed by economic interests and lobbies' (McCormick 1991:46).

Admittedly, some groups had acquired a degree of influence by the end of the 1970s through the pursuit of reformist strategies: by obeying the laws, working within the system and refraining from confrontational strategies, they gained access to certain areas of decision-making, particularly in the field of conservation and land use (Lowe and Goyder 1983). Conservative groups like the National Trust retain that 'insider' status. The advent of the Thatcher administration reduced interest group access to government, forcing a shift in pressure group tactics towards a higher profile, populist and media-oriented style of campaigning on the lines pioneered by more radical and adversarial groups such as Greenpeace and Friends of the Earth, which emphasised the strength of their support and their ability to attract media attention. There is no doubt that the environmental lobby was very effective in raising public awareness of environmental issues and pushing problems onto the political agenda.

However, perhaps one of the most important achievements of the lobby during the 1980s was in contributing to the party politicisation of the environment which unlike other European countries did not occur through the breakthrough of a Green Party, and the need for established parties to respond to that challenge. Instead, it came about as mainstream politicians felt obliged to respond to the swelling ranks of the environmental movement and the campaigns of the more activist groups, and the growing volume of critical media coverage of environmental issues. By and large, party leaders had not shown any great, and certainly no sustained, commitment to the environmental cause. Indeed, the build-up of party interest was at first halting and fitful. Nevertheless, there was a growing recognition that the environment was an important area for parties to present a positive image in public debate and possibly a potential source of votes (Carter 1992; Flynn and Lowe 1992). After Mrs Thatcher made her now famous speech to the Royal Society in September 1988, for the first time giving her authority to the significance of global environmental problems, the battle was on. Thus, when the Green Party, which had been around since 1973 with negligible political impact, received a temporary surge of popular support in gaining 15 per cent of the overall vote in the 1989 European Parliamentary elections, the mainstream parties were already vying with each other to improve their green credentials by fleshing out their existing, rather thin, environmental policies. The Green Party's flash success indicated to the established parties that they needed to be on their guard over environmental issues but that they might not be the beneficiaries if such issues became a central feature of electoral competition.

This politicisation has contributed to the second major development [a *change* *growing openness within the policy process*, as the combination of concerned *in policy* public opinion, investigative media, pressure group activity and Opposition *style.* party probing has exposed an increasing number of decision-making processes to a wider gaze. Certain institutional changes have enhanced this effect. Throughout the 1980s, the growing role of parliamentary select committees helped force issues, such as declining river quality and filthy bathing beaches, onto the political agenda. But perhaps most important was the indirect and unintended consequence of Thatcherite institutional radicalism, associated with the imperative of 'rolling back the state' and the introduction of market-type relationships into the public sector, which opened up established procedures to critical scrutiny. Privatisation has required new regulatory frameworks and generated institutional upheaval. The Government has had to re-examine basic questions concerning who and what to regulate and protect. Inevitably, this process of inquiry has been carried out at least partly in the public domain, enabling environmental pressure groups to gain access to previously closed policy communities, thereby creating opportunities for quite separate and major reforms with (generally) beneficial implications for environmental protection. In particular, ministers have been receptive to such pressures in order to enhance the popularity of their own institutional reforms and to help guarantee a smooth parliamentary passage for privatisation legislation. In this respect the development of environmental policy has benefited from having a high public profile while not being a central feature of the Government's political programme. A prime example is provided by water privatisation which yielded a powerful new regulatory agency, the National Rivers Authority (which in 1996 was absorbed into the new Environment Agency), and significantly improved environmental safeguards. One of the spin-offs from this institutional radicalism has been demands from both within and outside government for a more precise specification of policy objectives and the development of instruments, such as sustainability indicators, to evaluate the performance of the public sector.

Another factor encouraging the publication of detailed and comprehensive data on many aspects of environmental performance has been the need to comply with EC directives. This introduces the third key development: <u>the growing importance of European political integration in advancing environmental protection.</u> Environmental policy has been at the forefront of the movement towards political integration (Liefferink, Lowe and Mol 1993). All national policy communities are having to respond to a new European agenda but, among the states of Northern Europe, the impact has probably been most profound on Britain.

Yet during the early years of Britain's accession to the EC, the impact of membership on domestic environmental policy did not seem that significant. Britain's entry in 1973 coincided with the first EC environmental programme which stressed the need for member states to harmonise regulation policy while minimising interference with free trade and competition. Initially, the clash of regulatory styles between the British preference for voluntary regulation and the continental system of legislation and standard-setting caused the British Government few major problems. With the important exception of the *Wildlife and Countryside Act* 1981, until the late 1980s the Government was able to implement most EC environmental directives under existing law with only minor administrative adjustments.

However, the gathering pace of European environmental policy making increased the pressures for the integration of British policy into a European framework. This integrative process has challenged a twofold insularity: an administrative insularity which in the past exhibited a certain self-satisfaction over a system of environmental regulation that had evolved in an ad hoc, pragmatic and piecemeal manner; and a geographical insularity which, through exploiting the capacities of the prevailing winds, fast-flowing rivers and surrounding seas, had pursued a 'dilute and disperse' approach to solve, or at least dispel, major pollution problems. The establishment of uniform air and water quality standards and the standardisation of pollution control procedures has upset the cosy, 'gentlemanly' British style. Specifically, the EC directives on bathing areas, drinking water and on pollution from large combustion plants, which finally persuaded the Thatcher Government to commit Britain to specific reductions in sulphur dioxide emissions in an attempt to halt acid rain damage (Lowe and Ward 1998), have all contributed to this changing approach to environmental management.

Commission initiatives have also probed the traditional basis of British procedures by obliging the Government to explain and justify them fully. One consequence has been to make explicit principles that were previously implicit or merely rhetorical, a further step towards greater openness in the policy process. Not only has this established a more definite framework in which the achievements of regulation can be assessed and challenged, but it has also inspired a wide-ranging debate about established practices and alternative strategies of environmental protection (Haigh 1984). European and international pressures have increasingly required the Government to justify its policy in abstract terms and by reference to principles that have international currency, such as sustainable development (see below), the precautionary principle and integrated pollution control (IPC).

Another effect of the European Community has been to catalyse the trend towards greater centralisation of UK policy making. It is, after all, the central government that is responsible for negotiating with the Community and for ensuring implementation of agreed measures and policies. The Government has therefore been obliged to take a much more active and prominent role in setting policy objectives and standards, and to adopt a more strategic approach to environmental policy. These pressures have also coincided with a period of rule by Conservative governments with strong centralising tendencies. The consequence in the environmental field has been both a movement of responsibilities from local government to quangos and a diminution of the discretionary authority of both quangos and local authorities. Against these tendencies, local authorities have increasingly looked towards Europe for leadership and partnership in environmental initiatives. The EC Environment Directorate has made a conscious bid to move into the field of urban planning and, in the 5th Environmental Action Programme, makes a specific play for local authority support with the concept of 'administrative subsidiarity'. Many British local authorities, reeling from the drastic reduction in their responsibilities resulting from years of assault from central government, have sought to exploit the potential opportunities offered by Europe for carving out new competencies and responsibilities (Ward 1995).

Membership of the European Community has also provided a new forum in which environmental pressure groups can seek either to enter or to circumvent domestic policy communities. Pressure groups can now direct their lobbying at the European Commission and European Parliament. In addition, the EC actively consults pressure groups in the planning stage of policy formulation. Pressure groups also have the new instrument of judicial review: they can appeal to the European Court of Justice to require the British Government to implement EC directives. British pressure groups have made more use of EC institutions than their counterparts in other member countries. This can be explained perhaps by the existence of relatively greater dissatisfaction with domestic environmental policy prompting British pressure groups, which have a stronger lobbying tradition than elsewhere, to regard Brussels increasingly as a means of outmanoeuvring their own national government. For example, in the original proposal to privatise the water industry, the Government intended to leave the responsibility for regulating water pollution control with the privatised water companies. It was only after the Council for the Protection of Rural England, a leading conservation group, obtained legal opinion pointing out that such self-regulation would run counter to EC law that the Government conceded defeat and agreed to the creation of the independent National Rivers Authority.

The pressures from politicisation, growing openness and European integration combined to stimulate a major reappraisal of policy and institutions. In Autumn 1989, soon after the European election, Chris Patten, the newly appointed Secretary of State for the Environment, announced the preparation of a White Paper (i.e. a considered statement of government policy) that 'will set out our environmental agenda for the rest of the century'. When it was eventually published, in September 1990, *This Common Inheritance* (DoE 1990) was poorly received by the media and by environmentalists. Despite being a substantial document of 296 pages printed on recycled paper, most of the 350 or so measures were also recycled. Over half were decisions that had already been announced or implemented and there were few firm new commitments. Nevertheless, the White Paper was the first comprehensive statement of British environment policy and, significantly, it provided a clear expression of the principles of sustainable development that should inform policy making.

Subsequently, it has been *the gathering momentum of the sustainable development process* that has constituted the fourth important change in environmental policy. It was partly stimulated by international obligations; British involvement in the 1992 Rio Earth Summit resulted in a commitment to the Agenda 21 process. A year-long consultation period culminated in the publication, in January 1994, of *Sustainable Development: the UK Strategy* (HM Government 1994b). Again this document disappointed environmentalists for it contained no significant new policies and largely reaffirmed existing commitments. Nevertheless, the process has given rise to some interesting developments, including a series of institutional initiatives aimed at integrating environmental decision-making across policy areas. *This Common Inheritance* focused on internal machinery of government reforms by establishing: two Cabinet Committees to oversee government policy; 'green ministers' in each department who were responsible for environmental matters; and a process of annual DoE and individual departmental follow-up reports to record progress in implementing sustainable development. *Sustainable Development* included three new external initiatives: a Panel on Sustainable Development consisting of five eminent experts who report directly to the Prime Minister; a Round Table on Sustainable Development made up of 30 representatives drawn from business, local government, environmental and other organisations; and a 'Going for Green' programme aimed at carrying the sustainable development message to local communities and individuals. Institutional reform is usually a slow-moving process, so it is too early to evaluate properly the impact of these initiatives, although the early prognosis on the White Paper machinery of government reforms is poor (Carter and Lowe 1998). Another output of the sustainable development process, married to a government commitment to targets and performance indicators in general, is the production of a new and more extensive range of sustainability indicators

(DoE 1996). This approach was particularly evident in the elaboration of *Biodiversity: the UK Action Plan* (HM Government 1994a). Together these changes represent at minimum a small step towards a more co-ordinated structure of environmental policy making, and could have considerable potential to redirect government if they were pursued with sufficient political will.

Furthermore, the commitment to sustainability has contributed to change in at least one important policy area – transport. It was acknowledged for the first time in *Sustainable Development: the UK Strategy* that unlimited traffic growth is incompatible with the Government's environmental objectives; a conclusion supported by the report of the Royal Commission on Environmental Pollution in October 1994 which recommended that the roads programme should be halved, petrol taxes doubled and public transport and cycling encouraged. It is clear that the process of producing the strategy document – by generating a confrontation between the then separate transport and environment departments – contributed significantly to this shift and to the resulting policy changes. Subsequently, parts of the massive road-building programme have been put on hold, public spending on roads has been cut dramatically, road-pricing schemes are being investigated and planning policy advice has been altered both to discourage the building of out-of-town shopping centres and to concentrate new residential and employment development so as to reduce the need for private travel. There is still no integrated transport policy, car emissions continue to rise, and many other policies (such as cuts in railway spending) work against a sustainable transport policy, but these shifts mark a reversal of the 1989 *Roads for Prosperity* White Paper that viewed traffic growth as a sign of prosperity that should be accommodated by the Government. There is an old lesson here concerning the unintended consequences of policy change. Although the Government may not be an enthusiastic convert to the sustainable development bandwagon, it has given it official encouragement. Consequently it has provided ammunition to environmental organisations, local authorities and also the DoE, to employ against proposals from elsewhere in government that fail to give due regard to environmental effects.

The significance of these four developments in environmental policy and politics should not be exaggerated. The party politicisation of the environment had already weakened by the 1992 general election, when the issue was virtually ignored, as it was in the 1997 election. The environment has not yet become a salient issue in national elections and, under the current electoral system which discriminates against the emergence of small parties like the Greens, it will probably remain so. For neither the Conservative nor the Labour party has yet perceived any great political advantage in implementing a radical sustainable development strategy. The Conservative Government generally paid only lip-service to sustainable development. While it introduced a number of institutional initiatives, the most important change, the creation of the Environment Agency, owed little in its origins or ultimate design to the idea of sustainability. Instead, the Government was persuaded to consolidate the regulatory functions of pollution control within one agency on the grounds of administrative efficiency and political opportunism, not sustainability (Carter and Lowe 1998). The changing direction of transport policy is also due only partly to the sustainable development process: the cut-back in the road-building programme coincided with the need to cut public expenditure and the increasingly co-ordinated (and popular) anti-roads protests in the mid-1990s.

Nor are the four identified developments unambiguously beneficial to the environment. For example, while privatisation may have opened up many procedures to public scrutiny, the current pricing systems for the gas and electricity industries provide an incentive for the utilities to persuade consumers to conserve energy.

Crucially, the core policy communities remain very strong. Although environmental pressure groups may have scored some successes in changing government policy, they remain, at best, only intermittent 'insiders'. European law has certainly forced concessions from the policy communities but they have given ground only grudgingly, protecting the close-knit group and shared consensus as far as possible. Thus even with well-publicised controversies, such as that surrounding the spread of salmonella in eggs, when the public and politicians have questioned the role of government officials and producers (in this case, intensive farming practices), the policy community has managed to stave off significant policy change (Smith 1991).

The ability of policy communities to organise resistance to changing conditions raises an important question about the potential for EC law to improve the British environmental record. EC environmental policy undoubtedly offers a more progressive, far-reaching regulatory regime than the traditional British approach, but it is uncertain whether the shift to a more formalistic style will result in the *implementation* of a more progressive policy (Lowe and Ward 1998). Vogel (1986), comparing UK and US environmental policy in the early 1980s, while not claiming that British environmental controls were particularly effective, argued that the emphasis on voluntary compliance had proven no more or less effective than the more adversarial and legalistic approach adopted by American policy makers (p.23). However, it is questionable whether the UK's reliance on the consent of the regulated is capable of producing a radical shift in terms of both progressive policies and tougher, integrated regulatory institutions. In short, the conservative bias in favour of corporate interests that is inherent to British voluntarism may be inappropriate for resolving contemporary environmental problems.

The 'greenest government yet'?

In many respects, British environmental policy is still characterised by continuity rather than change. Certainly, there has been little significant shift in policy substance, with the possible exception of transport policy, but even here the extent of change remains uncertain. Moreover, all five core characteristics of the administrative structure and policy process remain clearly identifiable. In particular, it is difficult to envisage sustainable development becoming a guiding principle of government as long as environmental policy making is dominated by closed policy communities that use the voluntarist regulatory style to defend producer interests. Although this style inevitably involves some compromise, it does not have to be as conservative or as sympathetic to corporate interests as it currently is in Britain. The Dutch government, for example, has encouraged the notion of self-regulation within Dutch industry, but as a means to implement ambitious pollution reduction targets negotiated with particular sectors. This approach allows individual sectors to choose the means but not the objectives of policy.

Yet there have undoubtedly been some important developments since the late 1980s. In particular, the various institutional developments; the emergence of a sustainable development process, the publication of sustainability indicators, annual reporting of environmental performance, and the machinery of government reforms, together suggest some weakening of several of the five core characteristics. Thus the creation of the Environment Agency represents a move away from the decentralised tradition of regulation. And probably of greatest long-term significance is the gradual unfolding of the sustainable development process because it implies a shift towards a more strategic and integrated approach to policy making. But the Conservative Government failed to provide the leadership, strategic planning or political will that the implementation of sustainable development requires. Consequently, it has become fashionable in Britain to praise the EU for its progressive role in promoting environmental protection, for example the encroachment of EU law has resulted in a more extensive use of standards and targets.

Yet, in this respect, optimism should be tempered because the momentum of EU environmental policy seems to have slowed down since the signing of the Maastricht agreement. Moreover, the one-way direction of influence has also altered, for the British government has taken a more active role in certain aspects of EU environment policy; indeed, it has proven to be a leader in pressing such issues as subsidiarity, cost-effective regulation and integrated pollution control (Lowe and Ward 1998). While British enthusiasm for the first two concepts may be for reasons that are not entirely beneficial to the environment, with regard to IPC, Britain is acknowledged to be leading her Community partners. In this changing context, it is likely that the continued momentum for the sustainable development process will depend more upon domestic political factors than on international pressures. Much may therefore depend on the attitude of the Labour government towards environmental issues.

Prior to its 1997 election victory, few observers would have expected a change of government to herald a significant shift in the approach to environmental policy. Although Labour had a wide-ranging and ambitious policy document, *In Trust For Tomorrow* (1994), the Labour leadership had avoided unequivocally endorsing the programme. The environment was not a key plank in its electoral strategy and the specific promises made were both cautious and modest. Yet the new Government has displayed a surprisingly upbeat attitude, with several ministers promising to place environmental considerations at the forefront of policy making. Robin Cook, in his first interview after being appointed Foreign Secretary, declared that one of the Government's four foreign policy priorities would be 'to place human rights and the environment at the centre of European policy'. The Prime Minister, Tony Blair, expressing a new-found enthusiasm for the issue, made a powerful speech at the follow-up UN Earth Summit in New York in which he promised Britain would play a leading role in promoting international environmental agreements and committed Britain to a 20 per cent reduction in its 1990 level of carbon dioxide emissions by 2010. And the new Secretary of State for the Environment, Transport and the Regions, John Prescott, claimed that Labour would be the 'greenest government' yet.

It is too early to judge whether Labour's words and aspirations will be turned into action. Much will depend on the Government's political commitment to its environmental agenda. Here there are signs that Labour now sees political advantages in being greener. With electoral reform firmly on the agenda – proportional representation will be in place for the 1999 election to the European Parliament and a referendum is promised for electoral reform of the House of Commons – the threat from the Green Party may become more potent. Labour strategists are increasingly of the view that a positive environmental record may help retain the support that gave Labour its landslide victory in 1997, particularly among two groups, young and women voters, where the swing to Labour was very large. The presence of several long-standing environmentalists within the leadership, such as Cook and Chris Smith, and newer converts such as Prescott and Michael Meacher, should make the government more receptive to the environmental lobby. With the balance of power within the party shifting away from the trade unions in general, and manufacturing unions in particular, traditional objections to environmental policies may carry less force than in the past. Several unions are aware of the job potential in environmental technologies, recycling industries and a shift of emphasis to public transport. The Government is concerned by Britain's current small share of the growing global market in environmental industries: as Tony Blair put it, 'There is money to be made and there are jobs to be created' (*The Times*, 4 December 1997). Another incentive to be greener might be financial. Having promised not to raise income taxes the Government needs to find alternative ways of raising revenue. Some eco-taxes offer the combined benefit of generating revenue whilst also establishing a greener image for the government.

So there are signs that the political profile of the environment is likely to become higher under Labour. Moreover, some of its policies will certainly contribute further to the broad developments identified above. For example, the proposed legislation on Freedom of Information will create new rights of access to environmental information that will continue to open up the policy process. The creation of the new super-ministry of Environment, Transport and the Regions should improve policy integration by combining two departments, parts of which, according to Prescott, 'have not been speaking to each other for years'. This reform may also help overcome the long-standing criticism of environmentalists that the DoE was marginalised in the Whitehall hierarchy and may provide the opportunity for environmentalists to enter previously closed policy communities in transport and regional policy arenas.

However, Labour is likely to be judged on its response to the big substantive environmental issues, such as global warming and the conflicts between countryside protection and the need for new housing, and between road congestion resulting from growing car ownership and the need to improve the urban quality of life by investing in a crumbling public transport system and reducing worsening air pollution. In this respect, early evidence is mixed. Prescott certainly played a significant brokering role at the Kyoto Summit in December 1997, but back home radical policies are needed if Britain is to meet its commitments to reduce carbon emissions. During the 1990s Britain has had notable success in reducing emissions primarily as a result of the 'dash for gas' whereby power generators have switched from coal-fired power stations to gas-fired units. However, the potential for further reductions in this area is limited. Indeed, in response to growing concern about the desirability of using gas in electricity production – as it is a very inefficient use of this flexible energy source – and in order to protect what remains of the coal industry, the Government introduced a moratorium on permissions for new gas stations (after first allowing four new proposals through). Moreover, carbon dioxide emissions from road transport and home energy use, which together make up over 40 per cent of total output, are rising rapidly. Only radical policy initiatives will reverse these trends. Yet, in early 1998, there was no sign of the massive home energy conservation programme that will be needed to reduce consumption; indeed, the Government had exacerbated the problem by reducing VAT on domestic fuel from 8 to 5 per cent. The prognosis for transport is more optimistic. With evidence that public attitudes towards transport are slowly changing, under Prescott's leadership the Transport White Paper due to be published during 1998 may herald some genuinely radical changes, notably in taxation and charging structures. On balance, Labour should be able to achieve further reductions in carbon emissions, but it is difficult to see how it can meet its optimistic 20 per cent carbon reduction commitment. Ultimately, the test of Labour's pledge to put the environment at the heart of government will depend on its success in addressing these high-profile problems and in its resolution of the contradictions with more pressing priorities such as economic growth, industrial competitiveness and job creation.

References

Ashby, E. and Anderson, M. (1981) *The Politics of Clean Air.* Oxford: Clarendon Press.
Baldock, D (1987) *The Organisation of Nature Conservation in Selected EC Countries.* London: Institute for European Environmental Policy.
Carter, N. (1992) 'The Greening of Labour', in J. Spear and M. Smith *The Changing Labour Party 1979-92.* London: Routledge, pp.118–132.
Carter, N. and Lowe, P. (1998) (forthcoming) 'Sustainable Development in the UK', in S. Young and J. van der Straaten (eds) *Ecological Modernisation.* London: Routledge.
Cox, G., Lowe, P. and Winter, M. (1986) 'Agriculture and Conservation in Britain: a Policy Community Under Siege', in G. Cox *et al* (eds) *Agriculture, People and Politics.* London: Allen & Unwin, pp.169–98.
Cox, G., Lowe, P. and Winter, M. (1990) *The Voluntary Principle in Conservation.* Chichester: Packard.
Department of the Environment (DoE) (1986) *Public Access to Environmental Information.* Pollution Paper No.23, London: HMSO.
Department of the Environment (DoE) (1990) *This Common Inheritance.* London: HMSO.
Department of the Environment (DoE) (1996) *Indicators of Sustainable Development.* London: HMSO.
Flynn, A. and Lowe, P. (1992) 'The Greening of the Tories: the Conservative Party and the Environment', in W. Rudig *Green Politics Two.* Edinburgh: Edinburgh University Press, pp.9–36.
Gray, T. (ed) (1995) *UK Environmental Policy in the 1990s.* London: Macmillan.
Haigh, N. (1984) *EEC Environmental Policy and Britain.* London: Environmental Data Services.
Hamer, M. (1987) *Wheels Within Wheels.* London: RKP.
Hawkins, K. (1984) *Environment and Enforcement.* Oxford: Clarendon Press.
HM Government (1994a) *Biodiversity: the UK Action Plan.* London: HMSO.
HM Government (1994b) *Sustainable Development: the UK Strategy.* London: HMSO.
Liefferink, D., Lowe, P. and Mol, T. (eds) (1993) *European Integration and Environmental Policy.* London: Belhaven.
Lowe, P. and Goyder, J. (1983) *Environmental Groups in Politics.* London: Macmillan.
Lowe, P., Clark, J., Seymour, S. and Ward, N. (1997) *Moralising the Environment.* London: UCL.
Lowe, P., Cox, G., O'Riordan, T., MacEwen, M. and Winter, M. (1986) *Countryside Conflicts: The Politics of Farming, Forestry and Conservation.* Aldershot: Gower.
Lowe, P. and Ward, S. (eds.) (1998) *British Environmental Policy and Europe.* London: Routledge.
Lowi, T. (1964) 'American Business, Public Policy, Case Studies and Political Theory', *World Politics* 16: 677–715.

Macrory, R. (1986) *Environmental Policy in Britain: Reaffirmation or Reform?* Berlin: International Institute for Environment and Society.

Marsden, T., Murdoch, J., Lowe, P., Munton, R. and Flynn, A. (1993) *Constructing the Countryside.* London: UCL Press.

McCormick, J. (1991) *British Politics and the Environment.* London: Earthscan.

Rawcliffe, P. (1995) 'Making Inroads: Transport Policy and the British Environmental Movement', *Environment* 37, April:16–20, 29–36.

Richardson, G. *et al* (1982) *Policing Pollution.* Oxford: Clarendon Press.

Richardson, J., Gustafsson, G. and Jordan, G. (1982) 'The Concept of Policy Style', in J. Richardson (ed.) *Policy Styles in Western Europe.* London: Allen & Unwin.

Richardson, J. and Watts, N. (1985) *National Policy Styles and the Environment.* Berlin: International Institute for Environment and Society.

Smith, M. (1989) 'Changing Agendas and Political Communities: Agricultural Issues in the 1930s and 1980s', *Public Administration* 67: 149–65.

Smith, M. (1991) 'From Policy Community to Issue Network: Salmonella in Eggs and the New Politics of Food', *Public Administration* 69: 235–55.

Vogel, D. (1986) *National Styles of Regulation.* Ithaca: Cornell University Press.

Ward H. and Samways, D. (1992) 'Environmental Policy', in D. Marsh and R. Rhodes *Implementing Thatcherite Policies.* Buckingham: Open University Press, pp. 117–36.

Ward, S. (1995) 'The Politics of Mutual Attraction? UK Local Authorities and the Europeanisation of Environmental Policy', in T. Gray (ed) *UK Environmental Policy in the 1990s.* London: Macmillan.

Weale, A., O'Riordan, T. and Kramme, L. (1991) *Controlling Pollution in the Round.* London: Anglo-German Foundation.

Chapter 3

Denmark: Consensus seeking and decentralisation

Mikael Skou Andersen, Peter Munk Christiansen and Søren Winter

Introduction

In terms of natural endowment, Denmark differs significantly from the other Scandinavian countries. Denmark is a small country, around 43,000 sq km, most of which is cultivated. Denmark has more than 500 islands and a coastline of about 7,000 km. The population is 5.2 million. Like the other Scandinavian countries, Denmark was a late industrialist, but unlike Norway and, in particular, Sweden, Denmark has only few large industrial plants and almost no heavy industry. Large-scale industrial pollution has thus been less of a problem compared to other industrial countries, although of course Denmark has also experienced heavy air, soil and water pollution. The manufacturing sector is relatively dominated by food processing industries which produce large amounts of pollution in the form of waste water. The many bays and fjords have been vulnerable to municipal and industrial discharges, as well as to agricultural run-off. The shallow waters that surround Denmark and its many islands are more vulnerable to nutrient leaching than coastal waters in general.

As a consequence, Denmark's environmental policy has, compared with other countries, to a considerable degree focused on the aquatic environment. Water quality policy was one of the most important concerns of the Pollution Council in the late 1960s (Forureningsrådet 1971). During the last decade it has become apparent that agriculture, and in particular livestock farming, is the most important source of pollution of the aquatic environment.

Denmark is only moderately affected by transfrontier pollution. Thanks to the lime layers settled in Denmark's underground, acid deposits are neutralised, and Denmark is in fact itself a net exporter of air pollution, due to its reliance on coal for energy supply. With regard to water quality, the interior Danish sea waters (the Belts and Kattegat) have predominantly been affected by point and non-point sources from Denmark itself. The water exchange between the Baltic Sea and the interior Danish waters is slow, and influx of transfrontier marine pollution comes mainly from the North Sea, although scientific uncertainties remain. Nutrient run-off from Danish agriculture accounts for two-thirds of the total run-off from the Nordic countries (Bernes 1993).

Denmark was among the North European countries who responded to the environmental challenges in the late 1960s and early 1970s. Although the environmental challenges were new, the Danish responses during this early period to a large extent reflected traditional traits of Danish politics as well as administrative traditions. A number of these traits are still visible in Danish environmental policy. Policy has tended to rely on broad framework laws, the implementation of which has been somewhat problematic. Standards are set in negotiations between the environmental bureaucracy and interest organisations, and local governments have substantial discretion in implementing environmental laws and decrees. New policies have been developed in the 1990s, but traits of 'path dependence' are also visible.

This chapter presents a brief overview of environmental policy and its administration in Denmark. Firstly, we briefly review the historical development of 'modern' environmental policy. Secondly, we present some main trends in recent environmental policies. The third part of the chapter assesses the actors and interests involved in environmental policies and their implications for policy formulation. Fourth, we focus on the basic traits of Danish environmental administration.

The historical development of Danish environmental policy

Danish environmental policy has antecedents in the nineteenth century, but it was not until after the Second World War that environmental policy slowly gained momentum, and a coherent policy was not developed until the early 1970s.

In response to the cholera epidemics of the 1850s, parliament drafted various laws on hygiene, and the first regulations to restrict industrial pollution appeared in the 1880s. In 1907, the Sewer Act called for the extension of public sewerage, and in 1926 the Water Supply Act regulated the provision of drinking water. Pollution was indeed causing concern, but in spite of the establishment of various commissions in the 1930s when other welfare regulations were formed, it was only in 1949, when the Water Course Act was passed, that the first real policy instruments to control pollution became available. The Water Course Act established the first licensing scheme for industrial pollution and mandated local agricultural commissions to license dischargers. However, until the beginning of the 1970s there was no coherent environmental policy in Denmark. Regulation was very fragmented and merely a by-product of health legislation. The environmental conception of the legislation was rather narrow with a strong focus on neighbour relations to ensure that firms and farms did not pollute their closest neighbours (Christensen 1987).

The Ministry of Pollution Control was established in 1971 and was replaced by the enlarged Ministry of the Environment in 1973. The first comprehensive piece of legislation, the Environmental Protection Act, was passed in 1973. The organisation of environmental policy was characterised by two central traits. The first was a sectoral principle. At the central level, the main environmental responsibilities and competencies rested with the ministry. At the regional and local levels, the sectoral principle was followed less clearly since the scarce administrative resources were integrated in the boards and administrations dealing with 'environmental and technical matters'.

Over the last decade, the organisation of environmental policies has slowly but steadily challenged the dominant sectoral principle. Environmental issues have increasingly occupied other ministries and they have been responsible for implementing various measures. The second trait was a profound decentralised political and administrative structure, not found elsewhere in Scandinavia (Christiansen and Lundqvist 1996). Implementation of environmental policy rested with municipalities and counties, and the Minister of the Environment and the two associations organising counties and municipalities agreed that the main instrument of the central administration should be optional guidelines as opposed to binding statutory orders. Within a number of core areas this decentralised model, leaving the decentralised level with significant discretionary powers, is still practised, although local resources have increased and measures have been introduced to increase compliance with environmental regulations (Christiansen 1996).

The act introduced the so-called *balancing principle*, which required that the costs of pollution control be taken into consideration when requirements to industries were set. At the same time a 'Polluter pays principle', although weak, was set up. Furthermore, an Environmental Board of Appeal was set up. The Board may be seen as the 'supreme court' of environmental policy, as it is able to review complaints from industries that disagree with the conditions set up in their permits. Although headed by a judge, the Board consists of representatives appointed by trade organisations and the Environmental Protection Agency (Basse 1987). In the early 1980s, environmental organisations were also allowed to make appeals to the Board, but they have not yet been allowed to nominate experts for the Board.

In spite of several amendments to the 1973 Act during the 1980s, its basic principles have only been modestly altered. The 'balancing principle' was removed in 1986 to underline environmental protection as the primary objective. However, the change was not as dramatic as it might seem, as a principle of due proportionality is still maintained. A reform in 1991 broadened the scope of policy instruments available for pollution control, putting more emphasis on voluntary agreements and on economic instruments. The 1991 reform also aimed at simplifying the number of pollution control laws by integrating more than 20 different pieces of environmental legislation into four main laws.

With regard to the content of environmental policy, its early focus was primarily on industrial and other point-source pollution activities. The act was carried through by a coalition between Social Democrats and the Conservatives and was even endorsed by the Danish Federation of Industries. During the 1970s the primary concern was water pollution and more than half of both public and private investments were directed to this sector. Public sewage plants were extended and extensive water quality planning was carried out by local and regional authorities. In the late 1970s the environmental debate focused more on the question of nuclear energy. Strong political and popular opposition to the introduction of nuclear energy culminated in the rejection of nuclear power. As a result, Denmark developed an energy policy that on the one hand promoted energy savings and the development of alternative energy sources, such as wind power, and on the other hand relied heavily on coal and later on natural gas.

In the 1980s concern for environmental issues increased again. It was now the plural sources of pollution that came into focus, and attention was turned to a much more diverse spectrum of environmental problems. It was revealed how agriculture's contribution to eutrophication exceeded that of industry, and agriculture became the object of more intensive regulation. Acid rain, hazardous waste sites, ground-water contamination and a number of other issues entered the political and administrative agendas. In the late 1980s and early 1990s environmental concern also shifted towards global and inter-generational environmental issues, such as the greenhouse effect and the destruction of the ozone layer. As a consequence, the focus on solutions also shifted towards international fora and the relations between national and international environmental goals.

Although popular attention and concern clearly peaked in the late 1980s, public concern for the environment has been quite significant since the mid-1980s. At the parliamentary level, party competition was very high in the mid-1980s, not least due to the establishment of an 'alternative' majority in the *Folketing* (the parliament) which constantly pressured the Cabinet into introducing regulations that they would not otherwise have pursued. The high level of party competition in environmental matters and the capacity to respond to popular demands is one of the explanations of why Denmark has not had a successful green party.

Recent trends in Danish environmental policy

The 1990s brought some major developments in Danish environmental policy. New goals have been set, new instruments developed, and internationalisation has had a dynamic effect on national policies. However, it should also be noted that Danish environmental policies have clear elements of continuity. Some of the basic traits, such as decentralisation and consensus seeking, often survives when new initiatives are taken. Thus, there is a clear 'path dependence' also in environmental policy (North 1990).

New instruments have gradually been introduced since the mid-1980s. A couple of breaks in this process can be identified. In the late 1980s steps were taken for a major reform of environmental policy. Policies were to be based on simpler rules and economic instruments (The Ministry of Environment 1988). The reform carried through in 1991 was significantly less ambitious, but it did introduce new instruments. Another major step was the tax reform of 1993, which had as one of its basic principles the slight shifting of the tax burden from personal taxes towards taxes on consumption of natural resources. A third element has been the increasing EU regulation of environmental matters and other international initiatives.

To a large extent, Denmark still relies on traditional legal regulations. Some development of this instrument has occurred, such as the right of the authorities to demand that the least polluting technology be used. More spectacular is, however, the introduction of new instruments. In the second half of the 1980s, a number of taxes on toxic substances were introduced. The breakthrough came with the 1993 tax reform. From 1994 through 1998, taxes on water, non-renewable resources, and a number of other items will be gradually raised. It is difficult to calculate the amount of 'pure' green taxes. Denmark has had high taxes on motor vehicles and non-renewable energy sources for many years. They were originally introduced mainly for fiscal reasons. The government calculates environmental taxes at a level of DKK 36 billion in 1994. This includes all taxes having some relation to the environment, including motor vehicles and fuel (Christiansen 1996).

Traditionally, industry has been exempted from energy taxes due to their negative effects on international competitiveness. However, in 1991 a CO_2 tax was introduced which is paid by consumers as well as firms. The tax met with considerable opposition from business. In 1995, some additional environmental taxes were imposed on the business sector, but at a significantly reduced level compared to the original intentions of the government. Furthermore, a number of exemptory rules were allowed most of which require extensive bureaucratic control. The Danish experience as regards the use of economic instruments thus points to two major obstacles. Despite the widespread recommendations of economic instruments, they seem to be very difficult to introduce in the business sector. The second problem seems to be that when economic instruments are introduced, they do not live up to the expected quality of being administratively simple.

Voluntary agreements are another new instrument. The basic idea is quite similar to that of using economic instruments in terms of cost effectiveness. Under voluntary agreements a given goal is expected to be obtained at lower costs when the affected agents are free to pursue the goal in the way they want. As rational agents they are supposed to pursue the most cost-effective strategy in order to realise the politically set goal. Voluntary agreements, however, pose a potential enforcement problem: in cases of non-compliance, who is doing what and with what kind of authority? So far, environmental agreements concerning the manufacturing sector have been introduced in areas such as the use of PVC, emissions of NO_x from power plants, and the use of straw in power plants. In addition to the potential cost effectiveness of voluntary agreements, they also have desirable political values. The level of political conflict might be reduced by directly involving the affected trades. The Danish experience so far points to these political values as being more important than the 'technical' advantages of voluntary instruments (Christiansen 1996:43).

Besides the use of new policy instruments, environmental policy will move more in the direction of products and product regulation. The role of environmental labelling, such as the 'Ø' (O) seal for organic products, is likely to increase. Further national initiatives may come to supplement the EU labelling system and the Nordic Swan symbol. Life-cycle assessment, environmental management and auditing schemes and other measures which will allow not only consumers but also professionalised market actors to differentiate the quality of industrial products are likely to increase in significance.

The organisation of environmental policy has gradually changed from being primarily a sectoral matter, that is, an exclusive policy area, into an *inclusive* one in which the sectoral principle is less dominating. Environmental policy in the mid-1990s has clearly been affected by the advent of other ministries as players in environmental policies. The most obvious examples are agriculture, energy and transportation policies (Christiansen 1996). In 1993 it was decided that all new (relevant) laws should be assessed according to their potential environmental impacts and consequences.

Another significant development is the gradual internationalisation of environmental policies, which dates back to the early 1970s, that gained momentum from the mid-1980s. The national as well as international debates increasingly focused on transboundary and even global environmental problems, such as the debate on the greenhouse problem and the destruction of the ozone layer. It was recognised that a number of serious environmental problems can not be solved without international cooperation.

From an institutional perspective, the most important development has been the formal inclusion of environmental policy into the EU treaty. Denmark belongs to the group of EU members with well developed domestic environmental policies. Since the EU has taken over a number of earlier national competencies, not least in the area of product regulations, the Europeanisation of environmental policies poses a potential problem for Denmark. Denmark managed to get an environmental guarantee included in the treaty. According to the guarantee, national environmental regulations can be stricter and tougher than the regulations agreed upon in the Community.

In other international fora Denmark has pursued a pro-environmental course together with a number of other Northern European countries. Empty rhetoric is often used in such international fora. However, in a number of areas Denmark has actually set in motion national policies in order to fulfil some of the commitments agreed upon in international fora. For example, emissions of CO_2 are to be reduced by 20 per cent by 2005 compared to the

1988 level. Even though a number of measures have been taken, it is also clear that Denmark will have some problems meeting this target.

A number of significant changes in environmental policy have taken place during the last decade, mainly in the 1990s. At the symbolic level these changes may be summarised by key words such as sustainability. To the extent that it is possible to talk of a general policy strategy, it consists of a broadening of the scope and intensity of environmental policy, targeting towards a number of specific problems, and the exploitation of new instruments. However, one should not exaggerate the scope of changes that have taken place. Some of the basic choices made in the early 1970s are clearly recognisable in present policies. To understand the balance between continuity and change, we now turn to the arenas of environmental policy making.

Environmental politics: actors, interests and policy formulation

According to Wilson (1980), most environmental regulation is characterised by concentrated costs for polluting firms and diffuse and scattered benefits for citizens in general. Firms thus have strong incentives to organise and to resist regulation, while the environmental protection interest will be difficult to mobilise to such an extent that polluters are outweighed. Wilson hypothesised that in such a configuration of interests, regulation is unlikely to be introduced unless an entrepreneur is able to mobilise third parties, such as the media and public opinion, among the representatives of environmental interests.

This way of reasoning implies that the environmental regulation output can hardly be explained or predicted by looking at the simple distribution of costs and benefits of policy outputs. A study of the policy-making processes, their actors and interests, is also necessary. This reveals that public opinion has certainly been very important for the development of Danish environmental regulation and that the Danish manufacturing sector did not resist environmental regulation as strongly as expected according to Wilson's predictions. As mentioned above, the Danish Federation of Industries actually endorsed the first Environmental Protection Act. Danish environmental regulation of agriculture is better explained by Wilson's model.

Public opinion and the media

The environment did not become a political issue in Denmark until the late 1960s. Media coverage of serious pollution, as well as the student rebellion in 1968, probably drew the public's attention to the pollution issue. The political parties perceived that votes could be gained through environmental policy initiatives. The first environmental reform in 1972–73 had substantial media coverage, but considerably less than environmental media coverage in the mid- and late 1980s, and much less than the coverage of the negotiations on the housing and defence reforms at the same time in 1973 (Winter 1975). However, environmental concern was strong enough to push the political parties into putting an environmental reform on the agenda.

The environmental issue attention cycle has not followed the predictions of Downs (1972), according to which the environmental issue was likely to reach its peak relatively quickly followed by a gradual decline in public interest. On the contrary, public interest in the environment continued to grow in the 1970s and 1980s. In 1980, 21 per cent of the voters ranked environment protection as one of the top ten political problems, but this share had grown to 68 per cent in 1987. In fact, in 1985 Danish environmental consciousness seemed to be substantially higher than in Sweden, Finland and Norway (in that rank) measured as the percentage of voters assigning a higher priority to the protection of the environment than to economic growth (Andersen 1990:189–91). Public awareness of environmental problems decreased after the peak years in the mid-1980s. The share of the population that worried about pollution fell from its peak of 77 per cent in 1986 to 55 per cent in 1995 (Thulstrup 1996).

Interest organisations

One of the key elements of Danish environmental policy making and implementation is integration of affected interests in decision-making processes concerning environmental policies. The Danish manufacturing sector did not completely oppose the environmental reform proposal when it appeared in 1972. It was important for the Social Democratic minority government to reach some understanding with the trade organisations and especially the manufacturing sector and to depoliticise the environmental issue. Therefore, the government entered into negotiations with the Federation of Danish Industries. This inclusion of industrial interests mirrors a core trait of Danish policy making. It is a general norm that affected interests are integrated into political as well as administrative decision making when organised interests are affected by government intervention (Christensen and Christiansen 1992). Such negotiations and compromises are expected to facilitate not only the approval of a bill in parliament but also its implementation. Thus, Danish policy-making style including in environmental policy is consultative and consensus-oriented (Christiansen 1996:57ff.).

However, manufacturing and agriculture adopted very different strategies. Industrial pollution dominated the public debate at that time, and the Federation of Danish Industries anticipated the strengthening of pollution control. The strategy pursued by the Federation was effective. The Federation obtained substantial concessions in relation to both substance and form. Such concessions include the balancing and protection principles and the promise for financial support for environmental investments. More importantly, however, the Federation of Danish Industries had the bill changed to secure its influence in the implementation phase (Winter 1990; Knoepfel and Weidner 1983; Andersen 1989; Moe 1989). The Federation further secured that all administrative regulations, decrees and guidelines must be negotiated with the relevant trade organisations, and that the final administrative appeal be moved from the ministry to the Environmental Board of Appeal which would have representation from the manufacturing sector and agriculture. In Denmark the 1968 local government reform made local government the natural provider of services and regulation at the local level, and the Association of Municipalities was strong enough to ensure that the Ministry of the Environment would mainly use optional guidelines in accordance to municipal permit procedures and the like.

Agricultural interest organisations chose another strategy. Agriculture was not so much in focus when the environmental reform was discussed in 1972–73. Only a few farms had to obtain the environmental approvals which were mandatory for the highly polluting firms. Agriculture was only to be regulated by general guidelines. The agricultural organisations supported the concessions obtained by the Federation of Danish Industries, but did not endorse the final bill because the Federation disagreed with the 'polluter pays principle' and wanted compensation for environmental demands and investments. The agricultural organisations were even able to obtain a tacit concession: the environmental reform only applied to agricultural activities to a limited extent. Apparently, agriculture has obtained a much milder regulation by not entering into the same kind of negotiations and compromising as the manufacturing sector did.

However, in the mid-1980s, focus was again put on agricultural pollution of the aquatic environment. The agricultural interest organisations entered negotiations to meet the demands for further regulations, but their negotiation style was still much more conflictual than that of the Federation of Danish Industries. The defensive agricultural sector had enough political strength to avoid the most radical regulatory measures, implying that the means or measures adopted could not achieve the specified environmental objectives. Therefore, to a considerable extent, the plan has a symbolic character (Andersen and Hansen 1991). Since the adoption of the Plan for the Aquatic Environment in 1987, the environmental regulation of agriculture has followed a more consultative and negotiating style.

It is more difficult to assess the strategies of environmental organisations. There are a number of large and small environmental organisations pursuing different objectives and strategies. Their strategies are based on the exploitation of the media and the popular commitment to environmental goals as well as on direct negotiations with bureaucrats and politicians. The role of environmental organisations has changed over time. At the time of the first environmental reform in 1972–73, a few green organisations including NOAH (the Danish section of Friends of the Earth) tried to influence public opinion, but they were not invited and did not want to participate in the negotiation process. Later on a number of the 'activist' organisations, such as Greenpeace, have entered into negotiations with the ministry. These organisations typically try to pursue a double strategy: an 'activist' strategy in which the media are exploited through spectacular actions and a more traditional 'corporatist' strategy. During the 1980s and 1990s, a number of environmental organisations obtained the right to file complaints against administrative decisions, and they are listed as organisations with whom the ministry has to negotiate when introducing new regulatory measures.

The most important political role has been played by the Danish Conservation Society which was quite influential in 1986 in mobilising the media, public opinion and the political parties to demand more regulation and huge investments to protect the aquatic environment. The society functioned as a 'political entrepreneur' (Wilson 1980; Kingdon 1984) which was not only able to set the agenda, but also to define the objectives of the Aquatic Environment Plan. The Society has clearly lost momentum in the 1990s, maybe as a consequence of 'overdoing' their strategy against pollution from agriculture. The Society has also lost a significant number of members.

The political parties and parliamentary coalitions

When the political parties started paying attention to the environment, it was probably also because they had hoped to win votes on a popular issue. The initiative to form the Ministry of Environment and to create the first comprehensive act was taken by the Social Democrats. The Social Democratic minority Cabinet preferred to depoliticise the environmental issue by compromising with the manufacturing sector and one or more of the parties of the Right. The compromise with the Federation of Danish Industries made it rather easy for the Conservative Party, which has close connections to the manufacturing sector, to compromise with the Social Democratic government (Winter 1975). The Liberal Party (*Venstre*), with close relations to agriculture, followed the farmers' interest organisations in not recommending the compromise.

When in the 1980s the environment became an even more popular issue and a 'green party' was represented in some local government councils, the other political parties started competing in order to appear to be the most 'green'. In particular, the Socialist People's Party, the small Social Liberal Party (*Radikale Venstre*), and the Christian People's Party were eager to create a green profile. So was the Social Democratic Party, especially after losing power in 1982. During the Conservative-Liberal coalition Cabinets from 1982 through 1992, the Social Liberals supported the government in general and especially in economic questions, but they formed several alternative majorities in parliament on environmental (and security policy) matters. The threat of these majorities and the increased competition for votes made the government more willing to compromise, and the result was a number of acts that strengthened the environmental policy in the 1980s. The party competition on environmental issues continued after the Social Democratic-led Cabinets took over power in 1993.

Parliament has traditionally played a role in the administration of environmental policy. This has mainly to do with the position of the permanent parliamentary Environmental and Planning Committee. There is a tradition that the committee discusses with the minister different administrative measures. In the mid-1980s, the level of activity in the committee as well as in the plenary session was very high. Although the level of activity was reduced after 1988, it is still significantly higher compared to the 1970s (Christiansen 1996:40ff.)

Bureaucracy

The Ministry of the Environment plays an important role in the environmental negotiation system. The power of the bureaucracy depends both on the minister in charge and the parliamentary situation. The bureaucracy was reasonably influential when in 1982–87 and in 1990–93 an alternative green majority existed in parliament (see below). Especially in the first period, zealous officials forwarded proposals that were more 'green' than the Conservative–Liberal coalition government would be expected to accept. The Minister of the Environment from the small Christian People's Party found a platform to attract green votes and even seemed to be satisfied in cases when the green alternative made the government's proposals greener by using proposals from the Ministry of the Environment as private bills (Pedersen and Geckler 1987). Especially in the 1980s, the Ministry of the Environment had many conflicts with the Ministry of Agriculture, whose monopoly on agricultural matters was broken. The conflicts between the two ministries later developed into a strong conflict that split the political parties in the mid-1980s (Christensen 1987).

In concluding so far, public opinion has had a considerable though diffuse impact on environmental policy making in Denmark through the attempts of the political parties to attract voters. The first environmental reform in 1973 was rather depoliticised. The consolidation and implementation of this reform was characterised by consultation and negotiations between the Ministry of the Environment and the affected trades. In the middle of the 1980s, however, popular support for environmental protection and party competition revitalised the parliamentary arena and somewhat reduced the role of the corporatist arenas, although the consultative and consensus-oriented Danish policy style was not abolished (Christensen 1987; Andersen and Hansen 1991; Christiansen 1996).

Danish environmental administration: consultation and decentralisation[1]

The institutionalisation of Danish environmental policy proceeded in the early 1970s in the footsteps of, and under the influence of, the Swedish experience, but the structure of environmental administration turned out remarkably different from Sweden's, which had vested powers in a strong and centralised national authority. Danish environmental policy became profoundly decentralised due to conventional anti-state sentiments (rooted in the Danish farmers' movement, and reinforced in the aftermath of 1968) and has been a peculiar blend of traditional peak-level neo-corporatism and decentralisation. The 1973 Environmental Protection Act set no ultimate targets for environmental standards, but created a complicated system of negotiation among various actors.

The central level

Within the framework acts, parliament has vested the Minister of the Environment with a number of competencies. Danish ministries have two levels. The Department is considered to be the secretariat of the minister. The Department of the Environment employs around 80 persons, of which the majority holds a degree in law, economics or political science. The second level is the agency to which the minister typically delegates the competence to set guidelines and standards. The Environmental Protection Agency employs around 175 persons, of which a majority have a degree in the natural or technical sciences. Despite its administrative status and its technical staff, the agency has in some periods played a significant political role. Besides negotiating and setting rules, guidelines, etc., the agency also supervises the decentralised levels, but it has few hierarchical powers.

The central authorities issue optional guidelines on most of the tasks carried out by the local authorities. Furthermore, the Environmental Protection Agency reviews complaints over decisions taken by the local authorities on a case-by-case basis. The bureaucrats of the central environmental administration participate in a negotiation system of their own. Every guideline issued under the law is negotiated with the relevant interest organisations. With the 1991 reform the negotiation principle has been extended with the competence to enter 'voluntary agreements' with individual branches of industry.

Negotiations between the Environmental Protection Agency and the peak-level interest organisations about the exact guidelines have been a time-consuming affair. After the three basic guidelines on water, air and noise had been agreed upon in 1974, the manufacturing sector began to drag its feet. The Environmental Protection Agency was able to negotiate a number of branch guidelines with various branch organisations, but with the central guideline a revision was a lengthy matter. The revision of the guidelines on air pollution was put on the agenda several times, and when negotiations were finally initiated, it took several years before an agreement was reached. The corporatist traits of environmental policy might thus enhance consensus and legitimacy, but they also imply a strong conservationist element.

The negotiation system also includes the implementation process. The Environmental Board of Appeals seems to be unique, at least in the EU.[2] Besides its corporatist traits, the Board serves to protect the Minister of the Environment against highly politicised cases ending up on his desk. Although the Board rules on the basis of existing legislation and available scientific evidence, most environmental issues are marked by the typical 'crisis of proof' situation and are open for interpretation. Environmental organisations have not been allowed to appoint experts to the Board, and it was not until 1984 that they were given the right to complain. The Board to some extent disciplined the Environmental Protection Agency, since local decisions brought to the agency could be subject to a final review by the Board. Furthermore, since the procedure of the Board was often very lengthy, some companies began to speculate in complaining to the Board, simply to achieve another year or two's postponement in the demands set by local authorities.

In sum, the consultative procedure has proved complicated in practice. The consensus achieved between environmental authorities and the peak trade organisations has made resistance from the regulated less likely and has possibly enabled a smoother implementation of the guidelines. In retrospect, optimism was too high on behalf of decentralisation, and there was too much confidence in the consultation process. The act was passed just as the economic crisis, with its economic stagnation, replaced the previous decade's growth. Considering the character of this period, less faith in the consultation procedures would probably have been appropriate.

The decentralised levels

Denmark is a unitary state with two levels of decentralised authorities; at the local level, 275 municipalities (*kommuner*) and at the regional level 14 counties (*amtskommuner*). Two of the municipalities also have the status of a county (Copenhagen and Frederiksberg). The decentralisation principle regarding municipalities and counties is prescribed in the constitution. Within frameworks set by laws the popularly elected County Councils and Municipal Councils have the discretionary power to arrange their duties the way they prefer. The main duties of the decentralised levels are the provision of the major part of public services. Among the other tasks of counties and municipalities is the administration of environmental policy. It is generally recognised that the level of decentralisation of the Danish public sector is comparatively very high (Goldsmith 1992).

The 1973 Environmental Protection Act was passed under the influence of the coinciding reform of the structure of local authorities. Key elements of the reform were to make expanded local and regional entities more effective and to have them take over a number of tasks traditionally administered by the state or state-affiliated institutions. Within the framework of the Environmental Protection Act, authority was delegated to the local and regional authorities in virtually all practical matters.

The municipalities were made responsible for a number of tasks, such as the provision of public sewage treatment, waste and recycling, the granting of permits to polluting firms, and the general surveillance and inspection of the local environment. The Environmental Protection Act was based on an environmental quality principle that did not set specific standards to polluters. The logic of the decentralisation principle was to let the municipalities set the demands according to the capacity of local environmental quality, the preferences of their citizens and the economic situation of the firm in question.

The counties have since the early 1980s come to play a more important role; partly in response to the steadily growing list of issues to be managed, partly in response to the municipalities' lack of ability to fulfil their role. Among the tasks carried out by the counties are ground-water administration and protection, clean-up of hazardous waste sites, air pollution regulation and surveillance of surface waters, including permits to direct dischargers of waste water.

The implementation deficit

Ambitions were high as the Danish parliament passed the Environmental Protection Act in 1973. A common phrase among politicians was that Denmark possessed 'the world's best environmental protection act'. Fifteen years later, however, there was a widespread perception of failure, clearly expressed in the turmoil which led to the Plan for the Aquatic Environment. Surveys on the local environmental administrations were carried out in the mid-1980s, and it was found that the municipal environmental departments were typically understaffed, while the staff at hand did not have sufficient resources and education. For many municipalities, especially those with less than 10,000 inhabitants, the environmental staff consisted of a single part-time official (The Environmental Protection Agency 1985). This implementation deficit was related to the lack of political will and ability of the municipalities to meet environmental requirements, but it was also related to the administrative overload on local authorities, whose resources did not match the complicated guidelines drafted by the Environmental Protection Agency. Furthermore, it was related to the inability of the central actors behind the law to swiftly conclude agreements on new technological standards.

In practice, the authorities worked on a first-come-first-served basis. Firms received a license if they applied, but firms that did not apply were often not tracked, simply because the environmental authorities did not even possess a list of the relevant firms to control (Schroll 1985). The first national account of environmental licenses was presented in 1988 and showed that 15 years after the passage of the Environmental Protection Act, only 50 per cent of the Danish firms had obtained a license (The Environmental Protection Agency 1988). Legally, a license is a requirement to new firms, whereas existing firms only have to obtain one in case of changes or extensions in production. A survey showed that, on average, companies were visited and controlled only once every two to three years. Only the larger towns had enough personnel to meet the many requirements of the decentralisation principle. Many smaller municipalities had a rather pragmatic approach to the guidelines of the central authorities, as the following answers in a survey of municipal implementation of the control provisions revealed (DIOS 1987):

> If we had to read everything we receive from the Environmental Protection Agency, we would have to employ one person for reading only. He could not be interrupted with questions, and then the reading wouldn't matter anyway.

> Yes, we place it directly on the shelves, otherwise we would drown. We are not able to make use of everything they produce.

The local environmental inspectors were furthermore under close political supervision from the Municipal Council and its Technical Committee. Any legal step to assure compliance with licenses had to be approved by the politicians, and these were often softer in their approach, especially when local jobs were at stake.

The problem concerning the implementation deficit of the municipalities was in fact dual. Political will to support the intentions of the Environmental Protection Act had been lacking. Thus, the consensual policy style implied that ambitious environmental goals were to be balanced with other political aims of the local authorities (tax revenue, jobs). But as the interest in paying attention to the initial aim of the environmental protection act began to increase among municipalities, they found themselves confronted with a rather rigid regulatory scheme that required extensive planning and control, where even small changes in the production processes triggered the demand for new costly licensing procedures.

Due to some of the flaws in the regional and, in particular, local implementation of environmental policies, parliament and the Ministry of the Environment has pushed for improvements in the capacity of the decentralised levels to implement environmental regulation. From 1984 to 1993 the number of man-years committed to inspection and control were increased from 340 to 761 (Christiansen 1996:56). Increased popular attention to the implementation problems of the decentralised levels also has pressed for a reduction of these problems. The implementation deficit has also been reduced through the introduction of new instruments such as environmental taxes, increasing focus on products regulations and the like.

In general, Danish environmental administration is characterised by a consensus-seeking policy style and by a high degree of adaptation to local preferences. The principal role of the local authorities allows for adaptation to local environmental problems and preferences concerning the choice of solutions, and even the national system of standard setting is penetrated by a balancing of requirements with the acceptance of those being regulated. The development of new instruments and the increased EU influence on Danish environmental policy have pushed the overall balance of powers and competencies in favour of the central level and has challenged the consultative style. However, the emphasis on decentralisation and consensus-seeking is still an important part of Danish environmental policy.

Conclusions

Environmental policy and politics reflect a strong Danish tradition of a consultative policy-making style which involves affected interest organisations in policy formulation and implementation. The Federation of Danish Industries and the two organisations of municipalities and counties were involved in the compromise on the environmental reform in 1973. These organisations had considerable influence on the reform, later environmental laws and the implementation of environmental policies.

Environmental law has traditionally been more form than content in Denmark in the sense that environmental laws are framework laws with very broad principles which must be operationalised in administrative regulations which are always negotiated with the affected interest organisations.

Danish environmental policy is very decentralised compared to other countries. Municipalities and counties are responsible for inspection and issuing permits, and the *de facto* authority of the Environmental Protection Agency (EPA) to issue binding rules rather than guidelines for local governments is quite limited. One of the results of this decentralisation is a large variation in implementation among local governments.

While the environment only had a moderate saliency in the public opinion and the media in 1973, public attention grew in the 1980s to reach a peak in 1987 when it became one of the most important policy problems. Since then, public attention has declined but is still very high. Public opinion had a strong indirect impact on environmental policy making by promoting vigorous competition among the political parties which were trying to appear the most green. This strategy was very important in preventing a green party from entering parliament. The competition was reinforced by a peculiar parliamentary situation in the mid-1980s when an 'alternative green majority' was able to pass environmental laws against the will of the Conservative-Liberal coalition government.

The stronger attention of parliament to environmental policy implied a minor reduction in the influence of the interest organisations of the manufacturing sector and agriculture. In addition, the power of the local governments was reduced somewhat by increasing central control through more intense standard setting.

At the central level, a sectoral principle was followed by which the Ministry and its Agency were made responsible for issuing guidelines and co-ordinating policies. Dominated by specialists, the new EPA became a strong green policy advocate. However, more responsibility for environmental protection has been placed with other sector ministries in order to integrate environmental concerns with other policy concerns. This left the Ministry with more of a cross-sectoral co-ordinating role similar to that of the Ministry of Finance, but without the strong power of the latter.

The EU represents another institutional challenge to Danish environmental policy making. The creation of the Single Market and the policies agreed upon in the Maastricht Treaty are pushing and will further push the formulation of environmental policies towards EU standards. This trend offers interesting developments. Firstly, the EU's regulatory style seems to be different from Denmark's. A much more detailed style of regulation will clash with the tradition for framework laws and for the open implementation process. This clash will, however, be reduced by the national implementation of EU directives. Secondly, the political decision-making structure will change. The EU system is based less on institutionalised consultations than the Danish decision-making system. Interest organisations are potentially weakened in the decision-making process, but they seem to have maintained their position at the national level.

A number of significant changes in environmental policy have taken place during the last decade. It is difficult to detect a clear policy strategy, but a number of changes are clear. The development of new instruments includes a changing mix of traditional legal instruments and instruments relying on the 'logic of the market', environmental policy increasingly focuses on products and product regulation, more objects are the target of environmental measures and policies are increasingly formulated and implemented in other parts of the political and administrative apparatus than the exclusively environmental sector.

What then is the impact of environmental policy in Denmark? The evidence is inconclusive. Among the success stories has been the strong reduction in waste water pollution from industry and households. Also air pollution from SO_2 has been reduced by one half from 1972 to 1992 (Ministry of Environment 1995:19; Andersen 1995), while the emission of NO_x and NH_3 has decreased only marginally (Ministry of Environment 1995:19). The emission of CO_2 has been stable in spite of the increases in transport, production and energy use (Andersen 1995). The Ministry of the Environment concluded in a report in 1995 that the emissions from industries to air, the aquatic environment and the ground are by and large under control (Ministry of Environment 1995:389). Energy (Andersen 1995) and water consumption have decreased in private house-holds, and unleaded petrol has gained 75 per cent of the market (Moe 1993).

However, there are also examples of deteriorating environmental problems and policies which have not met the targets. There are still some industrial pollution problems. More than 10,000 severely polluted sites have been detected while, until 1993, only about 600 had been cleaned up. There are still problems in reducing the quantity and toxicity of industrial waste and the industrial consumption of energy and raw material, as well as the industrial use and sale of products which harm the environment (Ministry of Environment 1995:21, 389). Agriculture's nitrate pollution of the aquatic environment has only been marginally reduced, although it was decided in 1987 to reduce this pollution to one half before 1993 (Andersen 1995; Ministry of Environment

1995). Pesticides have been found in more than ten per cent of the tests taken from water drilling (Ministry of Environment 1995:21).

Available evidence is too sparse to reach any firm conclusions on how environmental outcomes can be explained by the policy-making and implementation processes. Some hypotheses can, however, be mentioned. In some environmental areas the consultative policy-making and implementation style in relation to the manufacturing sector seems to have been successful. This is related to generally positive attitudes of managers of industries where substantial support for environmental issues has recently been identified (Christiansen 1993). Business leaders seem gradually to have internalised environmental attitudes. In many cases this development is the result of market pressures.

However, the manufacturing sector has strongly opposed some environmental regulation and some instruments which would make the polluter pay. Denmark is one of the only countries where the clean-up of polluted sites is mainly a public responsibility (Church *et al* 1993). The manufacturing sector has also successfully opposed energy taxes on firms out of fear of international competition problems. The agricultural environmental problems can probably be traced back to the fact that the farmers' interest organisations, supported in particular by the Liberal Party, have opposed environmental regulation so strongly that the instruments adopted are insufficient to achieve the (symbolic) policy objective. Thus, the consultative policy-making style has been more successful in manufacturing compared to agriculture.

Modern environmental policies cannot rely solely on top-down regulation of producers and consumers. Policies must rest also on support from consumers and producers. Danish experiences show that it is difficult to hit the right balance between a *top-down* approach and a *bottom-up* one. Although many steps have recently been taken to increase the coherence between policy goals and the interests of producers and consumers, there is still a long way to go.

Notes

1. Unless otherwise stated, most of this section refers to Andersen (1989) and Christiansen (1996).

2. For a legal-political analysis of the Board, see Basse 1987; *Miljøankenævnet*, København: GAD. 523 p.

References

Andersen, J.G. (1990) 'Denmark: Environmental Conflict and the 'Greening' of the Labour Movement', *Scandinavian Political Studies*, Vol. 13, No. 2, pp. 185–210.

Andersen, M.S. (1989) 'Miljøbeskyttelse – et implementeringsproblem'. *Politica*, Vol. 21, No. 3, pp. 312–28.

Andersen, M.S. and Hansen, M.W (1991) *Vandmiljøplanen. Fra forhandling til symbol*. Harlev J: Niche.
Andersen, M.S. (1994) *Governance by Green Taxes. Making Pollution Prevention Pay*. Manchester and New York: Manchester University Press.
Andersen, M.S. (1997) 'Denmark' in M. Jänicke and H. Weidner (eds.) *National Environmental Policies – A Comparative Study of Capacity-Building*. Berlin: Springer Verlag, pp. 157–74.
Basse, E.M. (1987) *Miljøankenævnet*. Copenhagen: GAD.
Bernes, C. (1993) *The Nordic Environment – Present State, Trends, and Threats*. Copenhagen: Nordic Council of Ministers, Nord 1993:12.
Christensen, J.G. (1987) 'Hvem har magten over miljøpolitikken: politikerne, embedsmændene eller organisationerne?', in A. Dubgaard (ed.) *Relationer mellem landbrug og samfund*. Copenhagen: Statens Jordbrugsøkonomiske Institut, Report no. 36, pp. 65–77.
Christensen, J.G. and Christiansen, P.M. (1992) *Forvaltning og omgivelser*. Herning: Systime.
Christiansen, P.M. (1993) *Det frie marked, den forhandlede økonomi*. Copenhagen: Jurist- & Økonomforbundets Forlag.
Christiansen, P.M. (1996) 'Denmark', in P.M. Christiansen (ed.) *Governing the Environment. Politics, Policy, and Organization in the Nordic Countries*. Copenhagen: Nordic Council of Ministers, Nord 1996: 5, pp. 29–102.
Christiansen, P.M. and Lundqvist, L.J. (1996) 'Conclusions: A Nordic Environmental Policy Model?', in P.M. Christiansen (ed.) *Governing the Environment. Politics, Policy, and Organization in the Nordic Countries*. Copenhagen: Nordic Council of Ministers, Nord 1996: 5, pp. 337–363.
Church, T.W. and Nakamura, R.T. (1994) *Beyond Superfund: Hazardeous Waste Clean up in Europe and the United States*. Georgetown International Environmental Law Review, Vol VII, pp. 15–57.
DGXI (1990) *Schlussbericht*. Brussels.
DIOS (Dansk Institut for Organisationsstudier) (1987) *Tilsynsundersøgelse 1986*. Copenhagen.
DMU (Danmarks Miljøundersøgelser) (1991) *Vandmiljøplanens overvågningsprogram*. Copenhagen.
Downs, A. (1972) 'Up and Down with Ecology – the "Issue Attention Cycle"'. *The Public Interest*, Vol. 28, Summer, pp. 38–50.
Environmental Protection Agency (1985) *Miljøtilsyn*. Copenhagen.
Environmental Protection Agency (1988) *Miljøtilsyn 1987*. Copenhagen.
Environmental Protection Agency (1990) *Vandmiljø -90*. Copenhagen.
Forureningsrådet (1971) *Vand: En redegørelse fra målsætningsudvalget og hovedvandudvalget*. Copenhagen: Statens Trykningskontor.
Goldsmith, M. (1992) 'The Structure of Local Government', in P.E. Mouritzen (ed.) *Managing Cities in Austerity*. London. Sage.
Kingdon, J.W. (1984) *Agendas, Alternatives, and Public Policies*. Boston: Little, Brown and Company.

Knoepfel, P. and Weidner, H. (1983) 'Implementing Air Quality Control Programs in Europe', in P.D. Downing and K. Hanf (eds.) *International Comparisons in Implementing Pollution Laws.* Den Haag: Klüwer Nijhoff Publ. pp. 191–211.

Ministry of the Environment (1988) *Enkelt og Effektivt.* Copenhagen.

Ministry of the Environment (1995) *Natur- og Miljøpolitisk redegørelse 1995.* Copenhagen: Miljø- og Energiministeriet.

Moe, T.M. (1989) 'The Politics of Bureaucratic Structure', in J.E. Chubb & P.E. Peterson (eds.) *Can the Government Govern?* Washington: Brookings.

Moe, M. (1993) 'Implementation and Enforcement in a Federal System'. *Ecology Law Quarterly,* Vol. 20, No. 1, pp. 151–164.

North, D. (1990) *Institutions, Institutional Change and Economic Performance.* Cambridge: Cambridge University Press.

Pedersen, J.F. and Geckler, R. (1987) 'Græsrødder og embedsmænd', in R. Geckler and J.G. Christensen (eds) *På ministerens vegne.* Copenhagen: Gyldendal, pp. 103–134.

Schroll, H. (1985) 'Miljøgodkendelser af danske virksomheder', *Vand og Miljø.* Nr. 4.

Thulstrup, J. (1996) *Danskerne 1995. Holdninger, adfærd, planer og forventninger.* Copenhagen: IFKA-skrift nr. 1.

Vogel, D. (1983) 'Cooperative Regulation: Environmental Protection in Great Britain'. *The Public Interest,* No. 72, Summer, pp. 88–106.

Wilson, J.Q. (1980) 'The Politics of Regulation', in J.Q. Wilson (ed.) *The Politics of Regulation.* New York: Basic Books.

Winter, S. (1975) *En sammenlignende analyse af Det konservative Folkepartis beslutninger om deltagelse eller ikke deltagelse i forsvars-, miljø- og boligforlig i 1973.* Unpublished student's thesis, Århus: Institute of Political Science, University of Aarhus.

Winter, S. (1990) 'Integrating Implementation Behavior', in D.J. Palumbo & D.J. Calista (eds.) *Implementation and the Policy-Process: Opening Up the Black Box.* New York./London: Greenwood Press, pp. 19–38.

Chapter 4

France: Fragmented policy and consensual implementation

Corinne Larrue and Lucien Chabason[1]

Introduction

Compared with densely populated countries such as Germany, the United Kingdom or the Netherlands, the French environment has remained relatively undamaged. The average population density in France is around 100 inhabitants per square kilometre, but with an extremely unequal distribution over the country. The population is concentrated within cities (80 per cent of the population is urban) and mainly within the capital region which accounts for approximately 20 per cent of the population and for roughly 30 per cent of the national GDP. All this on less than seven per cent of the national territory. By contrast, some rural areas have been severely depopulated, some of them having less than 50 inhabitants per square kilometre.

In France there are approximately two to three hectares of open space per inhabitant, compared with an average of less than one hectare per inhabitant in Germany and the United Kingdom. As a consequence, many parts of the country have been able to maintain a wide range of flora and fauna. France supports 40 per cent of the flora species in Europe while only occupying 12 per cent of its territory. Unfortunately, many species of plants and animals are threatened, especially in the southern part of the country. Agricultural and urban development, particularly in coastal areas, are the cause of this unfortunate situation.

On the other hand, France has a positive record of accomplishment in some areas of environmental protection, such as air quality and industrial waste treatment. The quality of air in cities improved between 1970 and 1990 with regard to SO_2 levels, largely due to the use of nuclear energy to produce electricity. The fifty-five nuclear power plants that are in operation in France are the product of the strategic programme from the 1970s onwards to switch from imported fossil fuels to nuclear power. This has also led to a significant reduction in CO_2-emissions from energy production and has put France in an exceptional position among industrialised countries regarding

CO_2-emissions per capita. France has a great capacity for processing industrial toxic wastes. Half of the two million tons of toxic waste generated in the country is incinerated or otherwise treated.

Nevertheless, the recent National Plan for the Environment (Chabason and Theys 1990) made a cautious evaluation of the quality of the environment in France. While the most obvious sources of environmental damage have been eradicated, pollution from less definable sources has increased. Therefore, present and future policies are likely to be more difficult to implement.

With a few exceptions, the main areas dealt with by environmental policies have been regulated by means of laws and/or financial instruments. Important environmental problems remain in the area of water and air pollution as well as waste disposal, noise pollution and landscape degradation. Today, 50 per cent of waste is disposed of in landfills, without any further treatment; vehicle emissions of dust from transport have increased by 67 per cent since 1980; 43 per cent of the French population feel disturbed by noise at home; nine per cent of the French frequently receive inadequate tap water with unacceptable amounts of bacteria, while three per cent receive water too high in nitrate. More generally, 50 per cent of the population, according to a survey in 1993, feel that public authorities should do more to protect the environment.

Historical development of public action

In France, the public environmental consciousness is intricately linked with the creation of the Ministry of the Environment (ME) in 1971. Before that time, the environmental 'demand' was mainly generated by scientists, who were strongly influenced by American literature, and, to a lesser extent, by NGOs involved with nature protection (Poujade 1975; Charvolin 1993). The story of the establishment of the ME and its growing importance runs parallel to that of the development of public environmental awareness.

The ME was initially placed in the Office of the Prime Minister. Such an arrangement was an easy way to create a new administrative unit without unduly disturbing the already-existing overall administrative organisation. The ME was given an inter-ministerial role of stimulating and co-ordinating action in the areas of pollution prevention, water management, nature conservation, management of hunting and fishing, and the protection of nature sites and monuments as well as the management of landscapes. The ME was not given any executive powers and, consequently, it had no field administration. This new ministry was largely dependent on the sector ministries and their field administration in implementing policy, thus not threatening the existing ministries. In order to help the ME function, several structural changes have taken place:

- Some services were detached from other ministries and constituted as a new administrative branch subordinated to the new ministry. This was the case with the administration of the nature reserves and parks, the administration of

hunting, the administration of classification and the agencies of water management. These transfers brought together the organisational potential for a full-fledged administration, waiting to be born.
- The new ministry was also given authority over a number of secretariats, such as the secretariat of the Interministerial Commission of National Parks, of the High Commission of Sites (shared with the Ministry of Culture), and of the Interministerial Commission for Water.
- For local implementation of its policies, the ME was given the opportunity to use the local services of the Ministries of Agriculture and of Industry.

Within the ME, one can point out three main traditions which stem from different philosophical, practical and historical trends, and which also represent the main areas of concern around which the public environmental consciousness has developed.

The first is the hygiene tradition, which was represented within the administration by the service for classified installations, previously located in the Ministry of Industry. This service was created in 1810 in order to enforce the law concerned with industrial pollution control. The objective of the law was to ensure legal certainty to the plants regarding the conditions under which they could operate, while at the same time limiting the negative impact on the surrounding environment by means of a system of administrative authorisations given before the opening of the plant. This tradition was strongest during the period in which public health and hygiene became an issue. People such as Doctor Villermé, Pasteur and the architect Le Corbusier were prominent actors during this period. These concerns and the understanding of environmental problems dominated administrative thinking throughout the nineteenth century and up until the middle of the twentieth century. This hygienist thinking aimed to protect public health by improving environmental and living conditions, by fighting against emissions of smoke, dust, etc. and by improving the living conditions of the population by providing, for instance, electric lighting, clean water and sewage systems.

The second tradition is concerned with natural history. It stems from a venerable French tradition devoted to the study of nature species and natural space, topics very similar to those dealt with by modern-day ecologists. A good example of the institutionalisation of this tradition is the National Museum for Natural History (Muséum National d'Histoire Naturelle), created in 1791. Other examples are the local museums for Natural History and botanical gardens (Jardin des Plantes). This tradition was very strong during the French colonial period. At the beginning of the twentieth century, the first voices were raised in favour of nature conservation, the creation of natural parks and reserves, and the protection of endangered species. In 1950 the Ministry of Agriculture took over the responsibility for the implementation of this nature protection policy.

The last tradition is an aesthetic tradition which concerns the protection of cultural sites and landscapes. Before being transferred to the ME, administrative responsibility for these areas was located in the Ministry of Culture. It was inspired by the historic buildings service, which was created during the Revolution, and which was quite prominent during the Second Empire with Vivet, Merimée and Viollet le Duc. The law enacted in 1930 introduced a system for the protection and preservation of such sites: the sites and the landscapes are listed in a survey and classified in terms of their picturesque, artistic or historic value. Once included in the list of protected sites, these objects cannot be altered without a special authorisation from the Minister. In 1971, a distinction was made between natural sites, the management of which was given to the ME, and cultural sites, produced by human activity, which continued to be managed by the Ministry of Culture. The French environmental movement also reflects and is a product of these three traditions. This explains the diversity of the composition of this movement.

With regard to the ME, two other areas of concern were introduced during the 1970s, as well as special administrative services to deal with them: the first is the rational management of natural resources and the second concerns public participation, especially by means of environmental impact assessment (EIA) and by environmental information.

At the beginning of the 1980s, the conjunction of several factors led to a decrease in the development of environmental policy: the economic crisis, linked with the second fuel crisis, relegated the environmental concerns to a subordinate position; the ecologist groups, which were mainly against the nuclear programme, were not followed by public opinion, and thereby lost their credibility. The left parties, which acceded to power, had monopolised all the attention, and neutralised opposition. Thus, during this period, France was rather out of step with the development of environmental policy in other Western European countries and developed an environmental backwardness.

But after the Chernobyl accident, and the return of economic growth, environmental policy and consciousness reappeared. This ecological 'come back' appeared also in the political arena (see below). Such a political impetus helped the ME gain more competencies within the field of environmental protection, especially in the following fields:

- safety of nuclear plants;
- building of infrastructures;
- management of coastal zones and mountainous areas;
- landscape policy;
- jurisdiction over the National Office for Forests which manages four million hectares.

For the first time, the ME acquired proper local services at the local level (the DIREN) and now shares the Regional Service of Industry with the Ministry of Industry (see below).

But just recently, with parties of the Right coming back into power and with the economic crisis becoming more and more severe, environmental concerns are once again being downplayed. The environmental concern of the population has decreased as the latest survey shows (IFEN 1995): personal commitment is slacking, and the number of people who will accept tax increases for environmental purposes decreased by nine per cent between 1993 and 1994. Nevertheless, environmental concerns are still an important priority for the French, especially air pollution at local level.

Environmental politics: actors, coalitions and interests

The French political parties traditionally represented in Parliament (Socialists, Communists, Liberal and Conservative parties) initially paid little attention to environmental problems. They usually shared a positive point of view about 'productionism', especially with regard to the nuclear energy programme, which was carried out in order to ensure 'French energy independence'. More generally, the parties agreed on the programme of development that had been implemented after World War II. Consequently, they have tended to see concern for protection of the environment mainly as an obstacle to the modernisation of the country. The only steps taken in favour of the environment were those linked to pressure from public opinion. However, the strength of this pressure has varied greatly.

During the first period of Socialist government (1981–86), the state of the environment was alarming. The relative neglect of environmental policy at this time can be explained by the weakness of the environmentalist movement and ecologically oriented political parties. However, in the second period of its tenure in office, the Socialist government began to be more responsive to the revival of ecological parties and events at the international level, e.g. the Brundtland Commission report. Also, at the national level, there were a number of demonstrations and actions in favour of environmental protection, such as opposition to hydroelectric projects.

In the period 1988–92, the French government took a number of 'environmentally friendly' decisions: it cancelled important infrastructure projects, such as the dams on the Loire River; it developed a green plan; it strengthened the Ministry of the Environment with more financial and human resources.; and it took a number of international initiatives.

These actions were taken in a period of growing environmental awareness in the population and growing support for green parties. Before 1989, the green parties' performance in elections was in general relatively poor. The green movement was divided into several competing organisations and there was disagreement over whether the greens should participate in electoral politics at all. The national political discourse centred on the left-right dimension, and the green parties failed in articulating the non-environmental dimensions of their programmes. In addition, distinctive features of the French political system, such as the second-ballot electoral system and the centralised pattern of policy making are part of the explanation of the green parties' failure in the 1970s and 1980s (Cole and Doherty 1995:46–48).

A breakthrough came in 1989, when Les Verts (the French Green Party) performed well in the local elections, followed by a score of 10.59 per cent in the European Elections. In 1990 a new green party, Generation Ecologie, was formed by Brice Lalonde, who was at that time Minister of Environment. In the 1992 regional elections, these two parties together received 14 per cent of the votes (GE: 7.1 per cent, Les Verts; 6.8 per cent).

In other words: the electoral system for national elections normally makes the first ballot a contest between a small number of candidates who hope to win the necessary absolute majority. Where none manages this, there is a second ballot in which the victory goes to the candidate who gets either absolute or relative majority (Mény 1993:173). Such an electoral system favours the candidates of the big parties or the two big blocs of the parties of the Right or the Socialists. The candidates of small parties like the Greens are marginalised and filtered away. In the elections for the regional assemblies (the *conseils régionaux*), the representation is based on the proportional principle. Therefore, the green parties obtained a representation reasonably proportional with their success at the ballot box in these elections. Facing the possibility of overcoming the barriers of the second-ballot electoral system in the national elections in 1993, the two rival parties formed an electoral agreement in November 1992, the Entente Ecologiste (EE). The attempt failed, the EE captured 7.5 per cent of the votes and 'were left on parliament's doorstep' (Prendivill 1993:478).

The green parties failed to consolidate their gains and experienced a break in the upward trend experienced up until 1992. Nevertheless, French environmental policy has not been weakened since then. The stabilisation of its position can be explained, in part, by both the weight of international obligations entered into, and the pressures from the directives of the European Community. In addition, the government has been sensitive to sudden changes in public opinion. Finally, the personalities of the last two Ministers for the Environment have also been important in keeping the environmental issue and policy in the political spotlight.

At present, there is a discrepancy between the parliament and most of the political parties, which still do not pay much attention to these questions, and the government, which does try to fulfil its environmental commitments.

The present state of environmental policy in France

French environmental policy is quite comprehensive; almost all environmental domains are covered by a law. Some of these have been amended recently.

The present objectives of environmental policy in France were spelt out in the National Environmental Plan presented in 1990. The formulation of this plan was a consequence of the elections over roughly the last decade, which revealed an increasing importance of green parties. At that time, the environmental movement had recovered some legitimacy in the political arena within which policy making and implementation occurred.

The ecological wave of the nineties found its main manifestation in the formulation of this National Plan, which may be seen as part of a new offensive in environmental policy (Chabason and Theys 1990). New ambitious objectives together with new means for national environmental policies were proposed. At present the Plan is gradually being implemented, although to a lesser extent than had been intended in the Plan. Many changes have taken place within the environmental administrative sphere which will be discussed below.

The National Plan for the Environment of 1990 defines quite precisely objectives that are to be fulfilled, as well as the means to be applied in each sector. These explicit definitions are partly set within recent reforms of old environmental laws such as the Water Law or the Waste Law reform which were both amended in 1992. Presently a reform of the air pollution law is about to be enacted.

The main instruments used in connection with French environmental policy are command and control instruments and economic instruments. Moreover, the French environmental administration has developed various informative instruments such as the publication of annual environmental reports.

Command and control instruments provide the most important means of action for the central government. Emission standards, planning tools, protection zones, authorised permits and EIA were gradually introduced by laws and regulations, and today constitute the main tools at the disposal of the central and decentralised environmental administration.

But the French public authorities have also been developing economic instruments for a long time. For instance, since 1967, taxes have been imposed on industrial water discharges. These types of taxes are now imposed on air emissions, waste disposal and noise emission.

A further important characteristic of French environmental policy is the growing impact of EC environmental directives. As a member state of the European Union, France must fulfil EU requirements in the field of environmental policy as well as in other fields. In a study evaluating the impacts of EU environmental policy on French policy it was shown that although the EU and France have similar policies, in several environmental areas, European norms and standards are stricter than French ones (Larrue and Prud'homme 1993). Therefore, the EU policy serves to strengthen the regulatory pressure on France in favour of the environment. More importantly, many of the EU directives tend to reinforce the position of the ME, with regards to the industry and to the other often competing ministries. However, the above study also underlined that these benefits of the European environment policy are limited by the compromises produced by European decision-making processes. When policy negotiations at the European level do not incorporate the French government's position, the implementation of EU policies is made more difficult, because the position of the ME is weakened at home.

In sum, one can consider the French environmental policy as a 'mature' policy. On the other hand, it is still highly sectoral in structure, and difficulties mainly arise from the complexity of its administration and the quality of the implementation of these instruments.

The present state of environmental actors patterns: an overview

The main characteristic of French environmental policies and administration is complexity. This applies to the central level as well as to the regional and local levels, as Figure 4.1 indicates.

Figure 4.1: Organisational chart of the state administration (central and deconcentrated) in the field of the environment

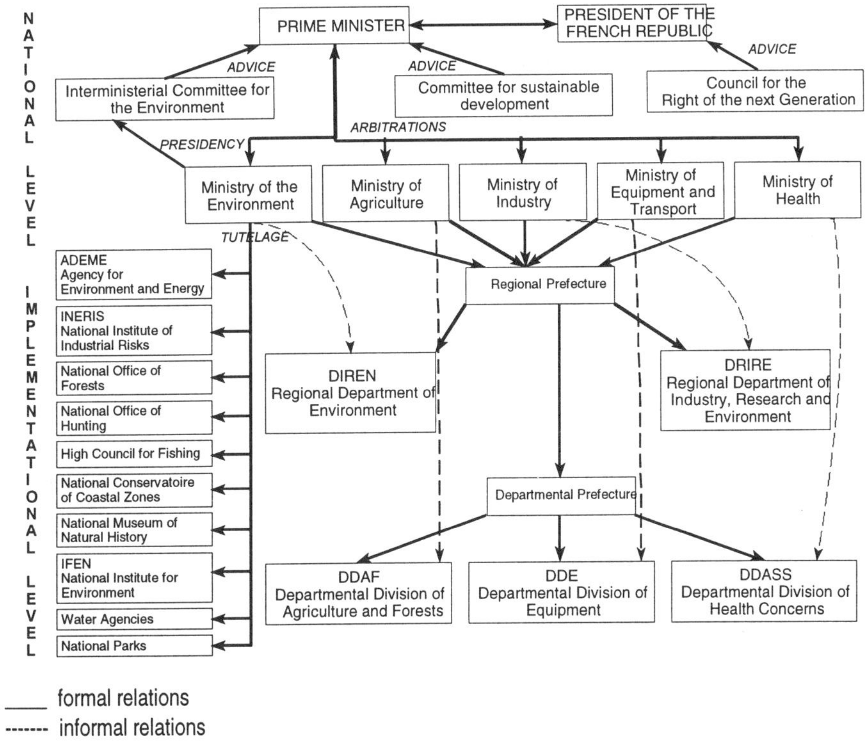

```
NATIONAL LEVEL

                    PRIME MINISTER  ←→  PRESIDENT OF THE FRENCH REPUBLIC
                    ↑        ↑              ↑
              ADVICE      ADVICE          ADVICE
    Interministerial    Committee for     Council for the
    Committee for       sustainable       Right of the next Generation
    the Environment     development
          ↑         ARBITRATIONS
    PRESIDENCY
    
    Ministry of the    Ministry of   Ministry of   Ministry of          Ministry of
    Environment        Agriculture   Industry      Equipment and        Health
                                                   Transport
    TUTELAGE
```

IMPLEMENTATIONAL LEVEL:

- ADEME — Agency for Environment and Energy
- INERIS — National Institute of Industrial Risks
- National Office of Forests
- National Office of Hunting
- High Council for Fishing
- National Conservatoire of Coastal Zones
- National Museum of Natural History
- IFEN — National Institute for Environment
- Water Agencies
- National Parks

Regional Prefecture
- DIREN — Regional Department of Environment
- DRIRE — Regional Department of Industry, Research and Environment

Departmental Prefecture
- DDAF — Departmental Division of Agriculture and Forests
- DDE — Departmental Division of Equipment
- DDASS — Departmental Division of Health Concerns

——— formal relations
------- informal relations

The actors at the central level

Generally speaking, environmental policies are matters for the Executive. The powers of the French National Assembly as well as those of the Senate are strictly limited by the Constitution. Consequently, the French politico-administrative system does not allow the parliament to play an important role in the field of environmental protection nor, for that matter, in any other fields. However, the authoritative objectives of environmental policies as well as the main instruments applied in the field of environment are based on framework laws that are discussed and passed by parliament. This general role in the legislative process provides parliament with a chance to introduce some changes which favour or disfavour environmental protection.

Within the Executive, the ME plays an important part in environmental policy. Originally the ME was intended as a light structure designed to influence other more traditional politico–administrative actors. The ME, which is made up of fewer than 500 persons, was to act as a trans-sectoral structure. In this sense, it falls into the category called in the French administrative vocabulary 'administration of mission', in contrast to the traditional 'administration for management'.

The choice of such a structure, which continues to define the ME's position today, means that the effective performance of the mission requires interaction with the sector ministries.

Consequently, interministerial arrangements, like the High Committee for the Environment, the Interministerial Committee for Quality of Life, the National Council for Noise etc., have been created to structure and co-ordinate these contacts. All these committees are made up of representatives of the involved ministries as well as of independent experts, and in some cases, representatives of affected interests. On the other hand, many controversial decisions in the field of environmental policy are made by means of arbitration at the level of the Prime Minister, a situation which does not always favour environmental protection.

Another consequence of such an organisational choice is that the resources of the ME have been limited. Such a limited capacity is a serious handicap today, since the instruments available to tackle environmental problems have become more and more complex and require more and more administrative skill. Acknowledging this problem, the authors of the recent National Plan for the Environment proposed a reform aimed at strengthening the environmental organisation at the central level.

Recently, and as a consequence of the implementation of this national plan, new interministerial bodies have been created: the High Committee for the Rights of the Next Generation, which replaces the High Committee for the Environment (which had not been functioning for many years); the Interministerial Committee for the Environment; and the Commission for Sustainable Development (which has the task of elaborating recommendations for improving environmental policies and for achieving better integration of sustainability).

This wave of reform also hit the ME itself. Since 1984, the ME has included four main directorates, one dealing with nature protection, one dealing with pollution, another one with quality of life and the last one with information and research. This organisational division corresponds to the main sectoral fields of competence given to the ME, but also to the different administrative *corps* involved in the environmental field. As is well known, at the upper level, the French administration is divided among a number of corporations, defined in reference to the higher professional schools where their members were educated (see Figure 4.2).

Figure 4.2: The classical operational chart of administration

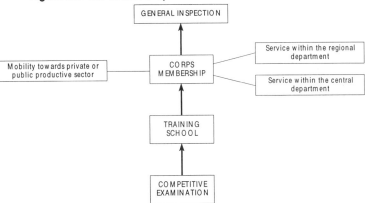

Three main *corps* are concerned with the field of the environment: the '*corps* of mining engineers', which deals with the industrial pollution aspects, the '*corps* of water and forest', concerned mainly with nature protection; and the '*corps* of bridges and roads' which is in charge of the question of the quality of life. The organisational structure of the ME is thus divided along both functional and socio-administrative tiers.

Furthermore, some of the sectoral ministries kept for their own use most of the competencies related to the environmental field. As a result, the ME is still very dependent, at the central level, and above all at the regional and departmental level, on four ministries in particular: the Ministry of Public Works (staffed with civil servants coming from the *corps* of bridges and roads), the Ministry of Agriculture (staffed by civil servants from the *corps* of water and forest, the Ministry of Industry (which is the central bastion for the *corps* of mining engineers); and the Ministry of Public Health (which is controlled, with regard to the environment, by sanitation engineers). Note that each important 'partner' of the ME has its corporate counterpart within the ME (except for the Ministry of Public Health). As noted, these partners have retained jurisdiction for various parts of the environmental field. For example, the National Plan for Highways, which was enacted in 1987, was formulated without any consultation with the ME; and likewise industrial and agricultural policies which have an important impact on the environment are often formulated and implemented without any dialogue between the responsible ministries and the ME.

Here again, a reform of the organisation of the ME was proposed in the National plan for the environment. A new organisational decree, promulgated in April 1992, expanded the four traditional directorates to five, by adding a directorate devoted solely to water problems. The jurisdiction of the ME has also been extended to include responsibility for landscape protection. In addition, the decree regulated the formal participation of the Ministry in the formulation of transport, rural development and urban planning policies.

Another important feature of the French environmental politico-administrative system is the existence of quite independent technical agencies, responsible for one particular field. These agencies exist in the fields of water management (since 1964), waste disposal (since 1975) and air (since 1980), but also in the fields of hunting and fishing, for special zones, like coastal zones, for forests or National Parks. These agencies can usually impose parafiscal levies, which are spent for controlling, promoting and investing within the field under their area of jurisdiction. Taken together, the budgets at the disposal of these agencies are approximately ten times larger than the budget of the ME. Some of these agencies do not fall under the jurisdiction of the ME, which further reduces the impact of this ministry on environmentally relevant decisions and activities.

Among these agencies, the case of the water agencies is particularly interesting. Each of these agencies manages one of the six great basins which divide French territory (Seine Normandie, Loire Bretagne, Adour-Garonne, Rhin Meuse, Artois Picardie and Rhône Méditerranée Corse). The agencies are governed by quite autonomous bodies called the 'Basin Committees'. These committees act as real 'Parliaments of water', inasmuch as they are composed of representatives of central and local governments, of professional interests and of water users, and because they define (within limits set by the Ministry for the Budget) the tax base and the rate of the charges levied on water consumption and water pollution.

Once again, the National Plan for the Environment led to reforms in the organisation of these technical agencies. The agencies for waste and for air have been merged with the Agency for Energy Conservation, which was created in 1982 in order to set up a new energy policy different from the one of Electricité de France. This new agency, called ADEME (Agence de l'Environnement et de la Maîtrise de l'Energie), was created in 1991 and has a staff of more than 600 persons at its disposal. Likewise, a new technical institute, called INERIS (Institut National de l'Environnement Industriel et des Risques), endowed with an extensive expert capability in the field of industrial environment and risks, was created in 1990 by merging two old laboratories. Finally, the IFEN (Institut Français de l'Environnement) was created in 1991, in order to provide the ME, and other actors, with statistical and informational support in the field of environment policy.

In short, under the present institutional arrangement, the legislative and rulemaking functions belong to the ministry whereas the technical and financial functions have been delegated to agencies.

The actors at a regional, departmental and local level

Traditionally, and until very recently, the French central administration did not have any real territorially 'deconcentrated' services at its disposal. The French regional and departmental politico-administrative system was composed of local governments together with representatives of central administration endowed either with general administrative powers, like the departmental or regional *Préfet*, or with specific ones, such as the Regional Directorate of Industry, Research and (since 1991) Environment (DRIRE), the Departmental Service for Agriculture and Forests (DDAF) of Infrastructures (DDE) or of Public Health (DDASS). These technical services 'belong' to their central ministries which they are supposed to represent in opposition to or in collaboration with local governments. They usually act under the joint control of their own Ministry and that of the *Préfet*.

Since 1978, the ME has had at its disposal a 'light' regional organisation called the Regional Service for Architecture and Environment (Direction Régionale pour l'Architecture et l'Environnement, DRAE). This service is in charge of many tasks (architectural control, site and monument protection, the implementation of Environmental Impact Assessment studies, the collecting and publication of environmental data and the link with environmental associations), but has very limited resources (less than 420 persons for 23 regions). The DRAE was under the joint control of the ME, the Ministry of Planning and the Ministry of Culture. Consequently, it has not been able, in practice, to implement environmental policies effectively.

At the regional and local level the ME was, and still is, dependent on the external or field services belonging to other ministries. In some cases these services were explicitly placed at the disposal of the ME (like the Regional Directorate of Industry, or the Regional Service of Water Management, belonging to the Ministry of Agriculture); others were more implicitly available, such as the Departmental Service of Infrastructures or the Departmental Service of Agriculture. Finally, some others, such as the Departmental Service of Public Health, play an independent role in the field of environment.

This state of affairs was critically commented upon in the National Plan, and the creation of a Regional Direction of Environment, directly under the control of the ME, was proposed in order to remedy the situation. These services, the DIREN, were established at the end of 1991. They were created by merging the old Regional Service of Architecture and Environment (DRAE) and the Regional Service of Water Management belonging to the Ministry of Agriculture (SRAE), and are now under the control of the ME. They are responsible for water quality management, and for nature, rural and urban landscape protection, and (as we will note below) they will probably change the way national environmental policies are implemented at the regional level. Notice, however, that the DIRENs are not responsible for industrial pollution, which remains under the control of the DRIRE.

Although the ME is in a better position at the territorial level, competencies as to public action in relation to the environment are still divided between different services. Last, but not least, the new decentralised actors have also tended to appropriate some of the environmental competencies. Traditionally, some of the jurisdiction over public environmental action has been granted to local, departmental and regional governments. Indeed, local government (*Communes*) was, and still is, responsible for water delivery, waste-water collection and treatment, and municipal waste disposal; and they usually fulfil their duties on an 'intercommunal' basis. Still, the powers of local governments are limited in environmental policy. For instance, the local Executive has more or less nothing to do with the pollution stemming from those industrial plants called 'classified installations', even if these factories have an important impact on the local environmental quality. The regulation of industrial pollution is the responsibility of central government.

Likewise, until recently, the government of the Department (called *Conseil Général*) mainly followed the central government initiatives, inasmuch as they participated in the financing of the purchase of equipment which was decided upon by the state. However, since the reform of 1982, even though such competencies were not explicitly transferred to them, regional, and above all departmental, governments have tended to involve themselves with the field of environmental and nature protection. They set up 'environmental services'; they define their own priorities; and they try to formulate their own policies with, or without, the participation of central government representatives. The new environmental law of 2 February 1995 tries to make clear the division of competencies between State and local governments in the fields of natural hazards, waste management and water quality protection. This law tends to strengthen the power of central government.

Finally, another kind of actor plays an important role in the management of environment: in order to compensate for its lack of means and staff, the ME has encouraged private organisations to involve themselves in the concrete management of the local environment, much as the case of social policy. For instance, the networks for monitoring air pollution are managed by local associations comprised of State representatives, local industrialists, and local governments. Another example can be found in the management of natural reserves: such reserves have often been given to non-governmental organisations, which receive subsidies in order to manage them.

After this brief overview of the different actors involved in environmental administration and politics, we now examine the main characteristics of the implementation of French environmental policy.

The implementation of environmental policies in France

It is beyond the scope of this section to examine in detail the implementation of environmental policies in France. However, on the basis of work done on implementation in the field of air pollution stemming from industry (Knoepfel and Larrue 1985; Maghalaes, Larrue and Darbera 1984), and water pollution stemming from agriculture (Larrue 1988a, 1988b, 1991), as well as on the basis of the general literature on the subject we will try to describe the main characteristics of the implementation of environmental policies in France. A new feature with regard to the administrative context of environmental policies must, however, be pointed out in order to understand the following description of implementation processes. The French politico-administrative system is well known for its alleged high degree of centralisation. However, since 1982, the country's four-tier system of government has undergone processes of decentralisation; some of the powers and competencies of the central administration have been transferred to local, departmental and regional governments. Decentralisation has occurred particularly in the fields of education and social policies, infrastructures, and urban and planning policies. This process has nevertheless had an important impact on the way in which public action takes place, even in those fields where the decentralisation has not been very extensive, like environmental policy. Local, departmental and regional actors now try to play new roles in politico-administrative arenas. This has changed the traditional context of policy formulation and implementation for central government.

A rather soft policy programme

The first elements concern the quality of the programme of environmental policies. Because the central government has implementation services at its direct disposal, the programme formulated at the central level is usually rather vague, leaving it to the implementation level to define many elements. This is also true for environmental policy despite the fact that implementation services are not always under the direct control of the Ministry. For instance, as far as the programme for reducing SO_2 is concerned, until the implementation of the 80/779 EEC Directive, no quality objective was really set in terms of milligrams of SO_2 per cubic metre of air, nor were national emission standards defined to be applied to individual firms. Emission standards were not defined at the national level but could be set by the regional administration, taking into account local conditions. Moreover, the programme left to the implementation actor (that is, the DRIRE) the task of granting delays of application of control measures and the spatial priorities within the region. A main consequence of such a soft programme was that implementation actors usually showed a capacity for innovation, which led them to create instruments not previously defined at the national programme level. For instance, in implementing the policy of SO_2 air pollution policy, the DRIRE of the Parisian region combined two instruments, alert zones and special protection zones, which had been defined by the national level as separate instruments. An alert zone is a zone with specific boundaries which have been previously defined and in which, when a given threshold of pollution is reached, main emitters must reduce the pollution they discharge into the air. The existing 11 alert zones are primarily located within industrial zones. A special protection zone, on the other hand, is usually an urban zone in which no one is allowed to burn fuel with a high sulphur content. By combining these two instruments, the Parisian DRIRE allows some big industries located within the special protection zone to ignore the fuel burning regulation, but obliges them to abide by the alert zone regulation, something not provided for through law. In many cases such soft programmes have provided a good basis for effective implementation, as extensive international comparisons of national practice in the field of air pollution show (see Knoepfel and Weidner 1986).

An implementation based on consensus

The main characteristic of the implementation of French environmental policy is that it is based on consensus rather than on imposition of constraints on the actors involved.

An example of such a consensual system is provided again by the case of air pollution alert zones, where the reduction of pollution is based on a general agreement reached with the companies involved. The negotiation process determines the level of the threshold to be fixed. In order to reach and to manage such consensus the DRIR, which is in charge of air pollution policy, has brought together, in a non-profit organisation, all actors whose interests are affected by the policy. Such independent associations have been created in more or less all the French regions. The members are usually local governments, and local and regional companies as well as environmental organisations and, of course the DRIR. Formally, these non-profit organisations are responsible for managing the air pollution network, which will be used in case of an alert. But they also constitute a framework within which the negotiation process concerning the threshold level takes place. Consequently, the thresholds set at the local level will vary from one region to another, taking into account local climate conditions as well as the 'goodwill' of local industrialists.

An implementation based on partnership

Environmental policy implementation in France is also based on partnership. The central environmental administration and its representatives at the regional and local level try to get target groups and/or local governments involved, as their partners, in the policy implementation process.

This partnership concerns, first of all, target groups. For instance, contracts, called branch contracts (*contrat de branche*), have been signed between the ME and the representatives of professional interests. These contracts set the schedule for the anti-pollution investments that are to be made by the companies belonging to the given branch. Subsidies from central government have been added in order to help the companies make these investments. Approximately ten branch contracts were signed between 1970 and 1980, for the most polluting branches. The results of these contracts have mainly been positive; the emission of pollution into the air as well as into water decreased during the same period, and although this reduction can partly be explained by the economic crisis, part of it stemmed from the branch contracts. However, since 1985, this instrument has no longer been used because the State Council (*Conseil d'Etat*) ruled that such contracts restricted local administrative agencies too much in making decisions regarding standards to be imposed on individual companies. The reason for this decision by the State Council was the characteristic of the above negotiation process. It did not include any external actors, and was, consequently, too secret and non-transparent.

Another example of such a partnership involving target groups is the water agencies and the charges they levy on polluting firms. As a matter of fact, a financial aid system is closely linked to the charges system applied to water consumption and pollution discharges into water. The revenues raised by means of effluents charges are redistributed to those firms which bear them, through subsidies on investments in waste-water treatment plants. Such a system of joint co-operation has also functioned as a strong argument for industrialists to accept the effluent charges system.

The partnership also concerns local governments. The so-called river contract, signed between the numerous public and private actors involved within a river basin, provides a good example. In order to tackle the pollution problem of a given river, contracts were signed for a five-year period, usually between the departmental government, the local governments, the water agency and the ME. Such river contracts led to a concentration of the purification financial efforts within a given time period. The 27 river contracts signed so far represent an implementation based on a partnership, which is intended to motivate local and departmental governments. Also in the field of water, approximately 200 contracts have been approved since 1984, between water agencies and local governments (called *contrat d'agglomération*) which determine a water purification investment plan for the participants. Such river or city contracts have greatly accelerated the improvement of the quality of French rivers, but remain insufficient with regard to the size of investments requested.

Such arrangements are currently used by the ME in order to encourage local and departmental governments to deal with the environmental conditions in an area as a whole. Local Environmental Plans and the Departmental Environmental Plans have been signed in which the problems of the local environment are described, and then the priorities for the local environmental policy, and the actions to be taken, are defined. The ME provides subsidies to carry out the preliminary studies, and also makes some financial resources available for the implementation of the plan (thus creating an 'environmental charter' instead of an 'environmental plan'). Here again, such policy implementation is based on partnership between the central government and regional and local governments.

Implementation based on negotiation

A particularly problematic aspect of environmental policy applies to the implementation of classified installation regulations. This legislation stipulates that each plant which is likely to disturb the surrounding natural environment as well as public health, must be given pre-authorisation by the administration, or in the case of the smaller factories, a declaration of pollution. During the authorisation or declaration process, the administration can impose technical conditions on the plant in question in order to reduce potential environmental impacts. But, as noted above, no specific emission standards have been set at the central level. The main regulation principle applied in the field of industrial pollution in France is BATNEC: best available technology not entailing excessive costs. The emission standards which are imposed on industries depend mainly on the financial resources of the companies. Such standards are usually fixed after a process of negotiation between the administrative agency and the individual company. It is safe to say that in France environmental constraints on classified activities are nearly always negotiated.

Two factors explain these characteristics of the environmental policy implementation. The first one is linked to the number of actors involved in environmental policy. As mentioned above, the environmental responsibilities are spread among many actors at the central level as well as at the regional and local levels. As a consequence, in many cases none of these actors is able to act alone to manage the implementation of the policy. Moreover, it is sometimes difficult to find any leadership at all within this implementation organisation. The administrative agency in charge of the implementation of an environmental programme usually needs to borrow 'legitimacy' from other actors. Negotiation and consensus are, therefore, required in order to implement every single policy.

But even in the rare cases when a single administration is responsible for the implementation of a particular policy, as in the case of the DRIR and industrial air pollution policy, the consensual aspect is present. In such cases there is a second explanatory factor which is linked to the cross-cutting character of environmental policy. This policy is often implemented by administrative actors already involved with the execution of other policies, and the way in which they implement environmental policy will be influenced by their traditional way of acting. They usually have previous relationships with the target groups of environmental policies, which will also influence the implementation of environmental policy. For instance, the DRIRs are, at the same time, responsible for industrial pollution abatement and for industrial development within each region. They are used to negotiating with companies in order to get them to locate in their regions. It is therefore not surprising that they use negotiations in the implementation of industrial pollution abatement.

One can then conclude that the consensual, partnership and negotiation aspects of the French implementation process appear to be an appropriate response to the complex nature of this process, and contribute to overcoming, at least partly, the inherent slowness that is usually a characteristic of such a system of action.

Varying implementation arrangements at the local level

Another feature of the implementation of environmental policy in France is that arrangements for implementation of policy at the local level vary from one region to another. The case of nitrate discharges in water stemming from agriculture is a good example of such a variation.

The diffuse water pollution caused by nitrogen fertilisers is a recent problem in France. At the central level it became an issue in the period 1980–84. There are different kinds of instruments presently available to tackle this problem:

- regulatory instruments which mainly limit liquid manure spreading activities and take into account the ability of the soil to absorb them;
- special protection zones surrounding a water supply plant, within which agricultural activities can be regulated. But here again, this instrument is not really effective inasmuch as it usually does not cover the water basin as a whole;
- stimulative instruments, based on a code of good agricultural practices. This code, defined at the central level, is then explained and communicated to individual farmers through professional agricultural services (mainly Chambers of Agriculture, *Chambres d'Agriculture*). This kind of instrument has been developed in the last few years;
- economic instruments, such as the charges on agricultural pollution which are about to be set at the national level (a general agreement was signed in 1991 between the Ministry of Agriculture and the ME); on the other hand, subsidies provided by water agencies and local or departmental governments are used to get farmers to invest in treatment systems (for liquid manure) or to encourage them to cover their land during winter periods.

The study of implementation in six French regions, representative of different agricultural contexts, revealed different configurations, varying from traditional ones to more innovative ones, according to the participants involved and the instruments applied.

The so-called traditional politico-administrative arrangement includes the Departmental Service of Agriculture and the Departmental Service of Public Health employing familiar regulatory instruments. The more innovative arrangement consisted mainly of the water agencies and the professional agricultural organisations (Chambers of Agriculture) which used social instruments.

In all the case studies, there appeared to be no permanent leadership. Depending on the local contexts, the leading actor has been the Departmental Service of Public Health (DDASS), or the Departmental Service of Agriculture (DDAF) or even the water agency. But in all cases, the action of the departmental government was decisive for the implementation of the policy. In those regions where the government of the department did not at least support the administrative actors, implementation failed.

The creation of the DIREN has not really reduced the uncertainty in this connection, even if they might be considered as the 'natural' implementation actors. Indeed, as the DIREN only act on the regional level, one can assume that the traditional actors like DDASS or DDAF will continue to play an important role in the implementation process.

Because local implementation actors belong to various administrations, and because no environmental 'corps' has been created, there is no common cultural background shared by, and integrating, those engaged in local implementation. Consequently, the local actors implementing environmental policies do not take into account a comprehensive environmental scheme. The use of such partial frames of reference leads to 'sectorialisation' of implementation: the implementation process is tied up like a sausage and the consequences of an environmental protection policy in a given field is not necessarily linked with the implementation of the policy in another environmental field. For instance, in the field of air pollution, the French policy has been effective in reducing SO_2 emissions mainly because of the ability of local implementors to manage the problem together with local industrialists, but the same cannot be said for NO_x emissions. Since this pollutant is emitted by industry as well as by vehicles, the DRIRE have not been able to manage this problem in the same effective way as with SO_2.

Growing competition between the central and decentralised government

A final and rather recent feature of the implementation of French environmental policy can be noted. As a result of the decentralisation process, local departmental and regional governments try more frequently to become involved in the environmental policy game. In some cases, they try to take over the leadership of the policy. Instead of contributing to the implementation of national policy, some departmental governments attempt to formulate their own policy, which will not necessarily be compatible with the national one.

The water pollution policy of the Department of Côtes d'Armor in Brittany is a good example of this. Considering that the environmental policy of the central government representatives was not severe enough, the government of this department (*Conseil Général*) decided, in 1988, to develop its own policy, particularly in the field of pollution from agriculture. It took the partition of the deconcentrated State Service of Agriculture (DDAF) which was carried out as part of the decentralisation process, as the opportunity to create its own environmental service.

With the creation of its new environmental service, the departmental government of the Côte d'Armor began to set its own policy in the field of water quality protection. It divided the departmental territory into water basins, within which priority actions were defined. But as the departmental government did not have any regulatory instruments at its disposal, it turned its policy towards stimulative instruments. Subsidies, information and advice were provided to farmers through the new departmental service. During that time, the central government representatives continued their own regulatory policies, but without any financial and political support from the departmental government. The result was that both policies were incomplete and more or less inefficient. After a two-year period of separate policies, a reconciliation between the two bodies took place (at the departmental government's initiative). The new politico-administrative set-up is now under the direction of the departmental government. The representatives of central government are more or less compelled to follow the departmental initiatives instead of being the initiator, as was the case before the crisis.

Such situations are continually developing in France. Many departmental or even local governments are creating their own environmental services. For the moment these services are mainly concerned with environmental data gathering and analysis, in order to be able to evaluate and to follow developments in local environmental conditions. In order to mitigate such a development, the recent law of 2 February 1995 tries to divide clearly the respective competencies at each decision level, and to put more precisely some clear limits on the power of local governments.

As a consequence of the growing decentralisation of environmental policies, the implementation structure in France has become increasingly varied. On one hand, we find local and departmental governments at the forefront of the environmental protection field and, on the other hand, local governments with little or no awareness of environmental problems. Such a situation could contribute to the creation of regional ecological disparities.

Conclusion

This brief overview of the implementation of environmental policies by the French administration stresses only the principal characteristics of the French situation as it is developing at present. Thus, implementation of French environmental policies appear to be highly complex in that it involves many actors, levels of action and authority, and many types of policy instruments.

But finally, one may wonder about the effectiveness of the French implementation apparatus. On this issue it is too difficult to draw any conclusions in this chapter. As mentioned in the introduction, the National Environmental Plan presents a moderately optimistic picture with regard to previous French environmental policies: while their results do not appear to be completely worthless, they are quite disappointing in terms of environmental quality. Consequently, the plan proposes many changes which are to be implemented. Comparative research efforts would then be interesting to undertake, not only in order to evaluate introduced changes, but also to point out ways of improving French and European environmental policies.

However, one can consider French environmental policy and its implementation as being incomplete and in need of improvement. In some areas the implemented policies seem fairly effective and efficient as in the case of industrial air pollution. In other areas, however, the policy pursued is clearly insufficient, as in the case of transportation-generated problems (i.e. noise, perturbations of natural life and landscapes by infrastructures, and last but not least, air pollution). Thus, French environmental policy can be seen as a piece of Gruyère cheese: depending on your perspective you can see either the holes or the substance.

Notes

1. The historical approach was written by L. Chabason, the description of the present state was written by both authors, and the implementation section was written by C. Larrue.

References

Billaudot, F. (1991) 'Les mutations administratives de l'environnement', *Revue Juridique de l'Environnement*, No. 3, pp. 333–353.

Chabason, L. and Theys, J. (1990) *Plan National pour l'Environnement, Rapport préliminaire en vue du débat d'orientation*. Paris: Ministère de l'Environnement.

Charvolin, F. (1993) *L'invention de l'environnement en France*. Grenoble: IEP (PhD Dissertation).

Cole, A. and Doherty, B. (1995) 'France', in Richardson and Rousset (eds) *The Green Challenge*. London: Routledge, pp. 45–66.

IFEN (1994) *L'environnement en France*, 1994–95. Paris: Dunod.

IFEN (1995) *Opinion publique et environnement*. Paris: Bordas.

Jeannot, G., Renard, V. and Theys, J. (1990) *L'environnement entre le maire et l'Etat*. ADEF.

Knoepfel, P. and Larrue, C. (1985) 'Distribution spatiale et mise en oeuvre d'une politique publique: le cas de la pollution atmosphérique', *Politiques et Management public*, 3 (2): 43–69.

Knoepfel, P. and Weidner, H. (1986) 'Explaining differences in the performance of clean air policies. An international and interregional comparative study', *Policy and Politics* 14 (1):71–92.

Larrue, C. (1988a) *La pollution des eaux d'origine agricole: comportements agricoles et politique publique. Deux études de cas dans le bassin Artois-Picardie*. Rapport rédigé pour le Ministère de l'Environnement (SRETIE) et l'Agence de l'Eau Artois Picardie, L'OEIL/IUP, Université Paris XII, Créteil, septembre 1988.

Larrue, C. (1988b) *La pollution des eaux d'origine agricole comportements agricoles et politique publique. Deux études de cas dans le bassin Seine-Normandie*. Rapport rédigé pour le Ministère de l'Environnement (DPP/Service de l'Eau), L'OEIL/IUP, Université Paris XII, Créteil, septembre 1988.

Larrue, C. (1991) *Impacts et conditions de la mise en oeuvre de la lutte contre la pollution d'origine agricole; Etudes de cas en région d'élevage intensif et en région viticole.* L'OEIL/IUP Université Paris XII, Créteil, août 1991.

Larrue, C. and Prud'homme, R. (1993) 'European Environmental Policies and Environmental Protection in France', in F.G. Dreyfus, J. Morizet and M. Peyrad (eds) *France and EC Membership Evaluated.* London: Pinter Publishers Ltd, pp. 68–79.

Lascoumes, P. (1990) *Un droit de l'environnement négocié: volet discret d'une politique publique: contrats et programmes de branches, programmes d'entreprise.* Paris: PIREN/CNRS.

Maghalaes, H., Larrue, C. and Darbera, R. (1984) *Politique nationale et mis en ouvre du controle de la pollution par le SO_2 en France.* WZB: coll. IIUG Report No. 84:9. Berlin: WZB.

Mayntz, R. (1979) 'Les bureaucraties publiques et la mise en oeuvre des politiques', *Revue International des Sciences Sociales,* XXXI (4). 677–90.

Mény, Y. (1993) *Government and Politics in Western Europe.* Oxford: Oxford University Press.

Poujade, R. (1975) *Le ministère de l'impossible.* Paris: Calman Levy.

Prendiville, B. (1993) 'The "Entente Ecologiste" and the French Legislative Elections of March 1993' in *Environmental Politics*, 2 (3): 479–86.

Vallet, O. (1975) *L'administration de l'Environnement.* Berger-Levrault.

Chapter 5

Germany: The engine in European environmental policy?

Heinrich Pehle and Alf-Inge Jansen

On the environmental situation

Germany is characterised by strong interdependencies, both in ecological and in economic terms. Located at the centre of the European continent, Germany shares borders with nine other countries, and has a long coastline stretching from the North Sea to the Baltic Sea. This means that many of the environmental problems arising in Germany are shared with other European countries. Another key feature is high population density. The two former German states (the FRG and the GDR) were united in October 1990, and the 'new' Federal Republic of Germany includes 80 million inhabitants on 356,854 square kilometres (FRG). The density of population varies from 256 inhabitants per square kilometre (km) in the former FRG and 148 inhabitants per square km in the former GDR (after the unification in Germany called the *Fünf neue Länder*). Additionally, a number of other environmental aspects are dissimilar in the two former German states. Therefore, one can speak of 'two Germanys in one state'.[1]

As to the utilisation of land, 53.7 (56.9 in the former GDR) per cent is used for agricultural purposes, while 29.8 (27.6) per cent is forested and 1.8 (2.9) per cent is covered by water. The land use of traffic, housing and development areas, 12.2 (7.9) per cent is increasing while agricultural use of land is decreasing. Protected areas comprise 13.9 per cent (11.9 per cent in West Germany, 18.4 per cent in East) (OECD 1993:218). Another key feature is the highly industrialised economic structure, with industry's share of GNP being 38.2 (34.9) per cent (OECD 1993:21). Germany is the largest consumer of energy in Europe and about half of the energy used is imported. In the former FRG, oil contributes 40 per cent, coal 27 per cent, gas 17 per cent and nuclear power 14 per cent to the total primary energy supply. In the former GDR, lignite coal's contribution of total primary energy supply was 69 per cent, while 14 per cent was based on oil.

The size of forests in Germany has been rather constant but the quality of the forests has been changing: 27 per cent of the German woods have been declared 'severely' damaged and 41 per cent 'slightly' damaged. In the area of air pollution there are considerable differences between the former GDR and FRG. The former GDR, a country that to a great extent produced its energy by burning brown coal, did not develop a clean air policy, while the 'old' FRG, since the middle of the 1980s, has been successful in reducing air pollution on a large scale. Nevertheless, the dying of the forests (*Waldsterben*) has not been checked.

A major cause is increased emissions from automobile traffic.[2] The number of private cars jumped from 17.8 million in 1975 to 30.6 million in 1990. Additionally, new cars are, on average, bigger and more powerful. Consequently, the impact of modern 'end of pipe'-technologies, such as the catalytic converter, has been more than offset by the increased number of cars. Automobile traffic provides more than four fifths of all passenger services, and according to all estimates, automobile traffic will continue to increase. For many years Germany has been called Europe's transit country number one, referring to the traffic running north to south. It is also likely to become transit country number one for the dramatically increasing east-west traffic, as a result of the unification and the 'opening' of Eastern Europe.

Although international comparison shows German waste generation per capita is average, the high population density makes waste management a considerable problem. Most waste deposits are full and new suitable places for landfill are hard to find. In the beginning of the 1990s, the whole German waste disposal policy was in danger of entering into a vicious circle: restricted space for waste deposits in connection with high environmental standards of waste treatment, the lack of public acceptance of waste incineration, and inadequate capacities for recycling certain problematic waste materials put high pressure on German waste management. Extensive export of toxic and problem waste has been used as the safety valve.

The environmental situation in the former GDR is a special case. Actually, no serious environmental protection worth mentioning was practised until the reunification. East German lakes and rivers have suffered a similar fate as the forests. At the time of the unification, there were few data available on the environmental situation. Within a few years it became clear that the situation was alarming: 42 per cent of the East German rivers and 24 per cent of its lakes are contaminated to an extent which made it impossible to purify them by means of water purification units. Only three per cent of the rivers and one per cent of the lakes in East Germany were considered healthy. In addition, the new FRG was forced to deal with the problem of waste sites (*Altlasten*) of the former GDR. Tens of thousands of deposits, many containing hazardous waste, had to be cleaned up. The total costs to clean up East Germany's environment has been estimated at between DM82 and DM321 billion, the largest share of the costs going towards waste water treatment (OECD 1993:91).

Outline of the historical development of public environmental action until the 1970s

Historically, public environmental policy measures originated mainly from public concern about public health. Cholera epidemics on several occasions generated heated public debate on the necessity of public measures in this area, e.g. keeping water clean. Since 1900 all drinking water has been subjected to inspections by health professionals under the provision of local *Gesundheitsämter* (Rüdig and Kraemer 1995:61).

Another important tradition in German environmental policy and politics originates from the naturalists and natural history. Scientists involved themselves in public debate and publishing for 'the good cause' – the protection of nature. They established associations for the protection of birds and other species and for national monuments as well.[3] In 1904 the German Federation for the Protection of the Homeland (*Deutscher Bund Heimatschutz*) was established in Dresden,[4] and during the following decade nature conservation associations were established in federal states like Bayern and Preussen. Their leaders enlisted the support of public authorities. Stimulated by these efforts, the Prussian government from early on became involved in the protection of national monuments; for example, in 1906 it set up an office that was assigned with this responsibility (Jahn 1996:3).

In order to understand the historical development of environmental policy we have to keep in mind some of the historical characteristics of the German state. The 1871 Constitution of the unified second German empire defined the pattern of government until 1916. The state of imperial Germany concentrated on core functions of the nation state, such as defence, foreign policy, finance and internal order and also set up offices for such matters as railways and posts. Public policies in most other fields were first and foremost policies of subnational governmental levels, and public measures in the field of environmental policy first emerged at the *Land* level. These measures were aimed at protecting people from the negative effects of manufacturing industry on their health and save their property from damage. Such measures had their roots in the General Trade Regulations (*Gewerbeordnung*) of Prussia, adopted by the *Reich* in 1869, which decided that all manufacturing facilities which resulted in disadvantages, menaces, or other inconveniences for the owners or the residents in the vicinity of the emitting premises had to be licensed by the responsible authority; i.e. the local authority (Wey 1982:31). The *Gewerbeordnung* was far from establishing environmental standards in Prussia. It left the issue to the discretion of the local authorities, and consequently there was much regional and local variation.

The federal character of the imperial state was also decisive for the jurisdiction of the environmental administration which was set up at the *Reich* level at the turn of the century. The *Technische Anleitung Luft*, TAL (Technical Instructions for Maintaining Air Purity) of 1895 for local authorities may be regarded as the start of a clean-air policy. This TAL consisted, however, only of non-binding suggestions on how to avoid excessive formation of smog. The bureaucracy at the *Reich* level only played the role of technical advisor. The *Reich* gradually did take the role as sponsor of data collection and research for both air and water pollution. As for ordinary administrative agencies, the role of The National Water Administration, based on the Water Law (1912), was typical. This agency was to act under *Land* law, rather than according to legal competence of its own (Weale, O'Riordan and Kramme 1996:38). The non-involvement policy of the central government and the great variations in degree of action and type of measures taken by local authorities were to remain long-lasting characteristics of German environmental policy and organisation.

Actually it is not surprising that this rule existed in the FRG until 1964. All political systems existing in Germany during the twentieth century (the Republic of Weimar 1918–33, the Regime of the National Socialists 1933–45, the GDR 1949–90, and the FRG since 1949) had to deal primarily with a great variety of economic problems. In the early years of the FRG, economic growth dominated all other topics, and economic policy was successful. The economic recovery soon became internationally proverbial as the German *Wirtschaftswunder*, for which Germany clearly had to pay a high environmental price: rivers were covered with foam, the landscape was spoiled by random deposits of waste, and the sky over the industrial regions, for instance the Rhine-Ruhr area, was invisible because of permanent smog.

There were some political reactions to these developments. The established political parties, the Christian Democrats (CDU/CSU), the Liberals (FDP) and the Social Democrats (SPD), cohabited in a comparatively harmonious relationship, especially around agreement on the necessity of pursuing economic growth policies in combination with stable prices, high employment and a balance of payments surplus. But already in 1952, members of parliament founded a working group on environmental problems (*Interparlamentarische Arbeitsgemeinschaft*). In 1953 this group prepared principles for environmental protection which indeed were 'remarkably modern' for that time (Müller 1986:51). The same applies to the *Green Charter of the Isle of Mainau* which had been passed by several environmental groups in 1961. In particular there was a lot of concern about air pollution in the densely populated *Land* of North Rhine-Westphalia which includes the most industrialised region in Germany, the Ruhr area. During the 1950s, air pollution was made an issue in the politics of this *Land*, and in 1961 the Social Democratic Party, in its election campaign, tried to draw people's attention to the bad environmental situation with the election slogan 'The sky over the Ruhr must be blue again'. The CDU government of this *Land* tried to seize the initiative by an innovative legislation aimed at reducing emissions at source and the system of regulation was given extra resources at the municipal level. The development in North Rhine-Westphalia was partly followed up at the national level. In 1959 a Federal Air Purity Act was passed, but the associated new *Technische Anleitung Luft* was not issued until 1964. With regard to water pollution, industrialists had in the 1950s started lobbying for increased availability of clean water for industrial purposes. As a result the Federal Water Management Act of 1957 was adopted. Although these developments represented the adoption of new standards, they were rather marginal and were not sufficient to put environmental protection on the political agenda of the 1960s (Müller 1986:51–55).

German environmental politics

The National Environmental Policy Act passed in the USA in 1970 and the preliminary work for the first international conference of the United Nations about the environment (1972) were important for the development of environmental policy in Germany, in particular at *Bund* level. The formation of environmental policy by the Federal Government, which passed its first environmental programme in 1971, cannot be explained by the pressure of the German public. In fact, one can argue that governmental action drew public attention to environmental problems.

The environmental offensive of the Social Democratic–Liberal coalition government (the SPD–FDP coalition) has been interpreted as an effort by the Minister of the Interior, Hans-Dietrich Genscher (FDP), to seize the opportunity to position his party as the *Reformpartei* in German environmental policy (Weidner 1991:14). These initiatives led to the establishing of general principles to be pursued in environmental protection policy as well as to the anchoring of these principles in a legal foundation for environmental protection. Furthermore, this offensive resulted in the setting up of an organisation to provide high-level technical and scientific advice and providing research background to regulation.

The economy of West Germany was highly dependent on imported energy and was therefore particularly vulnerable to the dramatic rise in oil prices in 1973 and to the concomitant falling off in world trade. The government responded by introducing a vigorous regime of deficit spending, and the budget deficit remained a problem as social spending steadily increased under the SPD–FDP government[5]: thus environmental policy suffered. There was massive pressure from representatives of industry and labour unions to weaken the implementation of environmental measures. Official attention to clean air also declined from the mid-1970s partly because the Federal Ministry of the Interior was focused on the growing difficulties of getting the nuclear programme socially accepted (Boehmer-Christiansen and Skea 1991:187). The period 1974–78 has been called 'the stagnation phase of environmental policy' (Weidner 1991:14). The implementation deficit (*Vollzugdefizit*) of environmental policy was obvious, and persuasively demonstrated (Mayntz 1978). Consequently, the contrast between the ambitious goals of environmental acts and realised environmental quality became explicitly striking. Numerous conflicts occurred in which environmental interest organisations and especially citizens' initiatives (*Bürgerinitiativen*) were increasingly involved.

The numerous citizens' initiatives were examples of a new environmental and social consciousness that was put into practice (Markovits and Gorski 1995:81–86). Through a network of alternative workshops, services and alternative presses as well as various kinds of environmental initiatives (in particular anti-nuclear initiatives) a fast-growing number of new groups came to life. Most of them can be seen as part of a new social movement. By the end of the 1970s there were about 11,500 alternative projects in which some 80,000 persons participated in the FRG and West Berlin. With friends and sympathisers the movement included 300,000 – 400,000 people (Huber 1980:29–30; Markovits and Gorski 1995:99–110).

Some groups gradually mobilised at elections. Green candidates scored breakthroughs in local elections, which gave the impetus to launch *Land* parties. In 1978 the Greens won 3.9 per cent of the vote in Lower Saxony and 4.5 per cent in Hamburg. In 1979 an alliance of various groups that called themselves the Greens (*Die Grünen*) won 3.2 per cent, almost one million votes, in the European Parliament elections. In the same year the Greens had won a seat in the state parliament (*Landesparlament*) of Bremen, and by the end of 1982 had won seats in six *Länder*. Through these successes the Greens gained national media attention and built up financial resources and political credibility. In the national election in 1983 the Greens won 5.6 per cent of the vote and 27 seats in the *Bundestag*.

The German Green Party had different historical roots, some leading back to the peace movement of the 1950s and the New Left student movement of the late 1960s, and had its basis in the growing support for postmaterial values in a growing 'new middle class' that primarily was recruited from the growing quasi-meritocratic cadres of teachers, administrators, health and social workers in the rapidly growing and diversifying public sector (Markovits and Gorski 1993:4–14). *Modell Deutschland* was viewed as having led to the disintegration between the rationality of processes and the desirability of outcomes, and new social movements and the Greens can be interpreted as representing a corrective force rather than adversaries of rationality and the project of modernity (Offe 1987:90). Doubtless indeed, by their electoral success, the Greens had significantly changed the party system and had even more changed the dynamics and pattern of environmental politics.

The Greens' success had greatly been at the SPD's expense. To counterbalance the FRG's dependence on external energy sources, Chancellor Helmut Schmidt had launched a vast long-term programme for the expansion of nuclear energy production. This demonstrated, that out of concern with West German competitiveness, he rejected the environmental position to impose costly environmental measures on German industry. During the *Land* elections in the end of the 1970s the SPD lost activists and voters to the Greens. When the growing concern about acid rain and forest damage led to public uproar after the cover story of *Der Spiegel* in November 1981, *Waldsterben* and SO_2 pollutants became major issues in national politics. It has been argued that the first step towards reversing official German environmental policy was taken by the foreign minister Hans-Dietrich Genscher in June 1982, when he committed the FRG to the '30 per cent club' whose aim was to reduce SO_2 emissions from 1980 level by 30 per cent by 1993. (Boehmer-Christiansen and Skea 1991:192) The FDP leader was well aware of the Greens as a rival to his party at the ballot box. Under the leadership of Franz Josef Strauss, the Bavarian prime minister, and strongly encouraged by CDU/CSU-oriented groups in the Southern *Länder*, the CDU/CSU majority in the *Bundesrat* demanded action.

Also in the federal government a dispute occurred regarding air pollution, in particular power station-generated pollution. The most salient issue was the requirements to be defined in the Ordinance on Large Combustion Plants (*Grossfeuerungsanlagen-Verordnung*, GFAVO). Several ministries as well as organised interest groups were involved. The Ministry of Economic Affairs, the defender of the energy industry, emphasised economic arguments against more stringent emission abatement. The SPD–FDP coalition had already begun to fray by the early 1980s, partly because of internal conflicts within the SPD and the subsequent erosion of the power of Chancellor Helmut Schmidt and partly because of disagreement between SPD and FDP on economic policy. On 1 October 1982, the government collapsed. However, one month prior to its collapse the SPD–FDP coalition adopted a far-reaching environmental protection programme in which the draft GFAVO was only a part. To a great extent this programme was included as important elements of the new government's subsequent environmental policy (Weidner 1991:15). It has been pointed out that there were small differences between this programme and what the Greens had proposed. Also, neutral observers have suggested that this case illustrates the phenomenon which is referred to as *Themenklau*; i.e. the theft of one party's policy themes by another (Boehmer-Christiansen and Skea 1991:193).

The SPD and its leading cadre were badly shaken by the electoral success of the Greens in 1983. The SPD had consistently pursued policies of economic growth and high employment. Thus, SPD policy was not only influenced by the labour movement, but also by organised interests of West German industry which constantly had pointed to the impact of environmental measures on industrial costs and competitiveness. During the remainder of the 1980s the SPD went through a significant 'greening'; most notable was its post-Chernobyl commitment to dispense with nuclear energy within a decade. The CDU/CSU was also up for 'greening'. In the new CDU/CSU–FDP coalition, the right-wing Friedrich Zimmermann (CSU) became Minister of the Interior. The new government coalition launched a campaign for clean air advocating strict requirements in GFAVO and the adoption of US type vehicle emission limits. After the re-election, Zimmermann ensured that strict emission limits were to characterise the content of the GFAVO. The new air-pollution control legislation became law in July 1983.

At the EU level, Chancellor Kohl's government pushed for an EU agreement on a reduction in acid emissions. A framework directive was signed in June 1984 and the subsequent EC directive on large combustion plants is largely modelled on German domestic legislation (Boehmer-Christiansen and Skea 1991:233–250). Similarly the government took the role as pacemaker, or *Schrittmacherrolle* (Weidner 1991:15), in another area when Zimmermann in July 1983 surprised the opposition with his announcement tightening the requirements on traffic-generated air-pollution. This announcement caused a sensation because the political weight of the German motor industry is well-known. When the EC Commission was persuaded by Germany to propose the equivalent of US standards for new vehicle types, this initiative failed due to vigorous opposition, particularly from Great Britain.

After the Chernobyl disaster in April 1986, the Minister of the Interior was not up to handling the situation. At first he minimised the consequences in public, and afterwards he was not able to co ordinate the other responsible authorities in order to limit the damage. In Lower Saxony the elections for the *Landtag* were underway and the CDU suffered because of the bad management of the consequences of Chernobyl. In this situation the Chancellor acted quickly and resolutely. On June 1st 1986 he set up the *Federal Ministry of Environment, Nature Protection and Reactor Safety* (ME).

The CDU/CSU–FDP coalition benefited from the reorganisation of the environmental policy organisation. The coalition remained in power in Lower Saxony, and, more important, was re-elected in the national election in 1987. It won a majority in Hesse, with the Minister of Environment, Walter Wallmann, as *Spitzenkandidat*. His successor as the Minister of Environment was Klaus Töpfer.

German environmental politics changed significantly during the course of the 1980s. The Greens had propelled environmental issues onto the public agenda and changed the terms of party competition. The environmental interest organisations had been strengthened and environmentally more radical action groups and movement associations had emerged as important actors.[6] The German people had become highly conscious of the threats and damage to the environment and the saliency of environmental issues in German politics had significantly increased. The relation between the governing coalition and its opposition had changed. The CDU/CSU–FDP coalition pursued environmental policies that mainly represented a continuation of those pursued by the previous SPD–FDP coalition.[7] But while the political opposition during the SPD–FDP coalition resisted more environmentally oriented policy measures, the opposition at the start of the 1990s was in favour of such measures. Moreover, there was much-increased media interest in environmental issues and subsequently a highly critical media coverage of the government's environmental policies and their results (Beuermann and Jäger 1996:192).

On one side the political opposition, the media and most environmentally informed commentators pointed to the weaknesses and shortcomings of the environmental policies of the CDU/CSU–FDP coalition. On the other side, representatives of the government, particularly the Minister of the Environment, pointed out that the Kohl government, in accordance with its policy strategy, had made the FRG a front-runner in the EU and in international environmental co-operation. As a result of this strategy, Germany had gained an international reputation.

In the 1990s, climate policy has developed into a salient policy area that has highlighted characteristics of German environmental policy and has given rise to new developments in environmental politics as well. The closing of this section will, therefore, be focused on climate politics.

In 1986, German physicists issued an alarming statement about global warming and the CO_2 emissions that drew wide attention. Similar statements by physicists, meteorologists and climatologists warning about the threats of global warming and ozone depletion were published in 1987. The ensuing public debate was greatly intensified as the concern with climate change was linked to the future of nuclear power in the FRG. The Chernobyl disaster in April 1986 had rekindled the divisive debate on nuclear power in FRG and the debate focused on an exit (*Ausstieg*) from nuclear power. The Greens and environmental associations demanded an immediate shutdown of all nuclear plants, and the Greens made significant gains at the ballot box. In January 1987 this party received 8.3 per cent of the vote in the federal elections (up from 5.6 per cent). The SPD in the autumn of 1986 adopted the position that nuclear power was to be phased out completely over a ten-year period (the nuclear power plants were to be partly replaced by coal-fuelled power plants). The issues raised by the above mentioned physicists were taken up by supporters of nuclear energy in Germany to counterbalance the arguments in favour of *Ausstieg*. The CDU had consistently been in favour of nuclear power and quickly took up the position that support of nuclear power made good environmental sense when confronted with the ominous threats posed by global warming (Hatch 1995:422).

Increasingly, public discussion centered on CO_2 and other greenhouse gases (GHGs) in conjunction with ozone depletion and nuclear power. The *Bundestag* in October 1987 unanimously approved the setting up of the Inquiry Commission on Preventive Measures to Protect the Earth's Atmosphere (*Enquete-Kommission Vorsorge zum Schutz der Erdatmosphäre*), to collect evidence on global changes in the Earth's atmosphere, up-to-date knowledge of the cause-effect relationships involved and to propose national and international measures of prevention and control in order to protect both man and the environment. The Inquiry Commission consisted of 11 members of the *Bundestag* and 11 scientists.

As a result of the setting up of the Inquiry Commission in December 1987, significantly more resources were made available to sponsor studies and hold hearings that drew on world-wide expertise (Hatch 1995:423). The commission's first interim report was based on hearings with scientists, representatives of industry, federal ministries and politicians as well as discussions with representatives of environmental groups. Recommendations for national targets and measures to be taken to contain the greenhouse effect were left for later reports, but the commission concluded that there was 'an extraordinary need' for action. It is noteworthy that the report was unanimously approved by the commission.

In its third report, released late in 1990, the commission proposed a 30 per cent reduction of CO_2 emissions from 1987 levels by the year 2005.[8] Although there were differing views on the role of nuclear power, all members agreed that energy conservation and improvement of energy efficiency should be given priority.

It has been widely concluded that the Inquiry Commission of 1987 was most decisive in these formative years of German climate policy. The commission filled a political vacuum left by the government and the political parties, which were not prepared to act on the global warming question, let alone provide leadership (Hatch 1995.425). After the broad consensus had been established in the commission, the parliamentary members promoted its views in their party and constituencies. Even as the focus was concentrated more and more on the question of appropriate policy responses, the views of the commission had the role of a baseline against which the positions of the government, its ministries and agencies, political parties and organised interests were measured.

In June 1990 the federal government agreed to a goal of 25 per cent reduction of the CO_2 emissions in the former West Germany based on 1987 levels, which the Minister of Environment had called for. In the former East Germany, larger reductions were expected, and in December 1990 the government stated that Germany's goal was to reduce CO_2 emissions by 25–30 per cent by the year 2005 based on 1987 levels. Within the government and the central administration there had been a great deal of politics in the process of setting this goal, and decisions as to the choice of strategy for fulfilling this ambitious target proved to be even more difficult and divisive.

After the federal elections in December 1990, in which the CDU/CSU–FDP was re-elected, the coalition partners agreed to work for the adoption of a CO_2 tax, but it was indicated that the government should make efforts to tie the German CO_2 tax to a European Community-wide climate-protection tax as soon as possible. Representatives of German industry interests, among them the Federation of German Industry (*Bund der Deutschen Industri*, BDI), the umbrella organisation for big industry, actively and persistently emphasised the necessity of adopting a CO_2 tax only as a part of an international framework; i.e. an EC-wide agreement was not enough to avoid competitive disadvantages for German industry in the world market.

With economic slowdown and increasing costs of German unification, the support for a German CO_2 tax decreased in CDU/CSU–FDP circles during the latter half of 1991. A CO_2 tax was also increasingly attacked by the opposition (the SPD and the Greens as well as environmental-interest organisations like the BUND and the DNR) because nuclear power would be exempted. The opposition recommended a tax on all energy sources with the exception of renewable ones. It is notable that the SPD also wanted to exempt coal. In December 1991 the federal government stated that for reasons of ecological effectiveness and German competitiveness an international agreement was necessary. The government supported the introduction of an EU-wide CO_2 and energy tax.

As widely reported, European industrialists were intensely lobbying against an EC-wide tax, and in a new proposal the EC Commission made any EC energy tax conditional on other OECD countries adopting similar taxes or measures (Liberatore 1995:64). In October 1992, a publicly disclosed letter from the Chancellor to the president of the BDI stated that a national tax was no longer anticipated, and that the Chancellor believed that it was necessary that an agreement with other OECD member states should be sought (*Die Welt*, 28 Oct 1992). During the German EU Presidency in the second half of 1994, Klaus Töpfer announced a new initiative to have the EU adopt a CO_2/energy tax. In December 1994, the Council eventually determined that there would be no tax set at EU level but member states were encouraged to develop their own taxes.

The general resistance of the business sector and trade unions to effective environmental policy has slackened, in part due to new markets for FRG-tested environmental products and technologies. There has been increased support for the argument that environmentally sound production and products would be necessary for German competitiveness in international markets. An ecological tax reform has received support from different NGOs and also from large well-known enterprises (e.g. the BUND's mutual statement with the Bundesverband Junger Unternehmer and the BUND's mutual declaration with 16 large enterprises, Weidner 1995:60).

It should also be noted that during the 1994 federal elections both the big political parties made efforts to link environmental protection with economic change and employment. However, the general pattern of German environmental politics appears to be stable. At the most recent elections the established pattern among the political parties from the 1980s prevails. Marginal gains for the Greens at the 1994 federal elections and later at some *Land* elections are of no significant consequence.

The environmental policy organisation: jurisdictions, policy-making process and policy system in practice

Environmental policy in Germany has not only been the result of the politics that has developed around these issues. The course of its development has also been strongly influenced by the dynamics between policy content and the policy organisation, mainly at *Bund* level.

Environmental policy and federalism

Germany is a federal state that consists of four levels or tiers of government: *Bund*, *Land*, *Kreis* and *Gemeinde*. A distinctive feature of its federalism is that the states (*Länder*) both participate in making and administering federal policies under a constitutionally assigned responsibility of their own. There are sixteen *Länder*, 543 *Kreise* (districts), 16,128 *Gemeinden* (communes) and 117 *kreisfreie Städte* (district-cities). To understand the making and implementation of environmental policy in Germany it is therefore necessary to understand the vertical distribution of responsibilities among *Bund*, *Länder*, *Kreise* and *Gemeinden*, as well as the horizontal division of labour at the respective political-administrative levels.

First of all, one should keep in mind that, with few exceptions, the *Bund* does not have its own administration for the execution of its laws, but depends on the executing administration of the *Länder* and *Gemeinden*. Secondly, it is important to know that the governments of each *Land* (*Landesregierung*) participate in the law-making process of the *Bund* through the *Bundesrat,* whose members are also members of a state government. In plenary sessions the delegation from each *Land* is obliged to vote as a group in accordance with the instruction of their *Landesregierung*. They have the right to propose laws and to comment on proposals of legislation from the Federal Government (*Bundesregierung*). All legislative proposals that directly concern the *Länder*, not only constitutional amendments, need the approval of the *Bundesrat* as well as the majority of the lower chamber, the *Bundestag*. The same applies to decrees and administrative rules, many of which are of great importance for environmental policy. The *Bundesrat* has a suspensive veto over the rest of the federal laws, which can be overridden by a renewed majority vote in the *Bundestag*.

This clearly demonstrates that the constitutional conditions can make policy making in FRG quite time-consuming. This situation is exacerbated by the fact that since 1949, the FRG has been ruled by coalitions. Moreover, political decision making becomes much more complicated if the party-political majorities in the *Bundestag* and *Bundesrat* are different, as they were during the time of the SPD–FDP coalition (1969–82) and again since 1991. The following sections will deal with various consequences for environmental policy resulting from this division of labour and the configuration of political forces in which it is embedded.

The national level (*Bund* Level)

The legislative powers of the **Bund** *in environmental policy*

Through an amendment of the Basic Law in November 1994, environmental protection was made, as a state goal (*Staatszielbestimmung*), part of the constitution. Previously competencies concerning the responsibility of the *Bund* for environmental legislation were only given implicitly in two enumerations in the Basic Law. They named those fields in which the *Bund* had the legislative power. The *Bund* is empowered with this right, if a certain subject cannot be regulated satisfactorily through the laws of the single *Länder,* or if a federal-wide rule is needed to ensure the 'homogeneity of living conditions' throughout the country. The implication for environmental policy was that the *Bund* was able to regulate two fields: nature protection and water management. This has been done through so-called framework laws, which have existed since the founding of the Federal Republic. These framework laws of the *Bund* are filled in by laws of the *Länder*; i.e. the framework laws have to be formulated in a way that leaves some room for the Länder's own regulations. Until the beginning of the 1970s, the two framework laws on nature protection and water management were the only environmental regulations made by the *Bund*, apart from the law for the peaceful use of nuclear energy.

The basic principles that in 1971 were launched in the *Bundesregierung*'s first environmental programme are still in force, e.g. the priority of the *precautionary principle* and the *polluter pays principle*. In the view of the *Bundesregierung* it was necessary for the realisation of this programme to provide further environmental political law-giving powers to the *Bund* (Genscher 1971:3). The *Bundestag* and the *Bundesrat,* with the necessary majority of two-thirds, changed the Basic Law. Since 1972 the *Bund* has been empowered to pass legislation to regulate waste management, and air and noise pollution. Subsequently, the *Bundesregierung* and its majority in the *Bundestag* were only partly successful *vis-à-vis* the interests of the *Länder*. Its demand for full regulative competencies for nature protection and water management were blocked by the *Bundesrat* (Müller 1986:83). In other words, the legislative reform in the establishment phase was influenced by one of the key features of the German political administrative system, the influence of the *Bundesrat* in the law-making process.

The centre of the environmental policy organisation: the Federal Ministry of Environment

As to the environmental policy organisation the federal government in 1971 set up the Council of Environmental Experts (*Sachverständigenrat für Umweltfragen*), responsible for providing scientific advice to the federal government. In 1974 the *Bund* established a new agency, the Federal Environmental Agency (*Umweltbundesamt*), responsible for environmental research, documentation and information. However, the questions concerning the organisation of environmental policy at the national level are primarily about the distribution of responsibilities within the *Bundesregierung* under the supremacy of the Chancellor.

During the SPD–FDP coalition, no strengthening of environmental policy through organisational reforms at the *Bund* level took place. No organisational changes were made by the CDU/CSU–FDP coalition when it took office in 1982 or when it was re-elected in 1983. For this reason as many as seven ministries had responsibilities for environmental policy of the *Bund* until 1986. The most important ministries were the Ministry of the Interior, which was responsible for 'technical environmental protection', and the Ministry of Agriculture, which was responsible for nature protection. Furthermore, tasks concerning environmental protection were carried out by the Ministries of Transport, Health, Regional and City Planning, Research and Technology and last but not least by the Ministry of Economic Affairs (Hartkopf and Bohne 1983:195). In the view of the concerned politicians, environmental policy had always been a field among others which was not very important. In particular, the Ministries of Economic Affairs, of Agriculture and of Transport were more concerned about polluters' interests than about environmental values.

Chancellor Kohl's resolute action in setting up the ME in June 1986 shows the crucial role of the chancellorship in German policy making.

The Ministry of Environment, Nature Protection and Reactor Safety (ME) was to be responsible for:

a) environmental protection, the security of nuclear plants and radiation protection (previously assigned to the Ministry of the Interior);
b) environmental and nature protection (previously assigned to the Ministry of Agriculture; and
c) health interests concerning environmental protection (previously assigned to the Ministry of Health).

The ME not only received these formal competencies, it also took over the staff working in these areas. Thus the work of the ministerial bureaucracy could continue easily.[9]

A significant additional assignment was given in 1988, when the Chancellor's office transferred the climate issue from the Ministry of Transport to the ME. Three central agencies report to the ME: The Federal Agency for Nature Conservation (1993) is responsible for planning and implementation under the federal nature conservation act. The Federal Office for Radiological Protection (1989) is responsible for implementation under the Act on preventive radiological prevention and the Atomic Energy Act. The importance of the third, the Federal Environmental Agency, should be emphasised. Although mainly a non-executive agency, it is considered to be the most important agency in the environmental policy area, through its role as an information centre and its influence on the public debate (Weidner 1995:31). Several advisory bodies are also attached to the ME (e.g. the Council of Environmental Experts, the Radiological Protection Commission).

It is difficult to evaluate the results of the establishment of the ME, and no definite answer can be given. Of great interest, however, are the answers of a number of ME officials who were interviewed about the new set-up one year later (Pehle 1988a, 1988b). These officials had previously been working on environmental questions in the Ministries of the Interior and of Agriculture, and could, therefore, compare the 'old' and 'new' solution.

These officials pointed to a disadvantage of the organisational changes that was related to the fact that the flow of information to a great extent follows the limited lines of the formal ministerial organisation. Consequently, the officials in charge in the ME often received information from other ministries about plans that were important for environmental protection at a very late date.

The officials saw it as advantageous that the Minister of Environment and his staff had been able to gain status through successful passing of environmental protection measures. Environmental protection was no longer the responsibility of ministries which also had to consider the polluters' interests. This was seen as very important because German ministries traditionally speak with one voice when they give their opinion to other ministries on, for instance, new laws that are being drafted. Previously several ministries had an environmental department, but this department seldom had spoken on behalf of its ministry because of opposition (mostly based on economic premises) from other departments of the ministry in question; i.e. the environmental department so often failed in its own ministry. This situation changed in 1986: the conflicts between environmental protection and other interests (mostly economic growth interests) were transformed from intra-ministry conflicts into inter-ministry conflicts. The conflicts had to be discussed and solved at a higher level, often top level, by the affected ministries, and could now also be discussed in public. To elaborate on this point: the responsible department of a ministry has to co-ordinate a law-draft with the other ministries involved before it is discussed and ultimately passed by the Cabinet. If the officials of the different ministries are not able to reach an agreement, the ministers are called in. If they also are unable to reach a decision, the proposal has either 'died' or the decision will be made in the Cabinet. Here the Chancellor's attitude will be decisive. It is in this context that the effects of the setting up of the ME in terms of whether or not the reform improved the environmentally-oriented politicians' prospects of success have to be judged.

On the relations between ministries in environmental policy

Before the establishment of the ME, conflicts between environmental protection interests and other societal interests to a great extent manifested as intra-ministry conflicts. When the ME was established, many of these conflicts were turned into typical inter-ministry conflicts. Illustrative of this are findings from the interviews with ME officials referred to above. Officials who, before they were transferred to the ME had been responsible for nature protection in the Ministry of Agriculture, said that their channels of communication had become very narrow after their transfer to the ME. They pointed out that competing ministries act in a tactical way towards another, and that the ME is informed rather late, if at all.[10]

In the area of waste management, such inter-ministry conflicts have also had demonstrable effects. Because of strong opposition from the Ministry of Economic Affairs, the Comprehensive Waste Management Act (*Kreislaufs-wirtschaftsgesetz* 1994) was on the government's agenda for more than four years (Weidner 1995:85). Inter-ministry conflicts and politics are highlighted in the case of German climate policy.

Based on the information from the Inquiry Commission of 1987, the Chancellor in early 1990 asked the ME to prepare a Cabinet Decision for a CO_2 reduction target (Beuermann and Jäger 1996:194). ME officials interacted informally with their colleagues at the Chancellor's office, worked on a feasibility study and consulted the Federal Environment Agency. As to the national target for CO_2 emissions, the government mainly followed the conclusions and recommendations of the ME, against strong reservations from the Ministry of Economic Affairs.

As the ambitious target had been set, the attention focused on the question of choosing means to reach the target. Ministries that are responsible for other policy sectors have indeed also interests in environmental policy, *in casu* climate policy. The Ministry of Economic Affairs as well as the Ministry of Transport, of Building and Planning, of Agriculture (Nutrition and Forestry) and of Research and Technology were important actors in interdepartmental working groups as well as in other efforts to develop a strategy for choosing instruments and developing measures. The ME and the Ministry of Economic Affairs, and the relation between them, became crucial.

The ME and its minister argued that Germany should take the role of pacemaker (*Vorreiterrolle*) in the international effort to curb CO_2 emissions, and proposed that a CO_2-levy should be introduced, regardless of whether such a tax was introduced in other countries or not. The Ministry of Economic Affairs and its minister, on the other hand, argued against such an instrument, emphasised the risk of jeopardising the competitiveness of Germany's economy and concluded that the minimum alternative was an EC-wide solution.

In 1991, as the costs of the unification and the economic slowdown were considered to be the most important issues, the Minister of Environment achieved less support from the Chancellor and other ministers. In December 1991, the government stated that an EC-wide solution was necessary. Michael T. Hatch has summarised the result of the process as follows:

> In essence, pressures from powerful interests, combined with the divisions within government itself, made it much more attractive to shift responsibility to the EC for a decision that was bound to impose heavy political costs. (Hatch 1995:31)

Furthermore, the process at EU level is equally interesting. In September 1991, the EC Commission proposed a combined CO_2/energy tax, which provided exceptions for energy-intensive industries. After intense lobbying by European industrialists against an EU-wide tax, the EU Commission in a new proposal made any CO_2/energy tax conditional on other OECD countries adopting similar taxes. In December 1994, the EU Council determined that no tax would be set at the EU level.

Although several other measures have been introduced by the government, the reluctance to introduce a CO_2-tax in Germany is interpreted as a defeat for the ME and its minister. It should be noted that the new Minister of the

Environment, Angela Merkel, at the beginning of 1995 made statements expressing optimism about achieving the national target.

The case of climate policy also shows the importance of the Chancellor in German public policy. Chancellor Kohl supported the ME and its minister when the national target for CO_2 reductions was to be decided upon. As the governing coalition became more vulnerable because of growing costs of unification and of economic slowdown, Chancellor Kohl transferred his support in favour of the position of the Ministry of Economic Affairs.

In short, the case of climate policy demonstrates the importance of inter-ministerial relations, and in particular the distribution of authority in environmental policy. These relations are embedded in a wider economic, social and political context, and the ministries' relations to different societal interests and in particular these interests 'weight' in the FRG political order is of great consequence for deciding German environmental policy. It is typical that while the initiatives of the Ministry of Economic Affairs

> ... are supported by intense lobbying of industry, BMU (the ME) has to compensate its lack of clout by intense public relations. (Beuermann and Jäger 1996:196)

Indeed, this process illustrates that it is not without reason that the MEA has been described as the 'attorney for German industry' (Paterson and Southern 1991:250). In addition, initiatives from the ME in this area have also had little impact on the Ministry of Transport. Without doubt, one of the main threats to German climate policy are the increasing CO_2 emissions from road traffic (Huber 1997:67, 79; ECMT 1997:47). German transport policy is still primarily concentrating on developing the infrastructure in order to satisfy what is considered to be the necessity to cope with the increasing traffic (Beuermann and Jäger 1996:210).

Environmental policy content: acts, instruments and principles
The federal government was able to use its powers consistently during the first half of the 1970s: based on its environmental programme of 1971, the Waste Disposal Act was passed in 1972, followed two years later by the Federal Air Quality Protection Act (*Bundesimmissionsschutzgesetz*). The framework laws for nature protection and water management were amended in 1976, and the Waste Water Charges Act (*Abwasserabgabengesetz*) was passed the same year. The federal government also took initiatives through its policy of internal reforms (*Politik der inneren Reformen*). By these measures the Federal Government had laid down its goals and principles of policy content and applied them in legislation in several areas of environmental protection.

At the beginning of the 1980s the development of the GFAVO was a watershed in German air pollution control policy. It was applied to all combustion plants rated over 50 MW thermal, and it was the first time that a time limit had been set for the reorganisation of such plants. In order to reduce emissions at source, strict emission limits were set for major pollutants like SO_2, NO_x, CO, halogen compounds and various metals. A significant reduction in air pollution by power plants followed; SO_2 emissions were reduced from 1.6 million tonnes (mio/t) in 1983 to less than 0.2 mio/t in 1989, and NO_x emissions were reduced from about 0.8 to less than 0.4 mio/t during the same period (Mez, 1995:175).

Since the establishment of the ME, important initiatives regarding content of environmental policy have been taken in various areas. Several laws and decrees have been passed, such as the Waste Avoidance Act (1986) and the comprehensive Waste Management Act (1994), the Environmental Impact Assessment Act, the Environmental Liability Act (1990) and the Packaging Ordinance (1991). As to policy content, the establishing of national targets for the reduction of pollution, of which the reduction of CO_2 emissions by 25–30 per cent by 2005 (using 1987 as the base year) is the most ambitious, are significant decisions in German environmental policy.

Despite numerous measures, waste management continues to be a severe problem: the new federal decree which went into force in June 1993 (with transition periods between eight to 12 years) prohibited the direct dumping of nearly all substances which are contained in 'normal' household waste in order to protect soil and ground water. This meant that most of the waste had to be treated in an incinerator before the remaining material could be deposited in a landfill. But construction and operation of incinerators was met by strong opposition from citizens throughout the country.

The reduction of packaging material was the declared aim of a subsequent federal decree passed in 1991. It originally aimed at reducing packaging material by forcing all enterprises to accept returned materials from the consumers, thus avoiding excessive packaging at the initial stage. This concept had been opposed by influential commercial and industrial interest groups. The political process resulted in a compromise. As a consequence of the federal decree on packing materials, a private system of packing-waste removal, besides the public (local) one, came into existence. Further, this decree mandated that certain quotas of the gathered material be recycled. After its initiation, the new private waste-removal system (*Duales System*) was strongly criticised because it had not yet been able to fulfil its obligations, especially for plastic materials for which suitable recycling technology did not exist. Consequently, these materials were either incinerated in Germany or exported to other countries. In short, one could observe a discrepancy between principles laid down in the legislation, and the problems arising in the implementation of these principles. In 1994 the parliament passed the comprehensive Waste Management Act which established a 'cradle to grave' responsibility for goods in the market.

This act, as well as other acts mentioned above, are (in spite of the implementation problems) considered to be an attempt to develop instruments that are active and cause-oriented, rather than reactive. Also other initiatives have been taken in order to establish alternative approaches in environmental policy. The traditional, regulatory approach has been criticised for its lack of flexibility and inefficiency, both by industry and various experts. The government is increasingly aiming at developing a strategy that puts stronger emphasis on information and negotiation as well as on voluntary agreements. Organised interests of German industry have most strongly argued that voluntary agreements are the most market sensitive and, therefore, most effective environmental policy instruments. Others, particularly environmental organisations, have, however, found reason to doubt this. They argue that voluntary agreements only work when they are backed by threats of economic loss or regulations. A definite answer cannot be given, but a clear finding as to industry's behaviour in the case of reducing CO_2 emissions was that as recession mounted, the enthusiasm of industry to make binding agreements receded. In this context the pattern of behaviour has been described as

> ... reverse piggy-backing, consolidation of established interests around core positions, the absence of precaution and only cosmetic compliance on the part of industry when it comes to implementing voluntary restrictions. (Beuermann and Jäger 1996:211)

However, legislation and the regulatory approach continue to dominate German environmental policy. Economic instruments do not play an important role, although there are important exceptions (for instance, in water management, transport and waste management). The strong emphasis on legislation and regulation shows that the central principles laid down in the early 1970s, the *Vorsorge*-principle ('the precautionary principle') and the *Verursacher*-principle, still are central guidelines for German environmental policy.[11] The strong reliance on the regulatory approach and the reluctance to introduce economic instruments (such as environmental taxes) in German environmental policy, must (as the above discussion of the climate policy demonstrates) be analysed with regard to particular processes within the different environmental policy areas.

Furthermore, the above general principles are interpreted and operationalised through *Stand der Technik*, which by the Constitutional Court has been defined as 'standards generally accepted if specialists, who have to apply them in practice, are convinced of their soundness' (Boehmer-Christiansen and Skea 1991:169). This principle implies that both lawyers and other types of professionals, and their professional associations, in the various fields, have a crucial role in defining the acceptable standards, and norms of proper technical practice. As pointed out by Rüdig and Kraemer, the technical norms developed by these professional associations have no legal standing *per se*, but they define proper technical practice and are empirically followed as rules (Rüdig and Kraemer 1995:63). Thus the policy process relies strongly on technical expertise in the interpretation of what is to be regarded as acceptable standards and for translating them into practical decisions. For instance, in the policy process in the area of air pollution, the German Society of Engineers has traditionally played an important role in the setting of pollution control standards. The number and characteristics of these organisations which represent such technical expertise in the policy process varies from area to area. For instance in the water policy area, which is crucial for water management and pollution, there are several important organisations. In addition to the *Länderarbeitsgemeinschaft Wasser* (LAWA) which consists of civil servants from the Länder (see below), there are three important professional organisations at national level that play an important role in different aspects of water management; the German Association of Gas and Water Experts (concerned with water supply), the German Association for Water Resources and Land Improvements (primarily concerned with the agricultural aspects of water management) and the Association for Waste Water Technology (Rüdig and Kraemer 1995:63).

In addition to these professional organisations, those which represent industrial interests in different sectors are important actors in policy making; for example, in the case of water management, the Federal Association of Gas and Water Industries, the main lobbying organisation for the German water supply interests, co-operates closely with professional associations.

Illustrative of the co-operative and voluntary aspects of the water policy process are policies to limit German farmers' use of nitrate-based fertilisers and pesticides. As a result of an amendment to the Water Management Act passed in 1986, farmers can be compensated financially for their loss from such limitations. The implementation of these policies is dependent on the co-operation between farmers and water utilities and on voluntary measures. The role of government, *Land* and federal, is mainly that of advisor and educator (Rüdig and Kraemer 1995:69–70).

These key features of German policy making are closely related to the third governing principle for German environmental policy, the co-operation principle (*Kooperationsprinzip*), a principle requiring government to regulate only after having consulted affected parties. This principle, firmly anchored in German federalism, surely has an effect on the environmental policy process. However, being a principle that merely formulates the opportunity for a large number of non-governmental and governmental actors to participate in the policy-making process, the effect of the different actors on the policy output is highly dependent on other factors, such as the technical and economic resources they control and are able to mobilise. In other words: general principles are concretised through interpretative and negotiation processes between actors that participate in policy making and implementation in the different environmental policy areas and policy sectors. Consequently, these processes are shaped by the characteristics of the different policy areas and sectors. This is also the case when we consider the role of the *Länder* and the local governments.

Environmental policy making and implementation: the function of the *Länder*

The *Länder* of the FRG are territorial authorities with the characteristics of states. They have their own constitutions, parliaments and governments. Some of the *Land* constitutions explicitly oblige the governments to protect the environment. Many *Land* governments were years ahead of the *Bund* with regard to setting up a ministry of environment. The first ministry of environment of a *Land* was set up in Bavaria in 1971. But there is not much room left for the parliaments and governments of the *Länder* for making and enacting their own environmental laws. As seen above, in all important fields the *Bund* has the legislative power and has made use of it. Even if the *Bund* only has the right to enact framework laws (e.g. water management and nature protection), these rules are often very detailed. Therefore, in spite of Germany's federal structure, it is justified to speak of one German environmental policy. The *Länder* have their own political possibilities, in particular in connection with the execution of federal environmental laws. The only law which is executed by the *Länder* on order of the *Bund* is the Nuclear Energy Law. Within this area the Federal Government, represented by the Minister of the Environment, has the power to give orders to the governments of the *Länder*. That has happened several times.

The other environmental laws are executed by the *Länder* as 'their own affairs'. This means that they are allowed to decide independently on establishing the necessary administrative apparatus and defining its duties. It is a matter for the authorities of the *Länder* to supervise the quality of the environment, and to grant the required permission for construction and operation of industrial plants, power plants etc. In some areas, however, the *Bund* also provides regulations to guide the implementation of environmental legislation. That happens by enacting administrative rules. The most famous example is the *Technische Anleitung Luft*. However, the quality of the implementation is largely decided by the *Länder*.

In effect the *Bund* itself decides on the nature of environmental laws. Nevertheless, the *Länder* play an important role in the process of environmental policy making.[12] As the environmental laws are executed by the *Länder*, they can only come into existence with the agreement of the *Bundesrat*. The Federal Government is authorised by a great number of these laws to enact regulations and administrative rules. This procedure relieves the *Bundestag* from considering detailed technical questions. But again these regulations need the approval of the *Bundesrat*; in such cases control over the Federal Government is assigned to the representatives of the governments of the *Länder* in the *Bundesrat*.

This tendency is supported by another distinct feature of German federalism. A quite essential element of German political culture is the demand for a 'homogeneity of living conditions'. This phrase is also included in the Basic Law, but its roots in the political mind of the population is just as important.

In effect, political decision-makers are pressured to co-ordinate their efforts, because of constitutional reasons and because of the need to be accountable to their voters.

In order to establish a homogeneous policy, a rather complicated and fragmented network of different committees and working groups is used. Co-ordination needs to be horizontal as well as vertical. Horizontal co-ordination means that the *Länder* try to agree on common environmental standards in those spheres which they may legislate themselves.

This phenomenon is often referred to as the 'third level' of federalism. It is not provided for in the constitution, but it has two important functions. Most of all it is used for the establishment of homogeneous rules, water legislation being a prime example. The *Länderarbeitsgemeinschaft Wasser* (LAWA) is in charge of this. The responsible ministers of the *Länder* work together in this study group. The representatives of several federal ministries have the status of observers without the right to vote. The representatives of the *Länder* make their decisions according to the principle of consensus. The other important function of the third level is the definition of a coherent statement of the position of the *Länder vis-à-vis* the *Bund*. This is important for the subjects which are regulated by federal laws. The laws have to be executed by the *Länder,* and they expect the *Bund* to pass laws which are practicable for the executing authorities. Corresponding negotiations with the *Bund* are prepared, e.g. with the *Länderarbeitsgemeinschaft Abfall* (LAGA), which deals with waste-management. In this case, the horizontal and the vertical co-ordination intersect with each other. The latter takes place in the so-called *Bund–Länder* committees. Through these negotiations between governmental representatives the actual law-giving process is simplified. A blockade by the *Bundesrat* at a later date is, therefore, unusual.

The bargaining between the governments does not result in a binding law. *De facto* it leads to preliminary decisions, which most of all confront the parliaments of the *Länder* with *faits accompli*. The parliaments have increasingly become mere institutions for the ratification of the administrative agreements. The democratic element of environmental policy is, of course, affected by this fact, which can only be justified by the efficiency of the decision-system. However, this way of proceeding has often been criticised from 'the bottom', from the local governments.

Environmental policy making and implementation: local government

Approximately 80 per cent of all legislation given by the *Bund* and the *Länder* is executed by local authorities. Consequently, local government is very important for environmental protection (Pehle 1990). It does not make sense however to talk about *one* type of local government in Germany. There are three different types of *kommunale Selbstverwaltung* (local self government). Smaller municipalities (*Gemeinden*) carry out only a part of the local tasks. The rural districts (*Landkreise*) take care of matters which exceed the capacity of a municipality, e.g. waste management. The larger cities are *kreisfrei* and take care of all local matters themselves. In other words: the effect of environmental policy is to a large extent determined by the quality of policy implementation by local authorities.

As far as their organisation is concerned, the cities and rural districts initially reacted much like *Bund* and *Länder* to environmental problems. During the last decade almost all municipal councils have established a standing committee for dealing with environmental matters. As for the municipal administration the situation is similar. More and more cities and rural districts have set up environmental agencies. Problems that are analogous to those mentioned above regarding the organisation of the ME can also be found in the municipalities. However, ineffectiveness is often not the decisive motive for organisational reforms. Often it is of greater importance to demonstrate local efforts towards environmental protection of the public.

The choice of how to organise the local environmental administration is not crucial: local government has to act according to the laws. That may be regarded as a weak point of the environmental decision-making system in Germany. Those who play the largest role in implementing the environmental laws hardly participate at all in preparing them. The municipalities have mobilised to form interest-groups (*kommunale Spitzenverbände*). The rules of procedure of the *Bundestag* and the Federal Government stipulate that a hearing has to be given to these associations if the laws affect local government. With respect to environmental policy it has, however, been demonstrated that those rights are not very useful for the communities. Sometimes the municipalities succeed in correcting details (e.g. concerning the legal definition of the concept of waste in the waste law), but mostly interests other than environmentally oriented interests have more impact. The results of decision processes, which to a great extent exclude the local policy, are often unsatisfactory. The municipalities frequently have to deal with measures that they cannot implement, often because special local problems are not sufficiently taken into account by the federal legislator.

Conclusion

In this chapter we have seen that Germany's approach to environmental problem solving was deeply rooted in her political, legal and administrative institutions. These are the outcomes of long historical processes. The experiences related to the collapse of the Weimar Republic, of the Third Reich and of the devastation of modern war led West German political leaders to spell out the normative framework of policy to a greater degree than is typical in West European countries (Dyson 1982:17) The comprehensive environmental programme which the SPD–FDP coalition government proposed in 1971 was of this kind. Similarly Germany's co-operative federalism, her party system, her semi-public peak interest associations as well as elements of her political culture that have contributed to Germany's consensual and incrementally oriented policy-making style (Katzenstein 1987:349–351) have also shaped environmental policy in Germany. In other words, during the 1970s, environmental policy was integrated as part of routine politics, policy making and implementation in Germany, and was part of the institutional continuity that was a distinctive trait of the FRG (Katzenstein 1988:334). Since the 1970s, important changes have occurred, which have significantly transformed the environmental policy field.

The high level of consciousness concerning environmental degradation and the increased saliency of environmental issues were both a consequence of and, even more importantly, a condition for the electoral success of the Greens. This success changed both the party system and the dynamics of environmental politics. Marginal electoral swings have significant impact on a party system such as Germany's that favours government by coalition. As the parties and governing coalitions are engaged in continuous election campaigns at *Bund* and *Land* level, this also made the three other parliamentary parties very sensitive to the changing public mood in favour of environmentalism. All three of them underwent a *greening* process, i.e. they employed the symbols and rhetoric as well as substantial positions of more environmentally oriented policy (e.g. in the areas of air and water pollution, and waste management). Moreover, the success of the Greens made the system less consensual. For instance, the need to compete electorally with the Greens made the SPD opt for less consensual positions.

As to policy instruments, traditional regulatory instruments are still the most frequently used. This may seem surprising given that the EU since the early 1990s has increasingly based its policies on procedural measures to encourage enhanced public participation and on self-regulatory and voluntary measures by economic actors. Such a policy strategy is at odds with the institutionalised role of the German state. Here the authoritative role of the state is taken for granted not only in the making of procedural arrangements but also in deciding the substantive content of policy and its implementation through quality standards stated in regulatory law. In a number of cases the German government has resisted the new EU environmental policy strategy, and has only haltingly and partly transposed EU directives into national law (Pehle 1997:201–05). Nevertheless, the government has made efforts to develop a strategy that puts stronger emphasis on information, financial incentives and negotiations as well as on certain types of voluntary agreements. Such a strategy reaffirms the established principle of co-operation that also is characteristically anchored in German federalism and political culture. This strategy has in addition reaffirmed that environmental policy evolves out of the interaction of a complex web of actors. Some of these actors are politicians and officials at the federal, *Land* and local levels, others are members of professions (in particular lawyers and engineers) or representatives of environmental groups. However, in such a strategy a crucial and often preferential role is assigned to representatives of economic interest groups that are affected by environmental policy. Consequently, environmental policy is dependent on the political weight of these types of actors within the different environmental policy areas as well as within the different policy sectors (e.g. industry, agriculture, energy, transport).

Up till now industrial and transport lobbies have persistently co-aligned with the imperatives of the German state to pursue policies of economic growth and of those that are crucial to Germany's international competitiveness. Similarly, in recent years in the context of German reunification and the great economic burdens incurred in the aftermath of the reunification, no bold steps have been taken to choose policy instruments or measures in environmental policy.

Issues relating to global warming represent a new environmental policy area. The approach to develop a common definition of the problems and their solutions as well as the subsequent national policy target represents a significant change from the routine approach. Instead of consensual negotiations between corporate interests and the state, the non-partisan inquiry commission of 1987 saw to it that systematically collected data on the state of the global environment were applied in advanced scientific models and were incorporated into reports that were widely published and debated. With the recommendations of this inquiry commission as an authoritative baseline, the ME succeeded in getting the Kohl coalition to set the ambitious target of reducing CO_2 emissions in Germany by 25–30 per cent by the year 2005, based on 1987 values. However, when the government was to select instruments and measures to achieve this target, it was not able to develop an effective strategy to apply measures necessary to achieve the target.

In this context it should be noted, however, that there has been a substantial reduction of CO_2-emissions in Germany (Mez 1997:26–27). Although the reduction so far is mainly due to dramatic reductions in East Germany caused by decline in industrial activity, this indicates – together with the successful policies to curb power plant generated SO_2 and NO_x pollution – that German environmental policies have had an effect.

To achieve the national target for reductions in CO_2 emissions is, however, a new type of challenge, in many respects going far beyond other types of challenges in German environmental policy. One of these aspects is underscored by the fact that Germany hosted the first session of the *Conference of the Parties* to the *UN Climate Change Convention* during the spring of 1995.[13] One of the decisions of this conference was to establish the Secretariat for the Climate Change Convention in Bonn. These events stress what is also otherwise obvious: the credibility of Germany's participation in international efforts in this crucial policy area is dependent on Germany's ability to meet its own reduction targets.

Notes

1. In the following text, information and data for the former FRG is followed by the corresponding facts for East Germany in brackets (unless otherwise stated; all data from *Umweltbundesamt* 1993).

2. In 1975 traffic was responsible for 2.2 per cent of the total SO_2 emissions in FRG, and in 1990 this share had risen to 5.4 per cent. In the case of NO_x emissions the development has been even more dramatic. The total percentage of NO_x emissions caused by traffic increased to staggering 73 per cent in 1990. The CO_2 emissions, mainly responsible for the greenhouse effect, caused by automobiles increased from 1975 to 1990 by about 70 per cent.

3. One of the most influential was the botanist Hugo Conventz, who has been called the 'father' of nature conservation in Germany, and who also in 1904 and 1905 visited Stockholm, Oslo and Copenhagen and stimulated his Nordic colleagues to establish a national association for nature conservation in each of these three countries.

4. The leader was professor in music, Ernst Rudorff, who has been called 'the Romanticist' among the early leaders of the German nature conservation movement (Schonichen 1954:35–55).

5. In 1981 public expenditure amounted to 51 per cent of GNP, compared to 38 per cent in 1969 (Paterson and Southern 1991:228).

6. Significant data are: Greenpeace was formally established in West Germany in 1980. The association that was set up as a national umbrella association for the numerous citizen action groups, *Bundesverband Bürgerinitiativen Umweltschutz*, BBU, as well as the German Federation for Environmental and Nature Protection, *Bund Umwelt und Naturschutz Deutschland*, BUND, a split-off from *Deutsche Naturschutzring*, DNR (which in 1950 had been set up as a national confederation of nature conservation organisations) had had their successes (Jahn 1996:6).

7. In Helmut Weidner's phrase: *'Die Kanzler wechseln, die Umweltpolitik bleibt'* (Weidner 1991:14).

8. This was even more ambitious than the 'Toronto goal', referring to the conference in mid-1988 in Toronto that has called on all developed countries to cut their CO_2 emissions by 20 per cent from the 1987 levels by the year 2005 (Jäger and O'Riordan 1996:18).

9. The Bundestag followed suit by setting up its own single environmental committee.

10. This practice is, however, at odds with the rules of procedure of the federal ministries, which stipulate that the ME always is to be involved in the preparation of any new law that might deal with the interests of environmental protection.

11. The *Verursacher*-principle is often equated with the polluter-pays principle, but this translation does not catch the strong legal foundation which this principle has in the German context; the *Verursacher*-principle assigns responsibility for effecting a remedy to the legal person causing environmental damage.

12. *'Ohne die Länderregierungen geht in der Umweltpolitik nichts'* (Nothing concerning environmental policy can be done without the governments of the Länder). (Hartkopf and Bohne 1983:157)

13 At this session of the first Framework Convention, the government announced 1990 as the base year for Germany's 2005 target. As CO_2 emissions in 1990 were five per cent lower than those of 1987, this was a strengthening of the 2005 target (Huber 1997:71).

References

Beuermann, C. and Jäger, J. (1996) 'Climate Change Politics in Germany: How long will any Double Dividend last?', in T. O'Riordan and J. Jäger (eds) *Politics of Climate Change. A European Perspective.* London: Routledge, pp. 186–227.

Boehmer-Christiansen, S. and Skea, J. (1991) *Acid Politics: Environmental and Energy Policies in Britain and Germany.* London: Belhaven Press.

Dyson, K. (1982) 'West Germany: The Search for a Rationalist Consensus', in J. Richardson (ed) *Policy Styles in Western Europe.* London: George Allen & Unwin, pp. 17–46.

ECMT (European Conference of Ministers of Transport) (1997) CO_2 *emissions from transport.* Paris: ECMT/OECD Publications service.

FORSA (1990) 'Das Land als politische Handlungsebene', in *Bericht der Kommission Erhaltung und Fortentwicklung der bundesstaatlichen Ordnung innerhalb der Bundesrepublik Deutschland – auch in einem vereinten Europa.* Düsseldorf.

Genscher, H.-D. (1971) *Umweltprogramm der Bundesregierung. Einführung des Bundesministers des Inneren.* Bonn.

Hartkopf, G. (1986) 'Umweltverwaltung – eine organisatorische Herausforderung'. Vortrag auf der 27. beamtenpolitischen Tagung des DBB Januar 1986 (unpubl.)

Hartkopf, G. and Bohne, E. (1983) *Umweltpolitik, Bd. 1. Grundlagen, Analysen und Perspektiven.* Opladen.

Hatch, M.T. (1995) 'The Politics of Global Warming in Germany', *Environmental Politics*, Vol.4, No. 3, Autumn 1995, pp. 415–440.

Huber, J. (1980) *Wer soll das alles ändern?* Berlin: Torbuch Verlag, pp. 29–30.

Huber, M. (1997) 'Leadership and Unification: Climate Change Policies in Germany', in U. Collier and R.E. Löfstedt (eds) *Cases in Climate Change Policy. Political Reality in the European Union.* London: Earthscan Publications Ltd., pp. 65–86.

Jahn, D. (1996) 'Pressure Group Politics and Party Competition: Organizing Environmental Concerns in Sweden and Germany'. Paper, Nottingham, May 1996.

Jäger, J. and O'Riordan, T. (1996) 'The History of Climate Change Science and Politics', in T. O'Riordan and J. Jäger (eds) *Politics of Climate Change. A European Perspective.* London: Routledge, pp. 1–31.

Katzenstein, P. (1987) *Policy and Politics in West Germany: The Growth of a Semisovereign State.* Philadelphia: Temple University Press.

Katzenstein, P. (1988) 'The Third West German Republic: Continuity in Change', *International Journal of Foreign Affairs*, pp. 325–344.

Liberatore, A. (1995) 'Arguments, Assumptions and the Choice of Policy Instruments: The Case of the Debate on the CO_2/Energy Tax in the European Community', in B. Dente (ed.) *Environmental Policy in Search of New Instruments.* Dordrecht: Kluwer.

Markovits, A.S. and Gorski, P.S. (1995) *The German Left. Red Green and Beyond.* Cambridge: Polity Press.

Mayntz, R. (ed.) (1978) *Vollzugsprobleme der Umweltpolitik.* Wiesbaden: Rat von Sachverständigen für Umweltfragen.

Mez, L. (1995) 'Reduction of Exhaust Gases at Large Combustion Plants in the Federal Republic of Germany', in M. Jänicke and H. Weidner (eds) *Successful Environmental Policy*. Berlin: Edition Sigma, Rainer Bohn Verlag, pp. 173–186.

Mez, L. (1997) 'Klimaschutzpolitik als CO_2-Minderungspolitik – Dänemark und Deutschland in nationalen Alleingang', in L. Mez and M. Jänicke (eds) *Sektorale Umweltpolitik. Analysen in Industrieländervergleich*. Berlin: Edition Sigma, pp. 15–32.

Müller, E. (1986) *Innenwelt der Umweltpolitik. Sozialliberale Umweltpolitik – Ohnmacht durch Organisation?* Opladen.

Offe, C. (1987) 'Challenging the boundaries of institutional politics: social movements since the 1960s', in C.S. Maier (ed) *Changing Boundaries of the Political*. Cambridge: Cambridge University Press, pp. 63–105.

Paterson, W.E. and Southern, D. (1991) *Governing Germany*. Oxford: Basil Blackwell.

Pehle, H. (1988a) 'Das Bundesumweltministerium: Neue Chancen für den Umweltschutz? Zur Neuorganisation der Umweltpolitik des Bundes', *Verwaltungsarchiv* 2/1988.

Pehle, H. (1988b) 'Das Bundesministerium für Umwelt, Naturschutz und Reaktorsicherheit: Alte Politik im neuen Gewand?', *Gegenwartskunde* 2/1988.

Pehle, H. (1988c) *Analyse und Bewertung von Umweltverwaltungen*, (unpubl.).

Pehle, H. (1990) 'Umweltschutz vor Ort', *Auspolitik und Zeitgeschichte*. BG, pp. 24–34.

Pehle, H. (1993) 'Umweltpolitik im internationalen Vergleich', in V. von Prittwitz (ed): *Umweltpolitik als Modernisierungsprozess*. Opladen: Leske and Budrich.

Pehle, H. (1997) 'Germany: Domestic obstacles to an international forerunner', in M.S. Andersen and D. Liefferink (eds) *European Environmental Policies. The Pioneers*. Manchester: Manchester University Press, pp. 161–209.

OECD (1991) *The State of the Environment*. Paris.

OECD (1993) *Germany*. OECD Environmental Performance Reviews. Paris.

Roth, Reinhold (1987) 'Die niedersächsische Landtagswahl vom 15. Juni 1986', *Zeitschrift für Parlamentsfragen*, 1/87.

Rüdig, W. and Kraemer, R.A. (1994) 'Networks of Cooperation: Water Policy in Germany'. *Environmental Politics*, 3 (4): 52–79.

Schmidt, M.G. (1992) *Regieren in der Bundesrepublik Deutschland*. Opladen.

Schonichen, W. (1954) *Heimatschutz. Ihre Begründigung durch Ernst Rudorf, Hugo Conwentz und ihre Vorläufer*. Stuttgart.

Umweltbundesamt (Hrsg.) (1993) *Umweltdaten kurz gefasst*. Berlin.

Weale, A. O'Riordan, T. and Kramme, L. (1996) *Controlling pollution in the round*. London: Anglo-German Foundation.

Weidner, H. (1995) '25 Years of Modern Environmental Policy in Germany. Treading a Well-Worn Path to the Top of the International Field'. Paper. Berlin: WZB.

Weidner, H. (1991) 'Reagieren statt Agieren', *Politische Ökologie*, Nr. 23/Aug–Sep., pp. 14–22.

Wey, K.G. (1982) *Umweltpolitik in Deutschland. Kurze Geschichte des Umweltschutzes in Deutschland seit 1900*. Opladen.

Chapter 6

Greece: Administrative symbols and policy realities

Calliope Spanou

Introduction

Greece belongs to a group of semi-peripheral countries characterised by late industrialisation (Mouzelis 1988). During the second half of nineteenth century, it experienced a relatively precocious urban and State expansion and developed an important infrastructure (roads, railways), failing, however, to industrialise fully. It is only in the post-war period and especially during the 1950s, that intensive economic development began. High rates of economic growth were attained during the 1960s and the early 1970s. Industrialisation has then expanded with the help of multinational capital through a rapid and unplanned process; it has been accompanied by an equally impressive urbanisation process, internal migration and absence of adequate infrastructure to handle the effects of these changes.

The Greek environment exhibits an important variety in terms of landscape, flora and fauna. Mountains, the insular complex and an extensive coastline (16,000 km) are its main features. The frequent upheavals and wars often had detrimental effects on the natural environment. Nevertheless, most of the present environmental problems have their origins in the post-war period of rapid and unregulated growth. Unbalanced regional development accounts for the concentration of economic activities and population in relatively few areas and results in interrelated environmental pressures (KEPE 1991; OECD 1983). At present, the main environmental problems are air and water pollution, solid waste, land degradation, forest fires, threats to biological diversity and natural reserves, and noise (Ministry of Environment 1992). Major economic activities with important potential impact on environment include tourism, agriculture and industry.[1]

Since the 1950s, urban growth has been rapid, unregulated and disorderly. Lack of land use planning, fragmentation of landed property in urban areas, illegal construction and speculation constitute major causes for environmental degradation (Economou 1994). High population densities, insufficient public facilities and road access, as well as lack of open spaces, often characterise the urban environment, while air and coastal water pollution preoccupy major cities (e.g. Athens, Thessaloniki, Volos). The same pattern was followed by the rapid development of tourism which represents a very important sector of economic activity. Insufficient state regulation and the subsequent inadequacies in infrastructure resulted in an increasing number of problems (e.g. solid waste, sewage, illegal buildings, landscape degradation).

Greece has traditionally been an agricultural country. The disproportion between arable land and farming population was one of the features of the agricultural sector. An important part of cultivated land is located in mountainous or semi-mountainous regions, unfavourable to the mechanisation of production. The resulting difficulties for the rural population led to a wave of external but also internal migration. Employment in agriculture fell from 41 per cent of the active population at the beginning of the 1970s to 21.3 per cent in 1993 (OECD 1973 and 1995). The abandonment of the countryside had important side effects on the environment, such as land erosion and forest fires. Internal migration, along with the lack of regional development policy, resulted in the formation of a few big cities. Present-day Athens, concentrating one-third of the country's population, is the outcome of this process.

Regional imbalances are also characteristic of the location of industrial and other economic activities. The majority of the economic activities are concentrated along the axis Athens – Thessaloniki, extending towards Patras and Kavala, all of which are coastal cities. Surface mining and processing of mineral deposits (mainly lignite and bauxite) are often responsible for serious landscape and environmental damage.

The subordination of environmental concerns to the economic imperative along with fragmentation of landed property, a consumer-oriented society, and low environmental awareness account for environmental degradation in Greece. At the beginning of the 1970s, there was a growing awareness of the high concentration of population and economic activities. The military dictatorship (1967–74) allowed, however, for the continuation of the concentration process, thus intensifying the above mentioned problems (Hadjimichalis 1988).

Environment as symbolic politics

In the Greek context, the start of environmental policy is considered to be an amendment on environmental protection that was included in the 1975 Constitution. Greece is one of the few countries having a constitutional provision stating that the protection of the natural and cultural environment is an obligation of the state.[2] It should, however, be noted that this is not the first time that environmental problems drew the attention of the authorities. Though the term environment was not in use as a concept, aspects of the environmental issue had long ago given rise to basic legislation, initiated by responsible sectoral departments.[3]

After the war, the priority given to rapid economic growth did not allow environmental considerations to influence the location of industrial activities. During this period of intensive industrialisation, in the absence of effective physical planning policies and under advantaged conditions, the first important polluting industries were created which are in part responsible for today's environmental problems. The year 1965 marks an important attempt at rationalising economic development. New policy instruments were introduced, such as the creation of industrial zones, the institution of technical controls and of licensing procedures for sewage and industrial waste disposal. The advent of the dictatorship did not allow these regulations to produce the expected positive results, since between 1967–74 environmental concerns were openly neglected.

The nuisance approach

The terms and the approach prevailing at that stage, were those of 'danger' and 'nuisance' (odour, smoke, private nuisances) affecting human safety and health. Regulations and classifications were most often based on these criteria, protecting the environment in an indirect way and not because 'a natural resource should be administered for its own sake' (Henning 1974). The nuisance approach thus limits the policy matter to what might have a directly perceived impact on people living in the surroundings of industrial activities.

But this approach has further implications with regard to the governmental division of labour. Since the focus is on nuisances, competent government departments are a) the technical departments regulating the sources of pollution or the polluting activities (e.g. Transport, Industry, Agriculture, Merchant Marine); b) departments responsible for health and safety (Health, Public Order, Justice); c) departments responsible for infrastructure (Public Works, Interior). The gradual emergence of policies to deal with aspects of the environment, followed at this stage the existing map of administrative responsibilities. The trans-sectoral, horizontal character of environmental problems contrasted with the predominantly sectoral/vertical logic of the government services. In the absence of any unifying concept and mechanism, the environmental issue did not seem to exist yet as such. It was rather considered as a form of accidental side-effects of various state and private activities.

The access to the political agenda

The access of an issue to the political-administrative agenda, e.g. the list of issues which will be the concern of decision-makers (Cobb *et al* 1976), usually constitutes the starting point of any public policy. During the late 1960s and the beginning of the 1970s, when at the international level the environment was put on the political agenda, Greece was under military rule. The major environmental effects of unregulated economic development had already become more or less visible. Air pollution, especially in Athens and water pollution in the Saronikos and Thermaikos gulfs were already alarming, while landscape alteration due to quarries and fires continued. The authoritarian nature of the regime and low environmental awareness did not, however, allow for any social mobilisation. Still, certain segments of the state apparatus started to become active in environmental questions. This is all the more striking given the discrepancy observed at that time between the complete state indifference towards environmental degradation, on one hand, and the numerous organisational restructurings and the reports produced, on the other. The various signs point to a *symbolic* form of inclusion of the problem in the political agenda (Edelman 1985; Spanou 1995).

In order to break its political isolation, the Greek military government was searching for legitimacy. The emerging environmental issue represented such an opportunity, given the developments taking place at the international level. The year 1970 was dedicated to nature conservation by the European Council. Though no longer a member, the Greek government eagerly followed the spirit of the time, organising campaigns and other events in tune with the rest of Europe. Among them, there were a National Conference on Nature Conservation, organised by the Ministry of Agriculture, and a similar conference organised in 1971 by the Ministry of Culture, with Greek and foreign participants. Reference to environmental developments in other European countries and North America was continuously made.

The issues were addressed either in an abstract or in a highly technocratic way, with very few comments on their socio-economic origins or on existing policies. However, remarks were made concerning the implementation gap, the lack of physical and land use planning that gave way to speculation, and the lack of environmental sensitivity within society. The general conclusion was repeated over and again: 'Time has come for an 'environmental policy' to be formulated and implemented with continuity and consistency'.

These initiatives contributed to transferring the international sensitivity for environmental problems into the Greek political-administrative system. Devoid of any socio-economic and political contest, the issue was then used as a value of wide acceptance and as an instrument for (auto)legitimising the authoritarian government vis-à-vis the international community as well as the general public.

The conversion of the environment to an agenda item was due to an *inside initiative* and corresponded to the mobilisation model of agenda building: the issues are initiated inside government and then expanded outside it, in order to attract widespread attention or awareness (Cobb *et al* 1976). Social mobilisation was, however, only partly compatible with the authoritarian regime. On the other hand, the symbolic character of the process was quite consistent with minimal solutions. Both reasons explain why the awakening of the public was limited to an appeal to individual, humanitarian values, downplaying the socio-economic components of the issue. Environment was seen as a question of the 'whole of humanity', a 'modern civilisation problem', to be solved through the development of technology.

Although symbolic, the access of the issue to the agenda was related to organisational rearrangements. The role of technocrats within government services in this respect deserves mentioning. They seem to have played the role of a bridge between the emerging environmental awareness, that they had mostly acquired during their studies abroad, and the government. A Committee on Environment was set up by the government in order to study environmental problems; it operated within the framework of the Centre of Planning and Economic Research (KEPE) in connection with the preparation of the 15-year Plan.[4] Among the suggestions made was the creation of a 'unified environmental protection agency' to control environmental pollution. The ministries of Co-ordination and Planning and of Social Affairs proved particularly active in creating special environmental units. With the assistance of the World Health Organization (WHO), this latter ministry initiated a project for the control of air pollution in the greater Athens area. The project was approved in 1973 and started to operate in 1974 under the name of Athens Environmental Pollution Control Project (PERPA) (Lekakis 1984; PERPA 1980).

Defining the issue

During this process of gaining access to the political agenda, the environment received various competing but also interacting definitions. At the *political* (governmental) level, where legitimisation needs prevailed, the environment was presented as a consensus topic, devoid of any potentially conflictual dimension. Responsibilities were diffuse, along with the risks that needed to be faced in a unified way. At the administrative-technical level, however, the socio-economic and technical implications of the issue were more obvious and were translated in a variety of organisational arrangements. At the social level, the limited expansion of post-materialist values did not yet provide a specific definition, with the exception of an organisation voicing the interests of the modernising fraction of business[5] (ERYEA 1976). By accepting the conflictual aspects of the environmental issue and the responsibilities of business, this organisation attempted to preserve or increase the social legitimacy of capital. It is, therefore, not surprising that the distinction was made between 'good' and 'bad' industries, the latter being indifferent to environmental degradation (Spanou 1995).

The co-existence of competing definitions of a problem is part of any public policy. Far from being incompatible with this vagueness, symbolic politics explain the gap between policy implemented, on one hand, and the organisational arrangements as well as the rhetoric developed, on the other. Symbolic politics, however, produced a dynamic that went beyond the hesitant intentions to provide impetus in favour of the environment.

In search of an integrated environmental policy

The post-dictatorial period bears two major characteristics, both related to the emergence of environmental policy out of the nuisance approach and the fragmentation it implied. This search involves, above all, building the necessary organisational infrastructure and legal framework. The main actors promoting these demands were groups of scientists and civil servants within government services (i.e. Centre of Planning and Economic Research, Ministry of Co-ordination and PERPA) as well as scientific environmental societies and the Technical Chamber of Greece.

Far from being accomplished, this process still raises questions and dilemmas linked to the bureaucratic division of labour and power, and faces resistance from the status quo favourable to well entrenched economic interests and practices. The difficulties confronted in trying to reduce sectoral fragmentation, improve co-ordination and institutionalise the horizontal trans-sectoral dimension of environmental management are but the bureaucratic reflection of the wider process of value restructuring that brought environmental issues to the forefront.

Building environmental administration: a long and difficult process

Co-ordination, though necessary for any policy, is of particular importance to environmental policy. The ecological notion of interaction finds in co-ordination its organisational counterpart (Henning 1974; Molnar and Rogers 1982). However, the achievement of co-ordination can prove to be extremely difficult. The trans-sectoral character of environmental policy implies antagonistic values and conflicting priorities.

At the organisational level, an attempt was made to enhance planning and co-ordination as well as organisational networking for environmental management. The ideal of a unified environmental agency was already being promoted in 1972, and dominated the discussion at that time. Nevertheless, in 1976, the numerous public, semi-public and private bodies involved in the various aspects of environmental management, included 12 ministerial departments out of 20, not to mention the non-ministerial agencies (KEPE 1976).

The effort to increase co-ordination and organisation networking for environmental policy-making passed through three stages.

The very first attempt at a global approach was introduced by Law 360/1976 on physical planning and environment. It provided for intersectoral co-operation among the various departments under the general responsibility of the Ministry of Co-ordination, responsible for economic development and planning. The law created (within the Ministry of Co-ordination) a Council, a Secretariat and a Commission for Physical Planning and Environment.

The National Council for Physical Planning and Environment, chaired by the Prime Minister, included the ministers of 11 related departments. It was given overall responsibility for environmental policy and for the approval and supervision of all physical plans and programmes for environmental protection. Interministerial co-ordination at a lower level was to be ensured by the Commission, in which every agency had its representatives. This solution provided the framework for environmental planning, but left the implementation and monitoring to existing administrative structures.

Despite this effort to centralise the responsibility for environmental policy formulation, many other initiatives (i.e. passing of laws) have been taken by the various ministries in a quite independent way. By 1980, over 15 ministries had developed a special unit with responsibilities for environmental protection (OECD 1983). The institutions created by law 360/76 were thus gradually marginalized. The National Council met on environmental issues only once, while the Commission, which was to have provided for the interministerial co-ordination, proved impossible to set up. Environment is the policy field that makes most clear the nature of co-ordination as a value conflict resolution process (Lieber 1979; Andrews 1981) as well as the links between various economic interests and government agencies.

At a second stage, a move was made to establish a specialised department to deal with environmental issues. If horizontal co-ordination confronted major difficulties in ensuring the respect of guidelines by the various sectoral ministries, the success of co-ordination by a specialised department was even less obvious.

The very first attempt at ministerial specialisation took place in March 1980. A new Ministry of Environment, Physical and Town Planning was created; it was responsible, among other things, for the preparation of environmental policy plans and for the management and co-ordination of their implementation. It did not, however, centralise under its authority environmental services already existing within the various ministries. Its environmental services were elementary; it was mainly based upon the Directorate for Town Planning, which included a section for environmental protection. Furthermore, this ministry was only given residual responsibility, that is for questions lying outside the scope of competence of other ministries. No implementing or executive tasks and powers were assigned to it. The National Council, still located within the Ministry of Co-ordination, continued to have the overall responsibility for the co-ordination of environmental policy.

In 1982, under the Socialist government, a move towards strengthening this ministry[6] led to the transfer of PERPA, but also of other services and responsibilities. Until 1985, its internal structure reflected the heterogeneity resulting from 'a number of historical amalgamations and influx from other ministries' (Ministry of Environment 1984).

The effort to further clarify the scope of competence and to rationalise the allocation of responsibilities led in 1985 to the creation of the present Ministry of Environment, Physical Planning and Public Works (ME). The important aspect of this reform was the grouping with the ministry of Public Works. Environmental policy formulation is currently the responsibility of the ME.

However, its responsibilities are not exclusive, which limits its decision-making power.[7] This time, the National Council and the Secretariat were transferred to the ME. But they had been inactive since the beginning of the 1980s without having been replaced by any other arbitration mechanism.

The framework law on environmental protection: a list of wishful thoughts

The second aspect of the emerging environmental policy concerns the institution of a basic law (framework law) for environmental protection. It was expected to rationalise environmental management by laying down a global approach and general principles and guidelines.

Contrary to what one might suppose, the legal framework for environmental protection was thought to be sufficient. A constant concern was that environmental legislation did not come into conflict with economic development. Creating the appropriate technical, economic and administrative conditions and strict enforcement of the existing legislation were the priorities. A codification was nevertheless considered essential. This should be done by the preparation of a basic law on environmental protection which would define the distribution of responsibilities, the system of monitoring and sanctions etc., as well as establishing the 'environmental impact statements' (KEPE 1976).

The elaboration of such a framework law proved extremely difficult. The successive attempts and the final product of this process provide the measure of the powerful opposition against it, coming especially from industry and the technical ministries (Louloudis 1987; Damanaki 1986). A draft had been prepared and submitted several times to the Parliament, only to be later withdrawn (Lekakis 1984, 1991). In 1981, a Presidential Decree was drawn up by the Ministry of Industry; it introduced the environmental impact statements (EIS) assigning all responsibilities on industrial pollution to that ministry. Considered to be too ambitious and strict, it was hardly implemented for the control of industrial pollution.

When the Socialist Party (PASOK) came to power in 1981, it included the passage of the Environmental Protection Act among the priorities for 'the first 100 days'. The delays due to the opposition of the industrialists and the priority set on industrial development by the government itself, led to an intra-governmental conflict between the ME and the Ministry of National Economy. In 1983 a new bill came up; it provided for the creation of the 'unified environmental agency', with a co-ordinating role under the supervision of the ME. The ministry centralised a number of pollution control responsibilities. Once again, serious objections were voiced by industrialists and the technical ministries. A new version appeared in June 1985, after reconciliation between the ME, on one side, and the Ministry of National Economy as well as the technical ministries on the other. It seems that most of the suggestions made by the industrialists were incorporated in it, since the last version was much less far-reaching than the 1983 bill.[8]

The framework law (1650/1986) was finally approved 11 years after the Constitutional provision for environmental protection. It contained various definitions intended to clarify the policy matter (e.g. environment, health, pollution, environmental protection, natural resources, landscape); it specified the ways to intervene (zoning, for example); it also provided for EIS (taking into account the obligation established by a correspondent EU Directive of 1985), as well as for environmental quality controls at the prefectural level. Its clearly stated objective was to lay the foundations for a more coherent policy.

This law constituted an attempt at a comprehensive approach but avoided any concrete regulation, remaining at a very general level. A considerable number of regulations were necessary to operationalise it, and still to date, a great deal remains to be completed. By preserving the uncertainty[9] it only delayed the confrontation with powerful opponents, while transferring the bulk of implementation responsibility to the administrative level. The regulations fall under the jurisdiction of various ministries and, therefore, need joint decisions. Once again, it has been confirmed that what might seem to be a success at the parliamentary level is often eroded during the administrative stage of the implementation process. In fact, most of its important and long-awaited provisions were incorporated without any real intention of implementing them. This is particularly true for the 'Unified Environmental Agency' which, while limited to an advisory role, has not been created to date.[10]

It is worth mentioning that the first period of the Socialist government was marked by the presence of a very committed Minister of Environment (A. Tritsis), who tried to push forward environmental policy and developed contacts with environmentally sensitive circles. The dynamic strategy he adopted provided the environmental issue with social visibility and set the way to promote basic reforms which shaped the present state of affairs. Nevertheless, this approach had as a side-effect an open conflict with adversary interests and practices as well as with the sectoral ministries responsible for their regulation. The 1986 Act should, therefore, be viewed as a product of these dynamics. Its highly *symbolic* character is due to the effort undertaken by PASOK in 1981 to reconcile the satisfaction of social expectations with regard to environmental quality and the pressure coming from prevailing economic interests. This is why it has been qualified as 'a list of wishful thoughts' (Tachos 1987).

Administering the environment: persistent weaknesses

Greek environmental administration is rather weak. This is obvious from many different points of view: political weight, concurrent responsibilities, staffing, field services, enforcement and control mechanisms. Its limited implementation capacity is partly explained by these weaknesses.

The central level
Located in a ministry also responsible for Physical and Urban Planning and Public Works, environmental administration is under the influence of a dominantly technical approach. Though the combination with Public Works had been absolutely rejected in 1975, it was later argued that this would strengthen the ME and make it more influential.[11] No general restructuring has, however, taken place; the two components of the ministry preserved their own form and responsibilities, forming two separate General Secretariats (GS): one for Environment and one for Public Works. The GS for Environment is further organised in three General Directorates (GD): GD of Environment, GD for Urban Planning and GD for Management.

The GS for Public Works forms a rather independent group of services, usually under the authority of a junior minister. Relations between these two parts of the ministry are limited. As a result, a number of important issues have been pushed up to the ministerial level for arbitration. This shows the considerable importance of the personality of the individual heading this sub-unit of the ministry. Environmental priorities do not seem to occupy a central position in public works policy.

Staffing
The number and specialities of the personnel involved in environmental administration are worth considering. On one hand, they exhibit the order of priority and the dominant view of environmental issues. On the other, the perception of the nature of environmental issues as well as of alternative courses of action by the civil servants has a considerable effect on the policy output. A brief overview of the staffing of the ME shows that central services occupy less than 800 employees and about 3000 in total for central and local services (including urban planning, excluding public works).

In terms of professional expertise, the predominance of engineers is quite explicit even outside the public works services. Through upward mobility, these officials reach top management posts, which require much less techno-scientific competency than a broader understanding of the social, political and cultural setting for environmental policy. At the central and regional level, engineers have access to all directorial posts at the expense of members of other environment-related professions (e.g. biologists, agricultural and forestry scientists, chemists). This state of affairs supports the view that the public works services and the professionals involved (mainly civil engineers and architects) have successfully tried to take over and control the field of environmental policy. Represented by the Technical Chamber of Greece, the 'consultant of the State for technical matters', this professional group has been particularly active in environmental questions since the 1950s (Tachos 1983). They have gradually expanded their expertise to cover questions not directly linked to their initial field of specialisation. This is an illustration of the offensive strategy of a professional group trying to capture the environmental policy field, based on their technical expertise within society and the state.[12] It is worth mentioning that during the 1980s, most of the ministers of the Environment, Physical Planning and Public Works came from this professional group and had been distinguished members or leaders of the Technical Chamber.

Environmental administration may generally be regarded as the product of the type and quality of personnel involved (Henning 1974). The fact that it is dominated by a very specific professional group (engineers of various specialities) introduces a highly technical perception of the policy issues. This does not fail to provoke some protest and criticisms. The environment cannot be reduced to techno-economical considerations which dominate that profession; the overrepresentation of engineers in the field services tends to reproduce the technical approach, legitimising their expertise at the expense of other excluded specialists.

The field services

A major issue for environmental policy is the geographical distribution of powers and responsibilities. The building of a decentralised network for implementation, monitoring and control is one of the top priorities in the development of environmental administration. Yet, this has not been the case in Greece. Traditional centre – periphery relations do not favour the strengthening of non-central institutions; furthermore, fragmentation and overlapping of responsibilities at the field level still constitute the basic features of the present situation.

Two deconcentrated agencies with geographical competence for Greater Athens and Thessaloniki specialise in the particular problems of these two areas. The need for a decentralised administrative network has been belatedly acknowledged. Very little has been done, however, since the 1986 law stated that 'implementation of environmental policy lies with the prefects and the local government'.

The structure of the Greek State is based on the geographical division of the country into 13 regions (i.e. state administration units with planning responsibilities), 54 prefectures (administrative units headed by a prefect) and a large number of municipalities. At the prefectural level, a second tier local government, under an elected Prefect, has recently started to operate (1995).

Only in 1989 did a reorganisation of the field services give birth to the Directorates for physical Planning and Environment in each of the 13 regions of the Country. They include an Environment Unit, which has no direct implementing or controlling activities. This is partly a result of the limited role of the regional level of administration in the country, being primarily an intermediate level for economic planning.

The prefectural level traditionally consisted of the deconcentrated services of central government departments under the authority of a Prefect appointed by the government. Historically, it constituted the most important deconcentrated administrative level in Greece, but the environment was until recently nearly absent from its services and mission. The urban planning services constituted the main part of the field services. Since the beginning of the 1980s, officials responsible for environmental matters had been appointed as a weak substitute for the institutional presence of the ministry at this level. According to the 1986 law, environmental quality inspection teams *could* be set up by prefectural decision. But it was as late as 1990 that a ministerial act created the 'environmental units' within the prefecture, institutionalising the presence of the ministry at that level. Their responsibilities concern the EIS, the implementation of environmental protection projects and the monitoring and control of polluting activities. Their staffing is most often inadequate as are the technical means at their disposal.

Building a network of field services has been a slow and hesitant process. For a long time, environmental administration has been deprived of local antennas. As a result, it has often been cut off from the realities of the field and limited in its monitoring and control activities. The centralising tradition of the Greek political-administrative system is only partly responsible for the inadequacy of field services. The low priority of environmental policy issues is primarily at fault. The increased responsibilities in environmental policy entrusted to the second-tier local government at the prefectural level may serve to strengthen its administrative capacity. At present, it is too early to assess the impact of this reform on environmental management. The sharing and transfer of responsibilities between the state and the two tiers of local government will need more time to be clarified and to produce results.

Local government and the environment

The traditional mistrust towards any form of autonomous representation of civil society permeates the general framework for all political-administrative arrangements. The large number of small municipalities (5921), the great majority of which have less than 1000 inhabitants, along with financial inadequacies, increases local government's dependency on central government. The limited role of first-tier local government as far as environmental protection is concerned is thus not surprising.

The balance between the more general and the local dimension of environmental issues is difficult to achieve. The distribution of powers between state and local administration, based upon the constitutional principle and concept of local affairs, is not particularly clear nor helpful. A review of the environmental legislation shows that the local character of most environmental problems is not reflected enough in the distribution of responsibilities (Remelis 1989).

The traditional responsibilities of first-tier local government include activities directly or indirectly linked to environmental management. Water supply, sewage, waste disposal, parks and gardens, traffic and parking arrangements constitute its exclusive responsibilities. There is, however, a series of concurrent activities having to do with the environment, which can be carried out either by state or by local government services. These include environmental pollution control, protection of water resources, traffic regulation and granting construction permits. The municipal council thus has the possibility to activate these competencies and therefore to create the necessary local services.

In general, municipalities have been rather hesitant to extend the scope of their responsibility for environmental matters, even when that was possible. This is the case for the monitoring and control of environmental quality. Furthermore, the possibility of creating a municipal police body was provided for, but has not been sufficiently used by local government. On the other hand, the traditional mistrust between state and local government did not help the initiatives taken by the latter to report problems, since 'it was not taken seriously'. Apart from the above traditional responsibilities, environmental legislation until recently has practically ignored local government. It was not until 1986 that the possibility for transferring responsibilities to local government on environmental matters was provided for.

At present, local government is consulted on regional development programmes as far as environmental questions are concerned or whenever government authorities plan to adopt an act dealing with questions of environmental protection or regional planning. Local government is responsible for approving the environmental conditions of activities having a particularly small impact on the environment. It has no involvement in the licensing of activities with a more important impact on environment, which lie within the scope of competence of the Prefecture and the central services. Local government also has an advisory role when special measures are to be taken for areas with particular environmental problems, but does not participate in the monitoring of their implementation (Remelis 1989).

Co-operation between state and local government is generally difficult: the first tries to dominate the options of the second, which often acts as a pressure group, supporting or getting into conflict with environmental protection goals and groups.

Environmental policy between state and society

Environmental awareness in society

Social mobilisation is generally vital for the support of a public interest regulatory policy (Weale 1992; Rourke 1984). To put it in a different way, public opinion as well as environmental action groups constitute an important resource for environmental administration: they help to increase the social visibility of the issues, to strengthen politically its position vis-à-vis its rivals, and to further develop its action through non-governmental organisations acting as its local antennas (Spanou 1991).

Greek civil society is considered to be traditionally weak (Mouzelis 1988) and social mobilisation is generally low. Until the 1970s, the environment has been a concern for an extremely limited group of people. It mostly took the form of the Friends of the Forest or mountaineers' associations, that later gave birth to the Hellenic Society for the Protection of Nature (1956). Scientific groups such as the Hellenic Society of Ornithologists were also created; however, until the advent of the dictatorship in 1967, environmental sensitivity was present only within scientific or naturalist circles with no wider social impact (Sfikas 1987; Louloudis 1987).

Although the period of military rule did not favour the emergence of environmental sensibility and mobilisation, international concern had some echo in Greece, through the translation of books or articles on the subject.[13] Opposition to industrial investments in various parts of the country emerged, integrating environmental protection among other broader political and economic claims (Sfikas 1987; Louloudis 1987).

Environmental awareness started growing after the mid-1970s. Protest and mobilisation against development projects brought together already existing scientific or naturalist societies with various environmental alternative groups that began to emerge all over the country. It was, however, during the 1980s that the number of environmental groups rose dramatically. New environmental groups (often local and specialised) emerged but also national sections of international environmental non-government organisations (such as the World Wildlife Fund and Greenpeace), as well as scientific societies for environmental protection. Possessing more expertise, financial resources and networks of relations, they gradually formed the social counterpart to environmental policy-making officials, submitting well-documented proposals, alerting the government and public opinion, and performing other informative functions (Citizens' Movement 1991). However, national non-governmental organisations for environmental protection are still few in number and generally have rather limited membership, financial support, and influence on public policy.

Environmental issues were brought into the electoral arena in the 1980s. In the 1982 municipal elections, for example, environmental pollution was a major issue in the platforms of the principal parties in Athens and a central issue in several other cities and towns. The first time green parties participated in the local elections was in 1986; the results, initially disappointing, have gradually improved. Green parties participated in three successive general elections between 1989 and 1990 and managed to gain a seat in Parliament (0.77 per cent), thanks to a more permissive electoral system. In the 1990 European elections they just missed the threshold allowing them to have a seat (1.11 per cent instead of 1.36 per cent) while in 1994 they did not participate in an independent way. The political presence of the Greens is nowadays an important stimulus to which political parties have to respond, even if in a partial and *ad hoc* way. Political parties, therefore, tend to integrate, to a certain extent, the environmental issue in co-existence with their prevailing productivist logic.

The belated emergence of the environmental movement as well as its weakness are by-products of the current stage of economic development[14] and a materialistic culture. Environmental awareness is today present in Greek society; it co-exists, however, with contradictory and conflicting values and interests, having low priority and, consequently, little policy impact. In this respect, it is interesting to notice the low mobilisation of the Athenians, despite the acute problem of air pollution (*nephos*) (Citizens' Movement 1990).

Environmental NGOs and the ME

Public opinion cannot be seen in an undifferentiated way. If the favourable attitude of the public at large is important, the existence of attentive publics, that is of groups that have a salient interest in the agency's policy (Rourke 1984), is crucial to environmental policy making.

Mutual distrust constitutes the general framework of this relationship. Since 1976, laws have excluded environmental organisations (NGOs) from being represented in various committees and from the possibility of civil action before the courts, a right accorded to local government and the Technical Chamber of Greece. This shows the way NGOs are seen by the government. Though often asserted to be important, their role is considered to be confined to the development of environmental awareness and social mobilisation, and excludes participation in policy making.

However, environmental organisations are becoming more important in Greek society. Their role in sensitising and informing the general public is developing, along with their utilisation of institutional mechanisms to oppose environmentally harmful governmental decisions. Increasingly, NGOs use the possibility of mobilising EU procedures or the (Greek) Supreme Administrative Court (Council of State), which has proved one of the most committed actors in defending environmental interests.

The relationship between environmental organisations and the ME is not yet well developed. The ME relies very little on NGOs in order to strengthen its position politically and increase its possibilities for promoting and implementing its policy. Ecological groups are mistrusted and unreliable because 'they do not have arguments, they take excessive positions and present no alternatives' or 'they do not know how to negotiate'. But also from their side, NGOs have the feeling of not being welcomed by the ministry, which is not responsive to them; they feel that their proposals or reports on specific problems are not taken into account, nor are their suggestions for co-operation. It seems, however, that the ministry officials readily acknowledge one function these groups perform: the function of alerting the authorities and of contributing to the social visibility of environmental problems.

A closer look at the way NGOs interact with the ME shows the complexity of the issue but also signs of change. The relationships between the ministry and these groups may vary a lot, according to different factors: the overall conception of the environmental issue, its origins and causes, as well as the means of handling it, contributes to the development of selective affinities and of more or less close links between them. Even the personality of the minister and the political situation may encourage or discourage co-operation (Spanou 1991).

The emergence or the strengthening of already existing environmental organisations with techno-scientific expertise is a factor of change. A few large, especially scientific, organisations have become formally, and more and more, informally, the advisors of the ministry on particular environmental questions.[15] Partly because of the inability of public authorities to perform certain tasks, the ministry progressively has turned to non-governmental organisations for assistance; the latter are thus being entrusted with certain activities.[16] The technical perception of environmental problems prevailing within the ministry allows techno-scientific expertise to serve as a *bridge* between a new approach to economic development and the well-rooted socio-economic interests and practices.

There is certainly still a long way to go before close co-operation between the two worlds is actually established. Nevertheless, this is how environmental organisations gradually gain access to the state apparatus in their effort to influence policy formulation or implementation. The ME itself admits that 'NGOs are changing tactics from protest and denunciation to constructive proposals' (Ministry of Environment 1992a). An important condition for effective environmental policy is, therefore, the 'opening towards new allies', i.e. NGOs, representative professional groups and other citizens' organisations (Ministry of Environment 1992b).

Environmental policy: general assessment and concluding remarks

Various factors account for the difficulties environmental policy faces (Ministry of Environment 1992b). The prevailing economic priorities relegate environment to a lower rank on the Greek political agenda. The weakness of the ME is undoubtedly linked to the emphasis on economic development (industry, infrastructure etc.). As J.N. Lekakis (1991) notes, the wish to support the industrial sector leads government to avoid imposing strict regulations on the private sector. During the last years, financial incentives have tended to acquire a more prominent place among environmental policy instruments, along with traditional regulation methods (Lekakis 1990). The environment is often only a means to serve the goals of other policies (e.g. employment) or even the goals of private initiatives, which are not necessarily in tune with broader policies. The low priority placed on environmental policy is clearly reflected by the corresponding public expenditure, as percentage of the GDP (about 0.25 per cent) (Pelekassi and Skourtos 1992).

The overview of the development of environmental policy highlights its symbolic, fragmented and reactive character. Lack of a comprehensive approach, inadequate territorial and land use planning and deficient enforcement mechanisms characterise the general situation. The lack of effective implementation and control reflects the low commitment of successive governments to environmental preservation, which involves making and effectively implementing environmentally important but politically unpopular decisions. This becomes particularly obvious during pre-election periods. Electoral considerations give way to all kinds of pressures upon the political officials of various levels (central or local) who readily ignore environmental questions.[17] Even when strict legal provisions do exist, laws are not enforced in order to serve the interests and pressures of the moment. In many cases, the ME either is over-ridden or excluded from decisions, or even captured by opposite interests. Thus, environmental legislation presents no continuity and individual, particularistic interests shape the decisions, neutralising any effort towards rationalisation.

Many examples might be given – the protection of forests, the regulation of quarrying activities, the licensing of industrial plants. Among them, a significant example is the delimitation of wetlands according to an obligation stemming directly from the Ramsar Convention and, indirectly, from EU Directive 79/409. Most of the wetlands are still not regulated and thus not protected. The delay is due to interminable bargaining with local interests, leaving the door open to any kind of trespassing, either authorised or tolerated. Sometimes, the destruction of the wetlands occurs with the blessing of the state, which awards the licence and even ensures financing through Community funds (Citizens Movement 1990 and 1991)!

The institution of EIS is a further example. The previously mentioned 1981 decree, EU Directive 85/337 and law 1650/1986 successively provided for the obligation of submitting an EIS for public or private works with possible harmful consequences for the environment. This obligation was not respected, however, since it was only in 1990 that a ministerial Act was promulgated, transposing the Directive into Greek law and making possible the implementation of the 1986 Law. The decision was taken only after the Commission had threatened to take the Greek state to the European Court for non-compliance (Paleologou 1992). Moreover, the ministerial Act only partly complied with the Directive, since it practically postponed its enforcement for almost three years.

In order to identify the specific problems of environmental policy in Greece, further reference to the socio-political context is needed. The way social interests are articulated to the political system favours individual access to decision-making centres. Interest groups are primarily related to the state in a vertical, heteronomous manner and personalistic, particularistic politics still persist (Mouzelis 1988). A tradition of patronage tends to short-circuit formal procedures. State power is fragmented into a multiplicity of decision-making centres which tend to function in terms of particularistic criteria, leaving no room for planning and long-term concerns.[18] Not only powerful interests use their collective weight to block major environmentally beneficial decisions, but also individual interests of lesser economic importance succeed in being exempted from obeying laws and regulations. In this context, the constant dilemma between economic development and environmental protection often appears as artificial or false. It is mostly used as a legitimising argument of limited value, since the economic development benefit to which environment is sacrificed is in many cases quite doubtful.

Another way to describe these phenomena is the deficient steering capacity of the Greek state (Spanou 1996). Despite the highly legalistic and dogmatic Greek administrative culture, and generally abundant and often changing legislation, enforcement is inadequate. Some kind of goal displacement brings the focus of attention to the elaboration of laws but fails to ensure their enforcement. This reflects the weakness of the Greek state since it proves unable to introduce and steer change and to implement its own decisions. Laws tend thus to be a list of good intentions with no more than symbolic value. As a result, the implementation gap is considerable, just like the distance between political rhetoric and actual practices.

In the absence of a strong environmental movement, the EU represents a valuable potential ally. It produces regulation which is difficult to initiate in the national context while it encourages consequent implementation. To put it in the way a higher official of the ME did, 'the EU succeeds in imposing what the minister would never dare', and, one might add, what social pressure is not yet able to achieve by itself. Much of the initiative for promoting environmental policy is, therefore, expected to originate as a result of joint social and international pressure upon the government.[19]

Notes

1. The origin of GDP at factor cost was in 1993: agriculture 10.2 per cent, services 66.7 per cent, mining, manufacturing and energy 15.6 and construction 7.5 per cent (OECD 1995).

2. The Constitutions of all Southern European countries include some reference to the environment. (Pridham and Magone 1993).

3. For example, in 1912 an important law was passed, introducing licensing procedures for the creation and operation of industrial plants

4. This Committee was composed of higher civil servants who specialised in the different aspects of environment (geologists, agriculturists, engineers, scientists, medical doctors etc.), of members of the National Academy of equivalent specialities and a University professor representing the Hellenic Society for Nature Conservation.

5. ERYEA (Society for the Control of Air, Water and Soil Pollution) was created in 1970.

6. On the problems reported see KEPE 1986, p. 61ff.

7. A report prepared for the ministry of National Economy in 1993 mentions 50 agencies of various kinds involved in environmental policy. See *To Vima* 16.5.1993 and Ministry of Environment 1992a, pp. 91–93.

8. It is interesting to note that in the meantime, the Minister of the Environment A. Tritsis was replaced by E. Couloumbis, the former Minister of Industry.

9. As Hanf (1989) notes, these open-ended framework laws often leave a great deal of uncertainty as to what the law in fact 'means' for industrial pollution.

10. The Secretary of State for the Environment (K. Gitonas) made clear during the parliamentary discussions that this agency was not to be created immediately. *Parliamentary Proceedings* 18.9.1986:1601.

11. According to A. Tritsis, at that time Minister of the Environment (Vlassopoulou 1991:67).

12. When PASOK came to power, this profession seized the opportunity to penetrate the state apparatus, having already acquired a significant social presence and created close links with this specific political party. They also conquered the local government level. See Kaler-Christophilopoulou 1989.

13. Such as *Ecology or Death*, by R. Dumont, 1973, *Ecology*, by P. Samuel, 1973 etc.

14. At the beginning of the 1980s, the GDP per capita was roughly equivalent to that attained by most European countries in the mid-1960s, when environmental protection became an important issue (OECD 1983:11).

15. Expression used by a higher official in the ministry.

16. Such as the management of the ACE-biotopes EU programme, the project for the preservation and management of Pindos mountain, the establishment of the Greek Biotopes and Wetland Centre, the management of sea-parks etc. See Citizens' Movement 1990:7.

17. A particularly significant example is the legalisation of illegal buildings (without previous construction permit). See *Eleftherotypia* 25. and 26.9.1993 and *To Vima* 5. and 12.9.1993.

18. It seems that most of these deficiencies are common to Southern European countries. See Pridham and Magone 1993; Pridham 1994; La Spina and Sciortino 1993.

19. This belief is shared among Southern European countries, i.e. that the EU can somehow compensate for the deficiencies of their system of government. See Pridham 1994 and Pridham and Magone 1993.

References

Andrews, R.L. (1981) 'Value Analysis in Environmental policy', in D.E. Mann (ed), *Environmental Policy Formation.* Massachusetts: Lexington Books, pp. 137–147.

Citizens' Movement (1990) *The State of the Environment in Greece. An Evaluation – 1989.* Athens.

Citizens' Movement (1991) *The State of the Environment in Greece. An Evaluation – 1990.* Athens.

Cobb, R., Ross, J.K. and Ross, M.H. (1976) 'Agenda Building as a Comparative Political Process', *American Political Science Review* 70:126.

Damanaki, M. (1986) *Parliamentary Proceedings* 12.9.1986, 1501–02.

Demertzis, N. (1991) 'The Green movement and Green Party in Greece. Political and Cultural Aspects', Paper presented at the ECPR Workshop *Europe: the Green Challenge*, University of Essex.

Economou, D. (1994) 'Physical Planning in Greece', *Perivallon & Dikaio* No. 1, 41–86.

Edelman, M. (1985) *The Symbolic Uses of Politics*, University of Illinois Press.

ERYEA (1976) *The Greek Environment.* Athens.

Hadjimichalis, C. (1988) 'The New Greek State. Regional Policy and Social Control', in Th. Maloutas and D. Economou (eds) *Problems of Development of the Welfare State in Greece*. Athens: Exantas, pp. 115–148.

Hanf, K. (1989) 'Deregulation as Regulatory Reform: the Case of Environmental Policy in the Netherlands', *European Journal of Political Research* 17 (2): 193–207.

Henning, D. (1974) *Environmental Policy and Administration*. New York: Elsevier.

Kaler-Christophilopoulou, P. (1989) *Decentralization in Post-Dictatorial Greece*. PhD. Thesis, London School of Economics.

KEPE (1972) *Long-Term Plan on Environmental Protection against Pollution* (1970–1985), Final Report. Athens: Centre of Planning and Economic Research.

KEPE (1976) *Report of the Working Group on Environment*, Plan of Economic Development 1976–80. Athens: Centre of Planning and Economic Research.

KEPE (1986) *Report of the Working Group on Environment,* Plan of Economic Development 1983–87. Athens: Centre of Planning and Economic Research.

KEPE (1991) *Regional Policy. Reports for the 1988–92 Plan*. Athens: Centre of Planning and Economic Research.

La Spina, A. and Sciortino, G. (1993) 'Common Agenda, Southern Rules. European Integration and Environmental Change', in J.D. Liefferink, P.D. Lowe and A.P.J. Mol (eds) *European Integration and Environmental Policy*. London: Belhaven, pp. 217–236.

Lekakis, J.N. (1984) *Economic and Air Quality Management in the Greater Athens Area*. Athens: Centre of Planning and Economic Research.

Lekakis, J.N. (1990) 'Distributional Effects of Environmental Policies', *Environmental Management* 14 (4): 465–473.

Lekakis, J.N. (1991) 'Employment Effects of Environmental Policies in Greece', *Environment and Planning* No. 23, 1627–37.

Lieber, H. (1979) 'Public Administration and Environmental Quality', *Public Administration Review* 30 (3): 277–286.

Louloudis, L. (1987) 'Social Claims. From Environmental Protection to Political Ecology', in *The Ecological Movement in Greece*. Athens: Meta ti vrochi, pp. 8–21.

Ministry of Environment (1984) *Environmental Audit Project*. Ministry of Environment and Physical and Town Planning. Athens.

Ministry of Environment (1992a) *National Report of Greece* 1991, U.N. Conference on Environment and Development, Brazil June 1992. Athens.

Ministry of Environment (1992b) *Annual Report to the Parliament. The State of the Greek Environment and the Measures Taken for its Protection*. Athens.

Molnar, J.J. and Rogers, D.L. (1982) 'Interorganizational Coordination in Environmental Management', in D.E. Mann (ed.), *Environmental Policy Implementation*. Massachusetts: Lexington Books, pp. 95–108.

Mouzelis, N. (1988) 'On the rise of Post-war Military Dictatorships: Argentina, Chile and Greece', *Comparative Studies in Society and History* 28 (1): 55–80.

OECD (1973) *La politique agricole en Grèce*. Paris.

OECD (1983) *Environmental policies in Greece*. Paris.

OECD (1995) *Economic Surveys. Greece*. Paris.

Paleologou, N. (1992) 'The Implementation of the E.C. Directives Relative to the Environment', *Nea Ikologia,* March, pp. 36–39.

Pelekassi, C. and Skourtos, M. (1992) *Air Pollution in Greece.* Athens: Papazissis/ WWF.

PERPA (1980) *Technical Report,* vol.1, Ministry of Social Affairs, General Directorate of Hygiene.

Pridham, G. (1994) 'National Environmental Policy-making in the European Framework. Spain, Italy and Greece in Comparison', in S. Baker, K. Milton and S. Yearly (eds) *Protecting the Periphery. Environmental policy in peripheral regions of the EU.* London: Frank Cass, pp. 80–101.

Pridham, G. and Magone, J.M. (1993) 'Environmentalism and Democratization: Comparative Policy Perspectives on Regime Consolidation in Southern Europe', Paper for the Conference on *Changing Functions of the State in the New Southern Europe.* Bielefeld, 8–10 July 1993.

Remelis, C. (1989) *Environment and Local Government.* Athens: Sakkoulas.

Rourke, F.E. (1984) *Bureaucracy, Politics and Public Policy.* Boston: Little Brown and Co.

Sfikas, G. (1987) 'Going on, while looking behind', in *The Ecological Movement in Greece.* Athens: Meta ti vrochi, pp. 298–307.

Spanou, C. (1991) *Fonctionnaires et Militants.* Paris: L'Harmattan.

Spanou, C. (1995) 'The origins of environmental policy in Greece', in C. Spanou (ed.) *Social Claims and State Policies.* Athens: Sakkoulas, pp. 223–286.

Spanou, C. (1996) 'De la capacité de régulation de l'Etat hellénique. Amorce de problématique à partir d'une étude de cas', *Revue Internationale de Sciences Administratives* No. 2, pp. 269–89.

Tachos, A. (1983) *Environmental Protection as a Legislative and Administrative Problem.* Athens: Sakkoulas.

Tachos, A. (1987) 'The "Framework Law" for the environment', *Nea Ikologia,* May, p. 43.

Vlassopoulou, C. (1991) *La politique de lutte contre la pollution atmosphérique.* Mémoire DEA, Université d'Amiens.

Weale, A. (1992) *The New Politics of Pollution.* Manchester: Manchester University Press.

Chapter 7

Italy: Environmental policy in a fragmented state

Rudolf Lewanski

The state of the environment

Like many other Western European nations, starting from the 1950s Italy entered a phase of accelerated economic growth in terms of both production and consumption. Presently, it is among the world's most industrialised countries, with a gross national product comparable to those of France and Great Britain. The country, however, is also deeply split between a highly developed North (Lombardy, for example, is among the richest regions in Europe) and a more backward South. The central government's strategy aimed at promoting the Southern economy by constructing large industrial and infrastructure projects has, in the last analysis, been largely unsuccessful in overcoming these regional discrepancies.

In environmental terms Italy, renowned as the 'Garden of Europe' at the beginning of the century, has suffered a sharp decline in its environmental quality as a result of the impact of economic activities, urban sprawl and traffic. Many of the industrial sectors that grew rapidly in the aftermath of World War II are highly polluting (chemicals, steel, cement, refineries, paper mills). Even the great number of small and medium-sized factories, perhaps the real secret of the Italian 'economic miracle', spread across northern and central regions, are responsible for serious pollution due to the concentration of numerous activities of the same type in relatively limited territories. For example, in the case of the ceramic tiles industry, 250 factories producing 30 per cent of the world's total sales are situated in a 50-square kilometre area near Modena.

With a population of 57 million living on a territory of 300,000 square kilometres, Italy has an average population density of 191 inhabitants per square kilometre; in urban areas, especially in the Po Valley and along the coasts, this figure soars to 2000 inhab/sq km, making these regions among the most densely populated areas of Europe. Built-up land has increased from 120,000 square kilometres in 1961 to 260,000 in 1986, occupying something like nine per cent of national territory and destroying rich farmland and areas of natural significance. In the Po Valley, for example, 'consumption' of land for construction purposes has been proceeding at an annual rate of six per cent (*Lega per l'Ambiente* 1989:227).[1]

Construction of buildings and roads, excessive mining (7000 sites were counted in 1980), timber cutting without replanting, mechanisation of agriculture in hilly areas, lack of maintenance of water drainage systems in a geologically fragile land (approximately one-sixth of the total territory is highly unstable) cause serious soil erosion and landslides (estimated at about 3000 per year). Large industrial plants have often been built in areas of considerable scenic and natural value, in particular along the coasts. Emissions and vibrations produced by traffic in ancient cities, among other things, have contributed to the deterioration of historical monuments and buildings. Although there have been some improvements in water quality over recent years thanks to the expansion of waste-water treatment facilities, large areas of coastal waters have been declared off-limits for bathing due to consistently high levels of organic pollution. Organic and inorganic (toxic metals and agrochemicals, in particular) substances cause high levels of pollution of surface waters from domestic household, agricultural and industrial sources. Forty-one per cent of all Italian lakes are plagued by serious levels of eutrophication; the same phenomenon in recent years has caused algae blooms resulting in fish deaths in the Adriatic Sea. Oil pollution is also quite common along the peninsula's beaches; approximately 44 per cent of all oil and petroleum products transported by ship in the Mediterranean basin is directed to or from Italian ports.

Tourism itself is a cause of environmental problems. Excessive concentrations of seasonal visitors exert serious negative impacts on fragile ancient monuments and city centres (Venice being an extreme example), and on natural ecosystems such as the coasts and sea waters. For example, one third of the 3250 km of beaches is subject to erosion. This is often caused by human interventions connected to tourism, such as wave barriers and boat marinas, which lead to modification of sea currents.

As a result of inadequate public policies, numerous environmental emergencies have occurred throughout Italy. Almost six per cent of the country (more than 17,000 square kilometres, inhabited by some 1 million people) has been declared risky areas due to high levels of environmental pollution. Intensive stock-breeding and cultivation methods relying on agrochemicals heavily concentrated in the four Regions of the Po Valley (Lombardy, Piedmont, Emilia and Veneto) are causing pollution of the soil and underground water reserves by nitrates and pesticides. As a consequence, water supply systems serving two million people in more than 300 municipalities in northern Italy had to be closed during recent years.

Finally, notwithstanding its relatively insulated geographic position, separated from the rest of Europe by the Alps to the north, and surrounded by water on the remaining sides, Italy is affected by pollution produced by other countries. The effects of long-range airborne pollution, in particular, are beginning to be felt here as elsewhere in Europe: about one-third of sulphur depositions and half of oxidised nitrogen come from abroad (UNEP, 1991:43). This transboundary pollution has caused serious decline in forests (along with more traditional forms of human intervention, such as fires, which destroy approximately 0.6 per cent of total forest area each year). In a 1989 survey 15–17 per cent of forests were classified as damaged by acid rains; in some areas a rate of 50 per cent has been registered.

Many of the developments mentioned above obviously are hardly typical of Italy alone; they can also be found in many other developed nations. A feature somewhat peculiar to the Italian case, however, is the 'mix' of rapid development of phenomena with negative impacts on environmental resources, on one hand, and the absence and delay, at least when compared to other industrialised Western countries, in adopting responses to the problem, on the other hand. To a large extent, public authorities have been incapable and often unwilling to regulate such market-induced processes and their negative consequences. Land-use planning policies, for example, have often been ineffectual in managing urban expansion. Construction without the required permits became a very widespread phenomenon during the 1960s and 1970s, especially in southern Italy.

The development of environmental policy in Italy

For heuristic purposes the development of environmental policy in Italy can be divided into three distinct periods.

Phase one: the birth of a national environmental policy (1966–76)

The beginning of an environmental policy in Italy dates back to the mid-1960s. The first explicit piece of legislation attempting to tackle a problem, namely air pollution, specifically in *environmental* terms (rather than for its health consequences or for its spill-overs on particular economic activities such as fishing and navigation) is Act no. 615, passed in 1966. The aim of this and other similar laws of the same period was to 'clean up' particular, geographically limited problem areas. Act no. 615, the so-called 'Antismog Law', was aimed primarily at the serious air pollution affecting large urban areas in the northern part of Italy, where frequent thermal inversions during the cold seasons are coupled with topographical conditions that hinder dispersion (for example the densely inhabited and highly industrialised Po Valley is surrounded by mountains on three sides). The reduction of the sulphur contents of fuels used for domestic heating formed the core of the Antismog Law's strategy. Only to a limited extent, depending on the actual will and capacity of local authorities, did Act no. 615 deal with emissions from industrial sources. Actual implementation and enforcement of air pollution control policies, however, began, essentially in a few areas of northern Italy only in the early 1970s when, after considerable delay with respect to the deadlines set by the Act itself, regulations were issued by central government, specifically the Health Ministry.

In this phase, environmental issues had a very low external or political visibility. Measures taken to deal with these problems came primarily from local authorities and individual judges (both of whom tried to tackle specific problems by means of the available antiquated legislation, such as the 1934 Health Laws). Politicians were hardly motivated to take up an issue that was of little or no relevance for inter- and intra-party power relations, and which at that time did not attract much attention from the media and public opinion. Consequently, the environmental issue promised very little political payoff in electoral terms, especially in a period in which, after the oil shocks, the economy was a high priority. Under these conditions, ecologists exerted very little political influence at all; the only well-established organisation (*Italia Nostra*) was mainly concerned with the protection of the national cultural heritage (monuments, etc).[2] Other, more environmentally specific associations (such as the Italian branch of the World Wildlife Fund, founded in 1966) were still in their infancy and had small memberships[3].

Phase two: filling the legislative toolbox (1976–86)

From the mid-1970s to the mid-1980s environmental policy underwent a substantial acceleration as a number of previously missing legislative instruments were added to the policy 'toolbox'.

A major piece of legislation, Act no. 319, concerning water pollution control, was finally passed in 1976 after more than ten years' debate. This Act introduced uniform national emission standards, i.e. it established identical limits to the quantities of pollutants allowed in industrial and domestic discharges across the entire country. Several other relevant laws, dealing with such issues as industrial and household waste disposal, sea protection and the biodegradability of detergents, were also passed in these years.

Access to the policy arena in this period was still limited to a rather small number of actors, although new ones, both institutional and social, began to appear. The gradually increasing political importance of the policy area was also reflected at the institutional level: although there still was no ministry with specific competence for environmental affairs, a first move in this direction was the creation of an Inter-ministerial Committee for Environmental Protection. This body was supposed to co-ordinate powers dispersed among 16 ministries. Important institutional actors were also created at the sub-national level, such as the Regions and the *Unità Sanitarie Locali* (USL), Local Health Units.[4] The ecological movement continued to grow slowly but steadily; a new group, *Lega Ambiente* was founded in 1980 with the support of leftist political movements, in particular the Communist Party[5].

The political visibility of environmental problems increased considerably in this period, as a result both of the emergence of the nuclear issue and of serious industrial accidents that attracted media attention and contributed significantly to growing public awareness of environmental problems. Particularly important in this respect was the accident that occurred in July 1976 at the ICMESA chemical plant in Seveso, a little town in Lombardy. As a consequence of dioxin emissions, thousands of persons were evacuated from the area and 220,000 people were placed under medical and epidemiological surveillance. Many suffered skin eruptions from which they have not yet completely recovered. The seriousness of the Seveso accident, together with the difficulties the responsible authorities met in handling this emergency, had a major impact on Italian public opinion. The following year, nuclear energy suddenly became a major public issue as the proposed construction of a new power plant at Montalto di Castro (just north of Rome) met strong resistance from the local population backed by the student movement.

Phase three: approaching maturity (1986–94)

The crucial turning point in the development of environmental policy in Italy was the creation in 1986, after a number of unsuccessful attempts during the 1970s, of a central authority specifically in charge of environmental affairs. The establishment of the Ministry for Environment 'officially' marked the upgrading of the status of environmental policy *vis-à-vis* other sectoral policies. However, the operational capacity of the Ministry remains far too weak (due to lack of personnel, logistic conditions etc).

During this period, the legitimacy of environmental policy grew considerably (though with strong regional variations, especially between the northern and the southern parts of the country). The level of environmental awareness among the general public became quite high. This was also due to environmental emergencies and disasters (Bhopal, Chernobyl) occurring elsewhere in the world and in Italy. Reported in the media, these played an important role in stimulating an increased public demand for environmental quality. Various surveys indicated that environmental quality was one of the public's main concerns. The ecological movement became increasingly active[6], though its capability to actually influence policy-making remained rather low. The good electoral results obtained by the Greens in local and national elections also showed the political potential of this issue. However, it should be noted that, especially in the late 1980s, the green vote represented not only environmental concerns, but in many cases a protest vote for a non-ideological party (at least compared to other traditional parties) in a context of growing disaffection with established parties (a role that in the last years has been taken up by the *Lega Nord* in northern Italy). Notwithstanding the appearance on the political scene of the green movements and parties, some parties (centre and right) continued to pay very little attention at all to environmental issues (which are hardly of any interest to organisations that collect their votes through clientelistic modes, as will be explained below).

Others have, however, attempted in earnest to incorporate some of the themes of the green culture, though such processes clashed with the traditional 'productivist' culture, as in the case of the Communist Party (now PDS, Democratic Party of the Left).

At the same time, the amount of resources made available by the Government to deal with environmental problems was also increasing consistently: by the end of the 1980s, total expenditures on environmental issues by central, regional and local authorities comprised one per cent of GNP (Ministero dell'Ambiente, 1992:332–337), a level considered by international experts to be the minimum for a developed nation. It should, however, be pointed out that a consistent quota of such financial resources are actually used to deal with environmental disasters (floods, landslides, etc.) often caused by environmentally unsound policies of the past (deforestation, construction of disrupting projects, etc.) or due to delays in cleaning up of heavily polluted areas that have been declared to be of 'high environmental risk' (such as the Lambro and Bormida rivers or the Venice and Naples areas), rather than for actions aimed at preventing further pollution. At the same time, it is interesting to note that the actual capability of the ministry to spend allocated funds proved to be rather low: it was able to disburse only approximately one-third of the funds that had been allocated to it. This was the lowest rate among all central ministries, which, on the average, are able to spend 50 per cent of their budgets. In 1992, the ministry's budget suffered a severe cut of 39 per cent as a result of the priority government gave to the reduction of state indebtment. It appears likely that this trend will continue for the next few years at least.

The new ministry was instrumental in providing previously missing legislation or strengthening inadequate legislation already on the books in such areas as industrial air emissions, landscape and soil protection, industrial waste and noise pollution. One of the ministry's priorities has been the protection of natural areas. After decades of neglect, the extension of protected natural areas has grown from 3.2 per cent of the total national territory in 1984 to 8.1 per cent at present. Eleven national parks, four marine parks plus hundreds of other protected areas (some managed by regions or environmental associations such as the World Wildlife Fund) have been set up. Recent legislation will increase the amount of protected areas to 10 per cent. Another important piece of environmental legislation passed in 1989, after some 20 or so years of discussions, is Act no. 183 dealing with soil protection and water management and instituting river basin authorities (national, interregional and regional).

Phase four: changing perspectives

With the break-out of *Tangentopoli* (the disclosure of extended corruption within the governing parties and the administration), the Italian First Republic, born after World War II, has come to an end; the electoral system has been changed from proportional to a 'first past the post' type, and the parties of the previously centre-left coalitions (Catholics and Socialists) have disappeared from the political scene. The period 1994–96 has been one of great uncertainty from which a new political order has not yet emerged. In this situation, environmental issues have hardly been a priority. During the Berlusconi government (1994–95) the ministry was headed by a representative of the extreme right who showed no interest in environmental protection and actually openly asserted to be in favour of nuclear energy, highways and hunting in national parks. In the following 'technical' Dini government, the ministry was headed by the Minister of Public Works, who did in fact manage to carry out several projects (institution of several national parks, increase of taxes on water and solid waste in order to provide financial resources for policies in such areas).

In May 1995 the situation changed substantially as a center-left government headed by R. Prodi took office. The Greens, as part of the coalition, obtained a number of posts in the Cabinet, among which that of the Minister of Environment, as well as in several important public agencies. Although economic (inflation, employment and public indebtment) and institutional (federalism, electoral rules) issues remain primary concerns of the government, the Prodi government has passed several new measures in the field of environmental protection, such as Acts no. 22/97 and 389/97, introducing considerable changes to past solid waste policies. These include Act no. 344/97, focusing on creating employment in environmentally-related fields; Act no. 137/97, updating previous legislation concerning hazardous industries; Act no. 179/97, aiming at reductions in emissions capable of damaging the ozone layer; and Act. no 39/97, granting freedom of access to environmental information. Other relevant measures concern the financing of national parks, the organisation of the environmental agency ANPA (Decree 335/97) and the reduction of noise pollution (e.g. in airports). The government also introduced economic incentives for owners willing to trade in their cars for new, less fuel demanding vehicles.

The increased attention of the present government to environmental issues was especially evident in connection with the Kyoto Conference in December 1997. Whereas previous Italian governments, generally speaking, hardly were involved in an active manner in the negotiation of international environmental agreements and even less eager to actually implement them, in this case the attitude was quite different. Besides taking active part in the formulation of the EU's position, several weeks before Kyoto, the Italian government approved a set of measures designed to reduce CO_2 emissions by 7 per cent. At present this programme is being translated into operational measures, such as promotion of energy saving house appliances, energy saving in the housing sector, promotion of solar energy, reduction of gasoline consumption by vehicles and promotion of vehicles running on alternative, less-polluting fuels (methane, electric, liquid gas). Although a start has been made, it has to be said that, in general, the official climate change strategy appears to be rather thin in content, for instance in terms of dealing with the increasing problems caused by developments in the transport sector (e.g. both increasing number of vehicles in circulation and distances travelled) and is to a rather limited extent based on realistic assumptions concerning the implementation of chosen measures (Silvestrini and Collier 1997:120–25).

It should also be noted that the decentralisation of powers from the central State to the Regions and local government, in the name of promoting 'federalism', has caused strong unrest within the environmentalist movement, which is concerned that this might bring about lower levels of environmental protection in many areas. An event strengthening this concern occurred in the summer of 1997, when several regions allowed hunting of a number of small bird species. Consequently the central government, under pressure from the Greens who threatened to leave the coalition, was forced to repeal the decisions of the regions. The environmentalists have also strongly criticised the government's policies in several other fields, such as a massive road construction program and a provision that would have 'pardoned' illegal constructions built along public coastlines (but the latter was repealed).

The institutional setting

Italian administration is essentially a four-tier system consisting of the central state, regions, provinces and municipalities.

Central government

The state's principal powers, generally speaking, are in relation to legislation concerning distribution of competencies, procedures, pollution standards and measurement methods, national sectorial planning and financial resources. The Italian central government is organised in terms of a traditional ministerial model. At present there are 31 ministries, which operate through both central and field offices. Until 1986, powers in the environmental field were dispersed among a number of ministries: Health, Public Works, Agriculture, Merchant Marine, Environmental and Cultural Goods and others. As the environment became a politically attractive issue, each of these ministries attempted to assert its hegemony over the topic. The Health Ministry promoted the first law on air pollution; Public Works tried to control water pollution policy (but was eventually pushed aside in the course of subsequent developments) and the Ministry for Scientific Research tried at one point to gain control of the policy, by publishing, for example, the first national *Report on the State of the Environment*. A first *ad hoc* ministry without portfolio was set up in 1973 to deal with environmental problems. It was, however, short-lived and its competencies were absorbed by the Ministry for Cultural and Environmental Goods, instituted in 1975 as the first attempt to concentrate environmental powers in this field in one administrative agency. In fact, the competencies of this ministry were limited to aesthetic and landscape aspects coupled with more traditional activities regarding monument conservation, museums and similar cultural activities.

The issue of administrative responsibility was not brought up again until 1983, when an Ecology Ministry without portfolio was established. Finally, in 1986 an actual Ministry of Environment (ME) was created after several years of discussions. The ME has been given responsibilities in the fields of air, water, soil and noise pollution, solid waste, mining, parks, as well as sea and coast protection. Its capacity for action has, however, been severely hampered by the procedural requirement that it consult with and secure the agreement of other interested ministries (such as Industry, Public Works, Health) in a number of matters. The initial objective of concentrating all powers in the field of environmental policy in the hands of one ministry has not been fully reached since pre-existing ministries were successful in protecting their powers with regard to other fields related to the management of natural resources. The Ministry of Agriculture, for example, was able to retain competencies over hunting, which is not only an activity of obvious importance for the protection of a major natural resource, i.e. wildlife management, but also a hotly debated issue in Italy and traditionally of great political interest since two million hunters/voters represent a substantial clientele and pressure group.[7] Other ministries, such as Merchant Marine, Transportation, Industry, Health, Cultural and Environmental Goods, Interior and Public Works (water management and land-use planning) have retained substantial responsibilities over matters of environmental relevance. Furthermore, in other areas of critical importance for the environment such as energy, industrial and agricultural policies, the ME possesses only very limited powers for ensuring that adequate attention will be paid to environmental aspects along with more 'traditional' technical and economic considerations. This has tended to weaken the effectiveness of the ministry's actions.

Furthermore, the ME has a rather small staff compared to other ministries (about 400 persons). The problem, however, is not only one of quantity. The Ministry was set up much along the traditional bureaucratic model, and personnel was taken over from other ministries, mainly Health and Public Works, together with the transferred competencies. These officials brought along attitudes, values and approaches (and, in some cases, the lack of professional competence) which were typically formal and legalistic rather than technical and problem-solving.

Scientific and technical know-how regarding environmental issues is provided by advisory bodies set up within the ME itself, by specialised services attached to the ME, such as the Geological Service, and by other agencies such as the National Health Institute (*Istituto Superiore di Sanità*), ENEA (the former nuclear energy agency, that was partially reorganised in the 1980s to broaden its field of interest to alternative energy sources and environmental issues) and the National Research Council (CNR). Nevertheless, the ME lacks much of the research and planning capabilities that the highly technical matters it deals with would require. A national environmental agency was instituted by law in 1994, but at present is still in the process of being set up. It will have a staff of approximately 600 (mainly coming from existing authorities such as ENEA). All in all, however, the in-house and external technical expertise available to the ME and other public authorities remains largely below the level needed.

As far as policy tools are concerned, Italian policy in this field mainly depends on a regulatory type approach based on emission and products standards. Ambient quality standards have been introduced only recently. These standards are to be applied uniformly throughout the country. What this comes down to is a considerable degree of formal rigidity that leaves little leeway for local authorities to tackle issues according to actual demand for environmental quality in their area on the basis of pollution levels, and the characteristics and uses of the specific environmental good to be protected. For example, in the case of water pollution, all industrial firms can discharge pollutants within prescribed concentration limits into a river, regardless of the cleansing capacity of that specific waterway, the number of industries discharging into it, and the various possible uses of the waters (recreation, fishing, drinking water supplies etc).

Since the creation of the ME, and especially during the period 1988–92, such policy tools as incentives and ecological taxes have played an important role in the new strategy of the ministry. Taxes had hardly been resorted to in the past: water pollution legislation of 1979 provided for effluents to be levied by municipalities, but the provision was in fact hardly applied (Lewanski 1986:71). The first actual example of an ecological tax on products was the charge imposed on plastic shopping bags in 1989 in order to provide consumers with an incentive not to dispose of them improperly. More recently, the proposal for imposing taxes on certain activities (pig-breeding, pesticide use, noise charge for aeroplanes and others) and products (pesticides like atrazine, and several plastic materials) has met opposition from the Ministries of Agriculture and of Industry, in defence of their respective constituencies, forcing the ME to set its projects in this area aside for the time being. Financial Acts of the last two years have introduced or consistently increased charges on environmental services such as refuse collection, water supply and depuration.

Another element of the ME's overall policy strategy is the creation of compulsory consortia, set up to promote recycling of selected materials, such as glass, plastic, aluminium cans, mineral oil and car batteries. The consortia are non-profit organisations whose members are producers, importers and user firms; they receive financial resources from charges levied on certain products. Minimum collection targets have been set for 1992; deposit-refund systems and other product taxes will be introduced should these targets not be met (Malaman and Ranci 1991).

Finally, an interesting component of the ME's strategy is represented by voluntary agreements stipulated with major industrial groups (such as petroleum companies, FIAT, ENICHEM, Montedison) in which such firms assure they will adopt clean-up measures in exchange for specific decisions to be adopted by the government. In the FIAT case, for example, the government pledged it would pass tax measures that would encourage customers to buy vehicles with catalyzers in exchange for FIAT's commitment to provide all its vehicles with such anti-pollution devices one year in advance of the EC deadline. In the end, the deal did not work out because of the Government's failure to honour its part of the agreement.

The ME has also been active in providing information to the public on environmental conditions: a *Report on the State of the Environment* was published in 1989, followed by a another one in 1992. The ministry gave financial assistance to local authorities to set up or upgrade monitoring activities, especially in the field of air pollution (Herman and Lewanski 1992). The ME at present is considering the possibility of setting up an Agency with responsibilities in the field of information collection and publication, much along the lines of the recently established European Environmental Agency.

The existence of a central government strategy in the environmental area appears to be tightly connected with the person who occupied the post for a long period (1988–92) shortly after the creation of the ME itself, Mr G Ruffolo. Although forced by the backlog accumulated under previous governments to carry out emergency measures, under his direction the ministry also started to pursue more long-term actions based on systematic plans. Furthermore, as we are now discovering, he strongly and successfully resisted pressures from his own party (Socialist) to take advantage of the construction of environmental protection projects to collect pay-offs for the party or himself (this aspect is further discussed below). After he left his post, environmental policy once again entered a period of stagnation, probably also because of the general political and economic situation in the country.

Regions

Regions were formally introduced into the 1946 post-war Constitution as one of the safeguards against future authoritarian backlashes. There was considerable delay in actually setting them up: with the exception of the five special regions – defined by particular characteristics such as the presence of strong ethnic minorities or geographical isolation[8] – set up just after the end of the war, the remaining 15 so-called 'ordinary' regions were constituted only in 1970. Reasons for the delay were partly political in nature, i.e. moderate parties in power were unwilling to let the Communists govern significant sub-units of the country where they could count on strong electoral majorities (especially in the central part of Italy: Emilia-Romagna, Tuscany, Umbria and sections of Liguria and Marche) and partly explainable in terms of bureaucratic politics (at stake for the national bureaucracy, obviously, are the loss of power, personnel and resources in favour of the regions). The resistance was so great that regions were born already 'mutilated' in terms of the powers and resources granted to them. Consequently, they had to fight bitterly in order to obtain the wider substantive (although still hardly satisfactory) competencies which were finally granted in 1976.

It should be noted that the legislative powers of the ordinary regions are strongly limited in terms of topics with which they can deal and the context within which they can be exercised; they are subject to the general policy principles fixed by national legislation. 'National interest', as this has been interpreted by the Constitutional Court in the cases of formal region/state conflicts brought before it, usually prevails. Consequently, the regions tend to be more bodies co-operating with centrally decided policies rather than autonomous policy makers in their own right. Residual powers (i.e. those not explicitly defined by the Constitution) belong to the central state.

Regions have been given numerous responsibilities directly or indirectly pertaining to environmental policy: nature and scenic areas protection, parks, hunting and fishing, pollution, health, land use, town planning, agriculture, mining. In fact, in the case of a few regions (especially in Northern Italy), though certainly not of the majority of them, environmental issues have represented a significant portion of their policy-making activity. On average, regions issue approximately 30 laws per year in the field of environmental protection. These regional laws are often quite similar, both because of the obligation to follow state 'framework' laws and due to a tendency to copy, to some extent, legislation passed by other regions. It should also be pointed out that within the public and the political system at present, a strong movement is emerging that advocates the transformation of Italy into a federal system. This would create a situation in which the regions would be upgraded in terms of legal powers, including those dealing with environmental policy and natural resources.

Provinces

The almost 100 provinces are a direct heritage of the brief period of Napoleonic domination. For years there has been much discussion of the fate of this institution. 'Rationalisation' proposals in favour of abolition of the province have been countered by the fact that the province represents both the territorial level at which both many decentralised state offices are typically organised and the basic level for political activities (trade unions, parties). Recently the province has made a comeback. General legislation in 1990, reforming the system of local government, granted them substantial powers, especially in such environmental policy fields as water and soil protection, waste management, natural resources (flora, fauna, parks) protection, pollution monitoring and public health. Thus, after considerable indecision and confusion during the 1970s on whether environmental responsibilities should be concentrated at the municipal level or at the provincial level, with competencies (for example in the field of water pollution) being shared, it is now quite clear that the provinces are emerging as one of the key future actors in this policy sector. After a referendum that abolished environmental activities (controls, monitoring, etc) belonging to the Health units, on the basis of Act no. 61/94, such responsibilities should now be taken up by the provincial offices of the new regional environmental agencies (at present in the process of being created; only five regions have passed legislation in this field).

Municipalities

The 8000 or so *Comuni* represent the basic bricks of the Italian local government system. Often dating back to the medieval period, they are general-purpose authorities, entitled to carry out practically any type of activity required to satisfy the needs of the local population, without the strict limitations constraining the action of broader provincial or regional authorities. Of particular relevance for the environmental sectors is the fact that the municipalities are responsible for such policies as public health, traffic, public parks, refuse collection and disposal, sewage water treatment, public transportation and occasionally energy production.

Municipalities also control another category of agencies that have important responsibilities in the environmental field, i.e. the above-mentioned USL created by the national health reform of 1978 and put into operation two years later. The 630 USL represent the basic organisational units of the public health service and are responsible for the management of hospitals and health services in general. In addition, they are also responsible for the quality of the environment in which people 'work and live' (Act no. 833/78). In order to meet these responsibilities, the USL organisation also includes technical apparatus (*Servizi d'Igiene e Prevenzione*, *Presidi Multizonali di Prevenzione*) which carry out the control activities aimed at checking environmental conditions in general (from the safety of household appliances, to the quality of drinking water and foodstuffs, and to workplace and environmental conditions). In practice it is the technicians working for the USL who carry out measurement and monitoring activities, and enforce regulation enforcement on the basis of their own work programs and requests from local authorities. Understandably enough, however, due to the interests and 'culture' of these medical personnel as well as pressure from their clients, more than 95 per cent of USL budgets goes to direct health activities. Environmental problems are hardly high-priority concerns of these agencies. According to the results of a referendum held in 1993, however, such responsibilities will be taken away from the USLs and probably passed over to field agencies directly depending from the Ministry.

Centre/periphery relations

Initially, for a long period, environmental policy in Italy was widely local in character. Even national legislation was often triggered by local inputs. To local authorities trying to deal with spreading pollution, state laws represented both a limit upon their own initiatives and a resource for effective action in dealing with these problems. The creation of the Ministry of the Environment, which controls substantial financial and legal resources, marks the beginning of a shift of policy responsibility from the periphery towards the centre. In recent years, stimuli towards the development of environmental policy have been coming more and more frequently from the central state, though implementation remains a local affair.

Regional and local authorities have had an important role in stimulating the development of environmental policy. This has especially been the case during the first two phases described above. Municipalities, being the institutions nearest to citizens, often are the first to register emerging social demands. In attempting to respond to these demands, they either become agents of political and administrative innovation, experimenting with solutions that can subsequently be adopted and extended by the central state; or act as pressure groups, pushing for legislation and intervention into new areas, such as the environment.

With very few exceptions, primarily in some parts of northern and central Italy, the performance record of local and regional authorities with respect to environmental management has turned out to be rather poor. This has been a powerful argument used by the Ministry of the Environment to justify its strategy of a 'centralisation' of the policies under its control (CINSEDO 1989:659).

Perhaps the biggest limitation on the ability of local governments to determine their own policies are the constraints imposed by Rome on the access of these authorities to basic financial resources. Practically all resources needed to finance the public sector are collected by the central government. Subsequently it, although often with considerable delay, redistributes a portion of these revenues to local authorities on the basis of a quota system reflecting historical patterns of expenditures, albeit with some corrections for other factors.

But even the resources handed out by the central government are, to a large extent, earmarked for very specific targets. These categorical allocations substantially limit the actual autonomy of local governments in making policy choices. Likewise, special state laws often specify the precise use to which the allocations are to be put. This situation has a political consequence as well, since it tends to break the 'responsibility circuit' between local administrators and citizens/voters. To a large extent, locally elected representatives cannot be held responsible for policy results since they do not control basic resources required to meet local needs. Nor can political parties differentiate their political programs in terms of more or less taxes *vis-à-vis* different packages of services delivered.[10]

Relations between actors up to now have been marked by a considerable degree of competition: the policy arena is crowded with an increasing number of actors who are more concerned with trying to stake out their areas of influence in this new field than with coping with the substantive problems. This can be seen in the conflicts both among central ministries and between the central state and local authorities. Notwithstanding the increase of functions and tasks required by the growing demand for environmental quality, the distribution of powers is perceived by actors as a zero-sum game. From this perspective, centre-periphery relations are seen in terms of a 'layer cake' rather than a 'marble cake'.

Yet none of the institutional actors appears strong enough to be able to 'carry' environmental policy by itself, neither at the local level nor at the central level. The Ministry of the Environment, notwithstanding its attempts to exert tighter control over policy, seems to be far too weak in terms of technical and administrative capacities, as can be seen in the large amounts of unspent moneys and the lengthiness of decision-making processes. The question is whether a virtuous rather than a vicious circle in intergovernmental relations can be started for the purpose of improving the performance of environmental policy.

Implementation

The picture that emerges from the elements outlined above may be summed up as follows: after a rather slow initial take-off in the 1960s, environmental policy in Italy has gradually been gathering momentum; by now most of the relevant policy instruments (legislation, financial resources, specialised organisations at various levels) appear to be in place; also, environmental issues have acquired both legitimacy in policy terms and a considerable public profile. All in all, in these respects environmental policy in Italy has caught up with other nations with similar levels of economic development. Yet, although the condition of environmental resources in Italy has suffered serious degradation, policy responses come later and are less effective than elsewhere. If this conclusion is correct, how can this pattern of policy developed be explained? Possible explanations are to be sought for in the peculiar features of the Italian political-administrative system involved in environmental policy making and implementation processes.

Perhaps the most serious problems lie in the implementation process. One of the main causes of the difficulties that policy encounters in the implementation phase is the high degree of institutional fragmentation that emerges from the description offered above. Powers are distributed in such a way as to require that activities be carried out by a number of authorities at various levels of government. This causes high co-ordination costs and time delays (Dente 1989:146). Furthermore, the administrative system is seldom able to develop adequate forms of co-operation to overcome such fragmentation.

With very few exceptions, the performance record of local, regional and national authorities with respect to environmental management has turned out to be rather poor. The low level of performance in the implementation of public policy can also be explained in terms of the specific characteristics of Italian bureaucracy. In quantitative terms, the size of Italian public administration is comparable to that of other similar European nations: public personnel numbers some 2.2 million units at the central level, plus 1.4 million at the regional and local levels. It is the qualitative aspects, however, that make the Italian case distinctive. The bureaucratic culture in Italy appears to be dominated by a legalistic orientation, rather than by problem-solving attitudes and values (Aberbach, Putnam and Rockman 1981:52), in the sense that the prevailing criteria shaping behaviour are respect for and strict application of formal norms, typical of last century's administrative model (Freddi 1989). The actual attainment of results is of secondary importance and, indeed, often remains completely out of the picture. Furthermore, technical competence, particularly relevant for environmental policy, is rare and has a low status in the eyes of the political and administrative actors. This tendency is reinforced by the character of national legislation, based as it is on rigid standards, absence of flexibility and administrative discretion, formalised and complexity of procedures[11].

Furthermore, bureaucrats, who are drawn largely from southern, less-developed areas of the country offering very few employment opportunities outside the public sector, enjoy the privilege of job security and adopt risk-averse, stick-to-the-rule attitudes. Moreover, there is little incentive for civil servants actually to pursue policy goals since career advancement is based on service seniority. This professional weakness of public bureaucracies is characteristic of all levels of administration (obviously with some exceptions).

Another set of causes of the deficits experienced in the implementation of environmental policy can be attributed to the political system. In the first place, the greater 'permeability' of parties and institutions towards particular clienteles, as will be discussed below, rather than to some form of broader collective interest (however this might be conceived) provides ample opportunity for various pressure groups to hinder implementation of unwelcome measures.

Furthermore, in an attempt to respond to the growing demand for environmental quality, the political system is exposed to the temptation to look for short-term answers mainly with symbolic or placebo effects, but with limited substantive impact on problems in the long term, and with little consideration of the problems involved in actually implementing the measures taken. This obviously occurs in other countries as well, but due to the relative delay Italy has experienced in developing its environmental policy, the phenomenon is more acute here.

Another feature of Italian policy style with consequences for the implementation of environmental policy can be seen in the tendency towards 'regulation without rules' or, to put it differently, in the fact that formal over-regulation is typically matched with substantive under-regulation. Even in those sectors strongly regulated by rigid and stringent standards, these are, in fact, not enforced (Giuliani 1992:89).

This basic *de facto* non-compliance is so well recognised that three different institutional mechanisms are frequently used to cope with it: (1) prorogation of deadlines; (2) *'condoni'* or remissions, which are very frequent in tax policy due to the state's incapacity to collect revenues; and (3) loosening up of parameters (e.g. when levels of pesticides above EC limits were found in drinking water, the limits were simply increased for several years). The inability to attain policy goals, at least within the declared time limits, causes uncertainty as to how serious the authorities actually are. It can also lead to a loss of credibility in the eyes of the target groups, who end up with the impression that deadlines will not actually be respected, or that, at the worst, some sort of prorogation or remission will eventually be passed. In light of this likelihood, it is considered, culturally speaking, a bit foolish to respect formal obligations and deadlines. The continuous modification of state legislation concerning deadlines and parameters has serious consequences for the legitimacy of policy, and therefore for its effectiveness. This situation sets off a sort of vicious circle where the less policies are implemented, the less they are credible, and, therefore, the less they are effective. Granting prorogations and remissions thus becomes the basis for further policy failures. Authorities responsible for environmental policy implementation therefore suffer repeated difficulties as a result of the legislature's decisions to postpone deadlines or grant remissions.

National policy style – consequences for environmental policy

A second order of causes responsible for the delays and inadequacies of Italian environmental policy are to be found in the specific features of the prevailing political culture (although this represents a variable somewhat harder to grasp) and the ensuing policy style.

The political culture of a nation may be considered to be the result of its collective historical experiences. In the case of Italy, political fragmentation of the peninsula until a century ago, government over the past century by despotic and often corrupt rulers, the lack of impact of the Protestant Reformation, the long secular power of the Vatican, the weakness of the national bourgeoisie, the dominant role played by the state in the economy and society (Pridham 1993:3) are all factors that have contributed in shaping the features of national political culture: individualism, familism, localism, clientelism, fatalism, a fragmented political culture, lack of trust in others and in public institutions in particular, absence of a 'sense of the state' and of public interest, and the frailty of public ethics (Cavalli 1992:393). In other words, one can speak of a predominance of pre-modern political values. For the purposes of this discussion, one particularly relevant consequence of this situation is that public goods, including environmental quality, have a position of low priority in the prevailing political culture. There are no collective goods as such, but only goods of the state of which each private faction tries to get its share (Galli 1992).

'Particularistic' interests of specific groups and clans, if not individuals, rather than wider perspectives of collective interests, dominate public policies. Such traits provide a fertile soil in which a clientelistic style of policy making can flourish: the political system in general, and especially the parties in power during the last 50 years, such as the Christian Democrats and the Socialists have typically generated consensus by granting immediate and direct benefits to specific groups or corporations in exchange for support and votes rather than paying attention to more general societal needs. Besides a tendency to pass legislation at the national (Di Palma 1977) and regional levels along such lines, political parties have been keen on putting their representatives in key positions (public agencies, state-owned industries) in order to be able to maintain tight control over decisions that in other countries are the domain of experts, bureaucrats or actors representing societal interests. The exposure of the so-called *Tangentopoli* (officially started in February 1992 and still underway) and *Mani pulite* (clean hands) judicial inquiries is bringing into daylight what many informally already knew, i.e. that political officials in power have obtained consistent payoffs on practically all public works, either in order to finance their parties or for their own personal benefit. Many private and state-owned firms interested in such public works, though somewhat blackmailed by decision-makers, made handsome profits themselves out of such practices.

If these were the criteria used in making decisions on public projects, one can now fully appreciate the actual reasons that lead to the construction of many infrastructures, often with highly negative environmental impacts which are occasionally completely useless even in economic terms (such as impressive highways built in areas that hardly have any traffic). On the other hand, projects badly needed (the protection of Venice, the clean-up of polluted rivers, the construction of mass transportation systems) were never approved, perhaps due to lack of interest from powerful groups or because such groups could not reach an agreement on how to split up the cake. Furthermore, even environmental protection projects (anti-pollution devices installed on power plants, sewage treatment plants, incinerators) have also been the objects of such practices.[12]

Thus, although many parties have paid lip-service to environmental issues in recent years in response to growing public concern, the activism of environmental associations, and the electoral successes of the Greens, the actual interest in such 'collective' issues remains low, and the governing parties seldom actively promote environmental measures.

The traits described above largely account for the difficulty collective issues encounter in getting on the policy agenda and for the largely reactive nature of environmental policy. Decision-making in Western democracies is generally characterised by consensual relationships. At least in part because of this, authorities tend to react to problems rather than to anticipate them. In the case of Italian environmental policy, however, the reactive nature of public policy appears to be especially pronounced. The 'collective dimension' is only expressed with great difficulty, typically under extreme or abnormal circumstances. Thus it is not surprising that policy measures are usually triggered by events that are 'abnormal' *vis-à-vis* the 'ordinary' policy process, such as environmental emergencies and disasters: it was only after the break-out of the eutrophication of the Adriatic Sea, the finding of atrazine and other pesticides in underground drinking water reserves in Northern Italy, the diplomatic crisis following discovery in Africa and Lebanon of illegal dumps of toxic waste produced by Italian industries or the alarming levels of urban air pollution that measures were taken in these areas.

The second type of input is 'external' to the domestic policy-making process, being represented by international environmental obligations, and especially by formal policy decisions taken at the EC level (directives, regulations etc.). There is little doubt that a very consistent portion of present national environmental legislation in such relevant fields as ambient quality standards, industrial air emissions and environmental impact assessment, only to mention a few, would not have come into existence without the pressure of complying with EC Directives. On the other hand, Italy has often proven to be both late in translating EC provisions into national legislation (with one of the worse records in the Community in this respect), and ineffective in actually implementing legislation once it is passed, as a result of its low implementation capabilities discussed above.[13]

Furthermore, it should be noted that the EC context provides significant informal inputs towards the upgrading of environmental policy, thanks to the complex dynamics and relations occurring in the economic, cultural and institutional arenas through integration with nations more advanced and active in this field.[14]

Summing up, it can be said that the role of EC inputs for the development of Italian environmental policy appears to be so relevant that one could legitimately wonder whether the real centre of Italian environmental policy lies in Rome or in Brussels. The question can, obviously, be posed for all EC members, and for many policy areas. However, this is even more the case for Italy, due to its weaker environmental policies and lower political-administrative capacities.

Notes

1. An indirect indicator of this process is represented by the per capita consumption of cement that is three times higher than that of the United States, Germany or Great Britain.

2. It should be noted, however, that in the Italian context environmental issues are often tightly interconnected with artistic and monumental goods; this link can be seen in the creation of a

Ministry for Environmental and Cultural Goods.

3. The WWF at present counts some 60,000 members in Italy.

4. These will be discussed further below.

5. The *Lega* plays an important role in keeping the public's attention on environmental issues, for example by carrying out monitoring activities on urban air quality (the Green Train carries out measurements in some 20 medium and large cities every year) or on bathing-water quality (the Green Schooner carried out quality tests along the coasts in order to control official data produced by local authorities and the Health Ministry).

6. Environmental associations count a membership of approximately 200,000.

7. Ecologists have promoted several national and regional referenda against hunting.

8. Val d'Aosta, Trentino-Sud Tirol, Friuli-Venezia Giulia, Sicily and Sardinia.

9. The latter activities are often carried out through 'controlled agencies' (*aziende municipalizzate*), of which there are approximately 500, concentrated mainly in Northern and Central Italy.

10. The following data give an idea of the distribution of resources among the various levels of government: in 1987 the regions spent 82,090 billion lire, the municipalities 66,011 and the provinces only 6588 for a total of 154,689 billion, compared to 465,395 billion lire spent by the State. During the second half of the 1980s, municipal and provincial annual budgets were equivalent to 5.2–5.4 per cent of GNP.

11. For example, the construction of an average-sized chemical plant requires 14 different authorisations, granted by nine different agencies, based on 20 distinct pieces of legislation.

12. For example, for the installation of desulphurization devices on state-owned power plants (as a result of Italian commitments to reduce SO_2 and NO_x emissions on the basis of international agreements) a payoff of 2 per cent of 5500 billion lire, approximately 370 million dollars, was paid to political parties.

13. An emblematic case is offered by a Directive that takes its nickname from an Italian town, Seveso: a total of 709 plants classified as risky industrial activities resulting from a census carried out in the early 1990s, in 1993 the necessary administrative procedures were under way only for 179, with only one having been completed up to now.

14. Italy is the member state that has least complied with Community Directives in time, and thus has been most often brought before the Court of Justice in matters related to environmental directives (Capria 1988:201).

References

Aberbach, J., Putnam, R. and Rockman, B. (1981) *Bureaucrats and Politicians in Western Democracies.* Cambridge: Harvard University Press.

Capria, A. (1988) *Direttive ambientali CEE. Stato di attuazione in Italia, Quaderni della Rivista Giuridica dell'Ambiente.* Milano: Giuffré,.

Cavalli, A. (1992) 'Un curioso tipo di italiano tra provincia ed Europa', *Il Mulino*, no.3.

CINSEDO (1989) *Rapporto sulle Regioni.* Milano: F. Angeli.

Dente, B. (1985) *Governare la frammentazione.* Bologna: Il Mulino.

Dente, B. (1989) 'Il governo locale', in G. Freddi (ed.) *Scienza dell'amministrazione e politiche pubbliche.* Rome: NIS, pp. 123–169.

Dente, B., Knoepfel, P., Lewanski, R., Mannozzi, S. and Tozzi, S. (1984) *Il controllo dell'inquina-mento atmosferico in Italia: analisi di una politica regolativa.* Rome: Officina Edizioni.
Di Palma, G. (1977) *Surviving without Governing. The Italian Parties in Parliament.* Berkeley: University of California Press.
Freddi, G. (1989) 'Burocrazia, democrazia e governabilitá', in G. Freddi (ed.) *Scienza dell'amministrazione e politiche pubbliche.* Rome: NIS, pp. 19–65.
Galli, C. (1992) 'La cultura politica degli italiani', *Il Mulino,* No. 3, pp. 401–406.
Gardin, P. and Pazienti, M. (1992) *L'ambiente in Italia: Problemi e prospettive.* Milano: F.Angeli.
Giuliani, M. (1992) *Giochi regolativi. La politica di protezione del paesaggio.* Doctoral dissertation, European University Institute.
Herman, A. and Lewanski, R. (1992) 'Environmental Monitoring and Reporting in Italy', in H. Weidner, P. Zieschank and P. Knoepfel (eds) *Umwelt Information.* Berlin: WZB-Edition Sigma, pp. 314–330.
Knoepfel, P. (1984) 'La tutela dell'ambiente nell'Europa occidentale', *Archivio ISAP* no. 2, *Le relazioni centro-periferia.* Milano: Giuffré, pp. 2289–2317.
Lega per l'Ambiente (G. Melandri editor) (1989) *Ambiente Italia,* Rapporto 1989: dati, tendenze, proposte. Turin: ISEDI.
Lewanski, R. (1986) *Il controllo degli inquinamenti delle acque: l'attuazione di una politica pubblica.* Milano: Giuffré.
Lewanski, R. (1990) 'La politica ambientale', in B. Dente *Le politiche pubbliche in Italia.* Bologna: Il Mulino, pp. 281–314.
Lewanski, R. (1991) *Progetto Istrice-Ambiente, Le politiche ambientali nel distretto ceramico.* ENEA, giugno.
Lewanski, R. (1992) 'Environmental Policy in Italy: The Case of the Ceramic Tile District', *European Environment,* vol.2, part 1, February, pp. 5–7.
Liberatore, A. and Lewanski, R. (1990) 'The Evolution of Italian Environmental Policy', *Environment,* Vol.32, No.5. pp. 10–15 and 35–40.
Malaman, R. and Ranci, P. (1991) *Italian Environmental Policy,* paper at the International Conference on Economy and Environment in the 90s. Neuchatel.
Mariani, S. (1991) *I processi regolativi in Italia: il caso di una politica regolativa.* Dissertation, University of Florence.
Ministero dell'Ambiente (1989) *Relazione sullo stato dell'ambiente.* Rome: Istituto Poligrafico e Zecca dello Stato.
Ministero dell'Ambiente (1992) *Relazione sullo stato dell'ambiente.* Rome: Istituto Poligrafico e Zecca dello Stato.
Pridham, G. (1993) *National Environmental Policy-Making in the European Framework: Spain, Greece and Italy in Comparison,* paper for the Workshop 'Environmental Policy and Peripheral Regions in the EC'. ECPR Joint Sessions, Leiden, 2–8 April.
Silvestrini, G. and Collier, U. (1997) 'Italy: Implementation Gaps and Budget Deficits', in U. Collier and R.E. Löfstedt (eds) *Cases in Climate Change Policy. Political Reality in the European Union.* London: Earthscan Publications Ltd., pp. 108–26.
Spaziante, V. (1980) *Questione nucleare e politica legislativa.* Rome: Officina Edizioni.
UNEP (1991) *Environmental Data Report,* 3rd Edition 1991/92. Oxford: Blackwell.

Chapter 8

The Netherlands: Joint regulation and sustainable development

Kenneth Hanf and Egbert van de Gronden

Introduction

The Netherlands is a small, densely populated country, cut through by a number of large rivers. Situated on the south-east shore of the North Sea, at the downstream end of three international river basins (Rhine, Meuse and Scheldt), the country's physiography has long been characterised by shifting balances between land and water. About 50 per cent of the total surface area lies below sea level, and, in the west and north, is protected from the sea by barriers of dunes and dikes. Despite occasional losses to the sea, the reclamation of marshes, lakes and tidal areas gradually expanded the total land area.

The Dutch have long been known for their forward-looking environmental policy. Concern for problems of environmental protection have been fed by the perceived pressures on an environment which is, given the particular characteristics of the country and its geographic location, exposed to impacts from abroad as well as of its own doing. The country's open economy made government reluctant to take actions that would place its industry and trade at a competitive disadvantage. At the same time, this factor worked to encourage the Dutch to push for international solutions to transboundary problems, so as to maintain a level playing field across countries. With an eye toward both obligations to the broader international community and its own well-considered national interests, successive Dutch governments have worked at developing a system of environmental management to protect the quality of the environment while at the same time protecting the economic interests of the country.

In developing its policy strategy and institutions, the Netherlands could build on a tradition of public planning, stretching back hundreds of years: wresting of land from the sea and inland waters to construct, in many cases, the physical foundation of the country; managing scarce resource of useable land through a system of regional and spatial planning; and applying the various tools of economic and social planning characteristic of its post-war welfare state. Environmental policy and management in the Netherlands has also been shaped by the consensus-oriented Dutch political culture which places a high premium on avoiding conflict and seeking negotiated solutions enjoying broad support from politically relevant societal interests. Most recently, the Dutch traditions of planning and consensus politics have merged with a redefinition of the environmental problem as one requiring the self-responsible participation or involvement of economic actors, traditionally seen as the cause of the problem now as part of the solution. For various reasons, this policy has been able to thrive because basic conflicts between economic and environmental interests have been avoided or, in practice, economic development interests have been consistent with environmental protection. There are, however, at present, signs that this happy state of affairs may be threatened by growing tensions between what business is able and willing to do voluntarily, on the basis of well-understood self interest, and the actions that are required if the politically defined quality objectives are to be achieved, thereby necessitating a stronger 'regulatory role' by government to define and police the constraints within which economic interests can be expressed/pursued in an environmentally responsible way. In this chapter we trace the way in which tradition and a growing awareness of the need to combat environmental pollution worked to shape the Dutch approach to environmental management, and suggest in what ways the present system of joint regulation is coming under increased pressure.

The economic and social structure of the Netherlands

Bounded by the North Sea on the west and north, by Germany on the east and Belgium on the south, the country covers an area of only 42,000 square kilometres. In this small area there lives a population of 15.3 million inhabitants (1993). With a population density of 407 inhabitants per square kilometre, the Netherlands is one of the most densely populated countries in the world. The Dutch population has always been unevenly distributed over the different parts of the country, with a heavy concentration in the western provinces of North and South Holland. Until about 1960 most growth took place in the big cities. Then, in response to government policy aimed at spreading the population over the whole country in a series of regional urban centres, people began moving to surrounding towns with more green space. Although this process of suburbanization came to a halt in recent years, it did generate and, indeed, continues to generate much traffic and has had a number of detrimental effects on the countryside.

The country is intensely exploited and, in particular, the western part of the country is highly urbanised. Here are found the large concentrations of population, industry and infrastructure. Much of the land is under cultivation or other human use, such as housing, transport and industry. About 65 per cent of the total area is used for agriculture, 12 per cent is covered with wooded and natural areas, while urban areas and roads occupy 14 per cent of the territory, and water covers 9 per cent (OECD 1995:19). Despite a declining birth rate, there continues to be a housing shortage, and the demand for living space per inhabitant has risen sharply.

The Dutch economy is very open, closely linked to those of its prime trading partners, especially Germany. Exports of goods and services now constitute almost half of GDP. Manufactured goods (especially relatively low-technology petrochemical and bulk chemical products) make up 64 per cent of total exports (1992) while agricultural and food products account for 25 per cent. This exceptional reliance on trade also makes the Netherlands very dependent on the environmental resources of other countries (OECD 1995:23). The Netherlands has major natural gas reserves and a small amount of oil, and exports large amounts of gas. Domestic coal reserves are no longer exploited. There are no other significant natural resources. As a result, the Dutch economy is based, to a large extent, on the processing of imported primary products which are then exported as end-products.

Despite its high population density, Holland exports a considerable quantity of agricultural products. The intensity of agricultural activity is particularly high, reflecting the economic development options in that sector. Among Dutch agricultural activities, livestock farming occupies a very important place. An indication of the intensity of these activities can be found in the fact that labour productivity is up at the same time that the number of farms, the area devoted to farming, and employment in agriculture as a percentage of the labour force are declining.

The geographical position of the Netherlands, at the interface of the North Sea and the European hinterland, and the economic development options available, have made the transport sector very important to the Dutch economy. Rotterdam is the largest port in the world. About 30 per cent of goods loaded or unloaded from ships in the EU pass through Dutch ports. Dutch carriers figure prominently in international transport via road and waterways. The development of the airport Schiphol to one of the 'main ports' for European air transportation has also been an important factor for the economy.

Agenda of environmental problems

With more than 20 per cent of the country reclaimed from the sea and more than half below the high water levels of the sea and rivers, it is necessary to protect the country against flooding. The need to protect the land from high water from rivers and sea, and the tradition of artificially draining low-lying areas, have combined to give the country a complex hydraulic infrastructure. But the regulation of water supply implies more than protection against excessive quantities; the water level also needs to be regulated in order to prevent land from drying out. Furthermore, in order to keep the country suitable for cattle farming and agriculture, salination of the soil by sea water must be minimised. For this, fresh water, primarily from rivers, is needed.

Of the three important European rivers passing through the Netherlands before entering the North Sea, the Rhine in particular is of enormous importance for Dutch water management. The upstream pollution of this river is one of the most serious environmental problems of the Netherlands. It has caused damage to fisheries as well as to the flora and fauna in the vicinity of the river, and it prevents people from swimming in the river. The most serious damage, however, is done to the supply of drinking water, since the Rhine is the most important alternative to the limited availability of groundwater abstraction. Though the water quality of the large rivers has clearly improved in recent years, despite the many continuing problems, long years of poor quality have caused severe pollution of the river bottoms' soil in many locations. Increasing levels of salt concentration in the Rhine also make its waters unsuitable for agriculture and horticulture.

Dutch bio-industry, and especially intensive livestock farming, also contributes to the pollution of water. Inadequate disposal and management of manure surpluses leads to pollution of ground water and surface water. The excessively high amounts of nitrogen from the extensive use of fertilisers and pesticides are particularly problematic. Dutch agriculture is also energy intensive. Agricultural activity itself, but also the claims on land from the consolidation of smaller holdings and for the physical infrastructure for this sector of the economy, has caused damage to the rural landscape.

Not surprisingly, given the heavy use of land for various kinds of economic activity, soil pollution is another serious environmental problem. So, too, in a densely populated country like the Netherlands, noise is inevitably a serious nuisance, while the processing of large quantities of solid wastes generated by the production and consumption activities of this population also constitutes a major problem. The national production of household refuse is higher than in the rest of Europe, calculated not only per square kilometre but also per inhabitant.

Acidification (due to road traffic, electricity generation, industry and agriculture) and depletion of water (desiccation due mainly to water extraction for agriculture) threaten the quality nature areas. Cities, industrial estates and roads steadily encroach upon the countryside with serious disruptive effects.

As a result of the different pressures on the country's environment, the Netherlands has produced domestic environmental plans which are considered by some observers to be exemplary in their willingness to take the issues seriously and to take action to address the matters at hand. The Dutch government has also taken numerous initiatives at the international level for dealing with the transboundary aspects of environmental problems, all this, at a minimum, on the basis of a well-understood self-interest in environmental co-operation. Nevertheless, even though the country has set itself ambitious targets and has pressed the environmental case hard at international gatherings, the Netherlands is itself a major polluter and, therefore, not only exposed to the pollution of several other up-river and up-wind countries.

Development of an environmental movement in the Netherlands

The Netherlands changed rapidly and radically in the period from 1900 to 1940. As a consequence of industrialisation and the resultant improvement of living conditions, there was a strong growth of population and urbanisation, concentrated in the conurbation stretching roughly from Amsterdam through The Hague and down to Rotterdam that has come to be called the Randstad. Although hygienic conditions continued to be quite poor in the urban areas, advances in medicine and health care did decrease the dangers of contagious diseases, and infant mortality declined. Nevertheless, public policy, especially at the local level, concentrated on improving social conditions and public health.

At the same time, there was some concern for what might be called 'spatial quality', that is, the relation between quality of the natural surroundings and human well-being. Both urbanisation and industrialisation made increasing claims on land and 'space', and, as a result of more intensive farming, mining and quarrying, and deforestation, there was considerable damage to nature. Although not perceived as a serious problem at the time, there was also pollution in some areas which resulted in a deterioration of environmental quality.

All in all, the perceived loss of nature values led to the establishment of organisations in support of nature conservation and the creation of parks and recreation areas. At the same time, a gradual process of 'cultural change' was set in motion as a result of growing awareness among some elements of the economically better-off public, of the costs of progress and especially the loss of the natural environment. Furthermore, a 'back to nature movement' was a prominent element of the socialist youth movement. It was during this 'romantic period' that the first Dutch environmental organisations were born: in 1899 for bird protection and in 1906 the still-existing Association for the Protection of Natural Monuments (*Natuurmonumenten*).

In the Netherlands there are now eight major national organisations for environmental protection, which work together through a regular national environmental consultation, co-ordinated by a full-time secretary. This consultative forum was established in 1976 and brings together organisations varying in size from 10,000 to 260,000 members. Three of the organisations involved were founded at the turn of the century, the others after 1960. These pioneers of nature conservation initially focused exclusively on the protection of nature from a recreational point of view. The organisations founded at the beginning of this century had operated, in the first instance, by raising money to buy natural and cultural monuments that were threatened with loss or damage. Since 1945, this activity has been subsidised by the national government. In addition, they also participated in government and provincial advisory committees, and sought to influence policy by means of government petitions and court proceedings against instances of environmental deterioration. Nature conservation organisations are generally stable, have a formalised internal structure and are directed by people with a lot of expertise.

Negotiations and consultations based on expertise, formal pressure and lobbying are currently the main elements in the strategy of the nature movement (Bressers and Plettenburg 1995:9).

In the period 1968–72 a series of environmental incidents prompted growing worries over the environment in the Netherlands. From 1970 to 1975 the Dutch environmental movement underwent a period of unprecedented growth. During this time, a number of new organisations were founded, although later on, at both the national and the provincial levels, there was a consolidation of the separate groups into larger, more comprehensive organisations (often funded, in part, by the government itself). One of the factors that played an important role in the emergence of the environmental movement was a general intellectual climate that nourished doubts regarding the dominant values and objectives associated with a faith in rationality, science, technology and progress.

Although this new movement was rooted partly in the traditional nature conservation organisations, it differed from its predecessors. In addition to its primary focus on combating pollution and the depletion of natural resources, these organisations also engaged heavily in expressive and symbolic actions designed to mobilise public opinion against the government's environmental policy. They clashed much more regularly, and sometimes even violently, with prevailing values and norms. On the whole, the movement was politicised, in part as a result of general developments and specific links to the broader movement demanding more democratisation and general socio-political criticism. This led, among other things, to increasing contacts with the parties of the left and the Social Democrats, but also, gradually, to the confessional and liberal parties. Although the newer environmental groups differed sharply from one another in ideological terms, they were all more strongly involved in political questions than had been the case with the more traditional organisations. Towards the end of the 1970s, as the popularity of 'extra-parliamentary' actions began to wane, many environmental organisations and action groups changed their strategy and were prepared to negotiate.

This change in attitude and tactics was greeted enthusiastically by the newly established environmental departments at the national, provincial and municipal levels, who were on the lookout for reliable allies and were, therefore, eager to recruit the environmental movement as a negotiating partner. In some cases, this new status was formalised by law. Today, the environmental movement has, on the whole, become a respected institutional and professional actor in the field of environmental policy.

A pivotal role in this regard is played by the Foundation for Nature and the Environment (*Stichting Natuur en Milieu*), which was established in 1972 and represents approximately 120 environmental groups. Similar federations of provincial and local environmental organisations were set up to serve as co-ordination points for the individual environmental groups, and to keep a critical eye on government policy and actions at these levels. The Foundation serves as the main think-tank for the environmental movement and supplies many of its representatives to committees and other consultative structures in and outside of government. The Ministry of the Environment considers the Foundation to be an important ally and has, therefore, always provided it with generous subsidies to guarantee the quality and strength of its arguments in the policy debate.

The Dutch traditions of co-operation and consensus has meant that the government tends to ask private organisations for advice on controversial issues. As a result of the politicisation of the environmental problems and the refusal on the part of the new environmentalist groups to accept government decisions which could lead to environmental deterioration, this accommodatory style was severely tested in the 1960s. Nevertheless, the politics of consensus and accommodation remains a key feature of the Dutch political culture. Government departments still tend to be surrounded by what is referred to as an 'iron ring' of advisory councils through which private organisations have an opportunity to influence policy decisions. As a result, government secrecy tends to be tempered by selective consultation with parties whose interests are germane to an issue and whose support is necessary or whose likely opposition needs to be taken into consideration. Over time, the representatives of the new environmental movement have been incorporated into the structures and practices of accommodation and consultation.

After the period of rapid expansion, by 1980 the membership of environmental groups had stabilised. It is interesting to note that figures show that those organisations that had appeared in 1970s pushing to 'change society' have remained relatively small. On the other hand, organisations such as *Natuur Monumenten,* the World Wide Fund for Nature and Greenpeace, without this commitment to fundamental changes in the social system, have grown rapidly. The different organisations now work together in the National Deliberative Council for the Environment (LMO) and are represented on various government advisory bodies.

Public perceptions of the environmental issue have fluctuated enormously over the last 25 years. The environmental issue was very prominent at the beginning of the 1970s and when most environmental legislation was passed. During the 1980s, on the other hand, general public attention was low, and it was the civil servants and interest groups who debated and negotiated over the development of strategies and measures. Survey figures show that during most years only a small proportion (one to five per cent) of the electorate sees the environment as the most important national problem. At times this proportion shot upwards, for example, in the early 1970s (about 20 per cent) and again around 1990 (45 per cent). The swings in public concern for the environment contrast sharply with public perceptions of the seriousness of unemployment. Throughout the period 1964–94, except for the early 1970s, unemployment has consistently been seen by the electorate as one of the most pressing social problems. There appears to be a kind of inverse relation between concern for unemployment and concern for the condition of the environment. In any case, the ups and downs of popular attention to the environmental problem seem to be relatively unaffected by the gravity of the actual problems. Apparently, many people still see the environment as a luxury problem (Bressers and Plettenburg 1995:14).

The political context of environmental policy changed significantly around the mid-1970s. The environmental euphoria disappeared, and with it, the most politicised environmentalist organisations also passed from the political stage. Other organisations began to act more pragmatically as they saw their membership decline.[1] In general, influence of the environmental movement seemed to decline, as a consequence of the growing social and political preoccupation with problems of unemployment and the slowdown of economic growth.

In the second half of the 1980s, the economy picked up and the environmental movement recovered as well. Membership figures began to look healthier and there was upsurge in publicity and media attention for and about the movement. From 1987, attention to the environmental problem grew, helped by the publication and subsequent discussion of the Brundtland Commission's report on sustainable development, and perhaps also due to the decline in the heretofore strong peace movement.

Dutch observers have concluded that the state of the economy seems to be very decisive for environmental policy and the power/influence of the environmental movement. In times of good economic conditions, attention shifts from material to immaterial issues and there is more room for an environmental policy. On the other hand, when times are bad economically, then the concerns of business and financial considerations prevail over environmental protection. Political power relations are also of consequence for the environmental movement. In the late 1980s and early 1990s, green parties and green issues gained considerable support. The Dutch political landscape consists of many political parties which display different shades of green and slight differences in ideology (Van der Tak 1994:25–26). These differences have to do with the amount of economic reform the parties favour, i.e. the extent to which they feel it is necessary to modify, radically, traditional patterns of production and consumption.

In the 1970s and 1980s the Labour Party (PvdA) and the left-liberal D66 and Green Left were more environmentally friendly than the Christian Democratic Alliance (CDA) and the conservative Liberal Party (VVD), with the VVD parliamentary groups being the least environmentally friendly, relatively speaking. But political power relations seem subordinate to economic factors. In the second half of the 1970s, when public support for environmental questions fell sharply as economic concern grew, the shift was made by all parties. Today, concern with economic recovery and questions of creating new employment opportunities, reducing the public deficit and controlling inflation continue to weigh relatively more heavily politically than do environmental issues. Still, government policy initiatives on these economic questions are constrained by the need to accommodate widespread popular demands for action on environmental quality.

It was during the first half of the 1970s that the 'first wave' of (modern) environmental concern crested in the Netherlands, leading to emergence of the environmentalist movement, the mobilisation of general public concern and the passage of a spate of governmental measures. But already in 1975, the tide had begun to recede. In the first half of the 1980s, economic and financial problems enjoyed greater political attention and priority than environmental questions.

Now there is talk of a second wave of environmental concern and action. There are signs that the political chances for effective environmental policy are now better. The environmental movement at present occupies an institutionalised position in society. At the same time, recognition is growing that the traditional solutions, which had been used in dealing with many environmental problems, i.e. passing them on to the next generation, are no longer a real option. Finally, support is growing for the proposition that environmental management is consistent with or even a precondition for economic growth. More and more economic actors and government officials agree that the very bases for economic production will be put in jeopardy if pollution and other abuses of the environment continue. In this sense, changes in the attitudes of both industry and agriculture in the Netherlands have opened new possibilities for an active role for these economic actors within the overall system of environmental regulation. In June 1997, the White Paper on Environment and Economy published by the government contained a number of concrete measures and financial incentives designed to bring the goal of a sustainable and environmentally-efficient economy closer to realisation. In this connection particular emphasis is placed on technology policy. At the same time, the environmental movement is busy redefining its attitude toward economic growth and its relations with the business community.

The development of Dutch environmental policy

Until about 1970 one could scarcely speak of 'environmental policy' in the Netherlands. Nevertheless, as in most other countries, policy in this field had its roots in earlier measures to control nuisances and promote public health. Around the turn of the century, people were primarily concerned with meeting the basic needs of life, such as food, energy, housing and dealing with the most annoying aspects or 'nuisances' of modern life. From 1900 until 1960, the Netherlands gradually industrialised and became more urbanised. At the beginning of this period, public attention was focused primarily on problems of public health, such as clean drinking water, sanitation, sewage and the quality of food, and on developing the physical infrastructure needed to make life in cities more bearable. At the same time, in the face of the pressure on land from both industry and housing, interest also grew in protecting nature and regional quality as well as providing adequate recreational opportunities.

More rapid economic growth, and more intensive urbanisation and regional development came to the Netherlands during the period from 1960–1980. The years 1945–1965 are usually referred to in the Netherlands as the 'post-war reconstruction period'. During this time, the material losses of World War II were worked away and the foundations laid for what became known as the modern welfare state. The primary concern of government was the stimulation of industrial development as a precondition for this material growth and social welfare. At the same time that the groundwork was being laid for material abundance, new 'scarcities' became apparent in the form of the declining quality of water and air, and the lack of peace and quiet. At first the seriousness of the problems caused by the very success of the reconstruction efforts was not widely recognised by society and its government. As they did become more apparent, however, public attitudes began to change and pressure was exerted on government to make care for the environment a visible part of public policy. Demands grew stronger to limit as much as possible the negative effects of these processes on nature and the environment.

As noted above, this was the period in which new environmentalist groups were formed and the more traditional organisations began to apply some of the tactics of the newer environmental lobby groups. Concern with public health was widened to encompass environmental hygiene, that is, those indirect threats to human health coming from deterioration and pollution of the natural environment. Not surprisingly, public policy at this time was still very much focused on cleaning up the mistakes from the past and in dealing with the most urgent problems in the different sectors of the environment. During this time physical planning was primarily concerned with allocating space to the different societal and economic functions.

In an important sense, there has always been a kind of 'environmental management' in the Netherlands in the form of water management (both maintaining the purity of the water and the water level). In addition, intensive use of limited surface area made strict physical planning necessary. Thus, the more modern environmental policy of the seventies found a nourishing basis in traditional concern for the physical surroundings. However, modern environmental policy in the Netherlands has developed to a large extent apart from water management and physical planning. This has been due to the fact that for environmental problems, new legislation was passed and, in part, new administrative organisations have been set up or existing structures adapted to handle the new programs.

Until 1960, the Nuisance Act was the only law in the Netherlands that enabled municipal authorities to implement any kind of environmental policy. Over the last 30 years, the number of environmental laws has increased dramatically. The 1960s saw the passage of the Pesticides Act (1962), the Nuclear Energy Act (1963) and the Nature Conservancy Act (1967). A second wave of 'modern' legislation, just now ebbing away, was inaugurated with the Surface Waters Pollution Act (1969).

In 1972 the government published a *Memorandum on urgent environmental issues*, which described the main environmental problems facing the country and developed both a long-term policy strategy for the environment and a set of urgent measures for a number of designated problem areas. Throughout the 1970s various pieces of environmental legislation were enacted to deal separately with the problems confronted in the individual compartments of the environment. During this period, laws were passed to cover the various sectors of the environment, i.e. air, water, soil, as well as certain forms of environmental hazards, (such as noise and hazardous substances. These included, among others, the Chemical Wastes Act (1976), the Waste Materials Act (1977), and the Noise Abatement Act (1979). Legislation on chemical substances and soil pollution cleared parliament in the latter part of the 1980s.

These laws took the form of so-called framework legislation which laid out general objectives to be achieved, and defined the broad lines of responsibility and range of instruments to be employed, but which needed to be filled in further with more specific regulations for implementing the general program objectives. The authority to issue such regulations by means of different kinds of executive decrees and orders is delegated in the law to the appropriate governmental authority. This environmental legislation aimed, primarily, at prohibiting actions or activities that threatened the environment. Exceptions to such prohibitions could be obtained by means of a licence, exemption or other dispensation. Licences were granted subject to conditions such as specified emission reductions or abatement control measures, or other restrictions on polluting activities. Other legal instruments available to the authorities were quality objectives, plans, general rules and financial measures. The laws usually provided for administrative sanctions in the case of non-compliance, such as the withdrawal of the licence, the imposition of a fine or the closure of the establishment.

All these acts are, then, examples of traditional direct regulation, based on general regulations and a system of conditional permits. Consequently, since each had its own set of regulatory procedures, companies and individuals had to apply for different permits and in most cases to deal with different regulatory authorities when engaging in activities that fell under the legislation governing the different segments of the environment. Permits were required for almost every activity, often even those with relatively minor environmental impacts. Under such a regulatory regime, it was, not surprisingly, difficult to provide a coherent assessment of the overall environmental impact of an activity. Starting in 1983, as part of the general policy move toward government deregulation, efforts have been made to reform, i.e. rationalise and simplify, this permitting system. A central focus of these efforts has been on the integration and harmonisation of the different regulations and permitting requirements presided over by separate governmental authorities.

Since approximately 1980, the concern for the environment itself began to take the centre of the stage in government policy as the primary goal became that of improving the quality of the eco-system by integrating human activities into the limits set by the objective of maintaining the integrity and quality of the natural environment as a precondition to human existence. This new perspective on the nature of the environmental problem meant a shift from a remedial to a preventive policy approach. It also had important consequences for the relationships among the various elements of environmental policy as well as for the relationships between this policy area and other areas of government activity. Most importantly, the commitment to preventive policy, as a precondition for the achievement of what came to be called sustainable development, was to have significant repercussion for both the policy instruments applied to reach these goals and the administrative organisation through which they were to be achieved.

The sectoral laws came in for increasingly sharper criticism from various sections of society: citizens alleged that the public participation and appeal procedures were biased against them, while industry, for its part, claimed that the licensing procedures were far too time consuming. Furthermore, they asserted that there was insufficient co-ordination between the various authorities involved in the licensing process. To remedy the shortcomings of this sectoral legislation, a new approach was begun in 1980 with the passage of the Environmental Protection (General Provisions) Act. This piece of legislation contained a uniform set of rules governing the procedures for applying for and being granted environmental licences; it also provided for uniform participation and appeal procedures. Furthermore, the Act served as the legislative vehicle for the development of a number of new instruments not included in the earlier laws, such as environmental planning, environmental impact assessments and modified financial regulations. The General Environmental Procedures Act was intended to provide the legal basis for developing a comprehensive system of multi-media management for an integrated attack on environmental quality problems.

But still it was felt that environmental legislation lacked coherence and pressure was exerted to expand further the scope of the new General Environmental Provisions Act at the expense of the sectoral laws. This culminated in 1993 with the incorporation of the general provisions legislation into the new Environmental Management Act (EMA). The new law, which radically simplified Dutch environmental law, was to be an 'all-embracing' general law for the environment. A key feature of the act is the provisions for a new integrated system of environmental permitting which will enable installations to apply for a single permit covering all operations. As a consequence, the five licensing systems from five separate environmental acts – nuisance, air, noise, waste materials, chemical waste – were transferred to the EMA. These changes were made to make environmental legislation more transparent and easier to enforce. A separate license is still required under the Surface waters Pollution Act, since the Ministry of Transport and

Public Works refused to give up its permitting authority over these activities. In January 1994 a regulation on Waste Substances came into force containing new requirements for the disposal of household waste, industrial waste, car wrecks and chemical waste substances. All provisions of the new EMA are aimed at increasing opportunities for conducting a consistent, integrated environmental policy, based as far as possible on controlling the sources of pollution (Matthews 1995:17). The EMA has also widened the opportunities for participation as well as for lodging objections and appeals. Authorities are now required to draw up a draft decision, for example, concerning a license application, before the actual decision can become legally binding. This enables individuals and organisations to object to the proposed decision at an early stage. Provision is also made for public hearings if demanded or otherwise deemed necessary. An important change is that those wishing to participate in, object to or appeal these proceedings no longer need to be directly involved in the activity under consideration.

The 'renewal' of environmental management

The foundation for the environmental management strategy being employed at present in the Netherlands were laid during the period in which Pieter Winsemius was minister of the environment (1982–86). In the period before him, the ministry of the environment had established itself within the administrative system of the Netherlands and had begun to expand its policy terrain to include a variety of environmental compartments and issues. There had also been a gradual shift from a strongly anthropocentric orientation (where public health considerations were central) toward a policy with a more eco-centric point of departure and objective. Furthermore, initial steps were being taken to confront the criticism that the sectoral approach to the problems was not sufficient, since environmental problems could not be treated separately, apart from developments and decisions in other policy areas.

Already late in the seventies and more so in the early eighties, critics pointed to the disadvantages and limitations of the sectoral environmental policy pursued up to that point (Leroy 1994:42). Efforts to counteract the substantive, procedural and organisational fragmentation of environmental management took a number of forms. Environmental policy was no longer formulated according to separate environmental compartments (air, water or soil), but according to themes (acidification, eutrophication, water depletion), geographical areas (vulnerable sandy soil, water protection areas), flows of materials (e.g. cadmium) and target groups (traffic, industrial sectors, agriculture). The new approach initially stressed the need for internal integration within the field of environmental policy itself. Later, more attention was given to the 'external integration of environmental policy', i.e. inter-policy co-operation. This means that policy and programs in other departments were to be fine-tuned or co-ordinated with the objectives of national environmental policy as well as with measures of provincial and municipal authorities.

Winsemius brought with him a vision regarding the way in which environmental quality objectives could be achieved most effectively. He changed the internal organisation of the ministry and its relations with its surroundings. Internally, the move from a focus on compartments of the environment to one emphasising comprehensive environmental themes embracing a number of sector-transcending environmental problems, e.g. acidification, eutrophication or water depletion, made a more project-oriented type organisation necessary. Such a substantive reorientation required the setting up of project teams drawn from the different bureaus and sectors to develop policies for each theme and the different target groups. Moreover, in moving beyond the sectoral approach, the new strategy also moved beyond the limited competencies of the Ministry of the Environment (ME) [2] It now became necessary to collaborate regularly and systematically with other departments, such as transportation and public works, agriculture and nature, and economic affairs. Within the jurisdictional lines of the existing ministerial division of labour with regard to environmental policy, new mechanisms for inter-departmental co-ordination were developed, including the use of environmental contact officials in other ministries. These officials were responsible for monitoring and co-ordinating environmentally-relevant activities within their own houses. Inter-departmental teams have been set up for regular exchanges of information and consultation among these officials.

The expansion of objectives meant that more attention needed to be paid the broader social and politico-administrative context within which environmental policy was formulated and implemented. Attention to the relations between the ME and the various actors in its external environment led to a rethinking of the way in which policy objectives can be achieved. This re-examination of the general management strategy of the ME has been part and parcel of the more general set of changes in the Netherlands regarding the capability of the state to intervene and 'steer' the course of societal development. In essence, there was less optimism regarding the steering capacity of the state. It was increasingly recognised that government had to be able to convince other actors of the necessity and 'correctness' of the proposed course of action and to get them to co-operate in its implementation. This meant that more emphasis would have to be placed on achieving consensus among the parties involved. In the Dutch case, this insight, combined with the focus on preventive policy, meant stressing the self-responsibility of target groups with regard to making their activities more environmentally friendly.

Changes in the approach of the ME to societal actors, such as the chemical industry, the power industry and other economic actors, reflected the minister's conviction, coming as he did from the private sector, that regulation from above was inadequate for achieving the objectives of environmental policy. Traditional top down regulation was not accepted by the target groups; they continued to try to evade their obligations or otherwise complied only half-heartedly. Instead of imposing quality objectives and the associated regulations from the 'outside', the target groups should, according to Winsemius, be encouraged to make the objectives of government policy their own goals. That is, they needed to internalise these objectives into both the value system as well as the organisational structure and operations of their own firms, so as to create the basis for internally-motivated commitment to the policy objectives. They needed to develop their own sense of responsibility for the environmental impact of their activities. In this way, concern for environmental quality could be effectively integrated into the decision making and behaviour of the firm, thereby promoting the realisation of both objectives of environmental policy and the economic goals of the enterprise.

According to Winsemius, this self-responsible commitment to the objectives of environmental quality could best be achieved by developing more occasions and structures for consultation and deliberation between the target groups and the government, by concluding private law and legally non-binding covenants or gentlemen's agreements between the ministry and economic organisations, where the target group accepts certain duties or obligations in exchange for the government's word that it will not come up with more regulations.

The different strands of Dutch environmental policy and management come together in the series of policy plans that has been produced since 1989, following upon the heels of earlier planning and programming activity. Since the early 1980s, the ME had operated with rolling three-year Environmental Programs, updated yearly, which set out the broad lines of the Ministry's activities and accomplishments. A more ambitious exercise in environmental planning, involving not only the national government but also provincial and municipal authorities as well as a number of target groups of environmental policy in society, produced the first National Environmental Policy Plan (NEPP) in 1989. As input into this plan, the National Institute for Public Health and Environmental Health (RIVM) had published a comprehensive survey of the state of the environment and trends up until 2000, entitled *Concern for Tomorrow*. This report provided the data and interpretation on achievements and likely trends to support the government's environmental policy plan. It was also instrumental in raising public consciousness and focusing political debate on the critical state of Dutch environment and the need significantly to strengthen environmental policy.

This policy plan, together with the different sectoral plans keyed to the objectives of the NEPP, was intended to be a comprehensive policy document serving as guidance to all levels of government as well as various target groups. It is to help redirect environmental policies, to associate all levels of government and all relevant societal groups with the implementation of these policies, and to integrate more effectively environmental concerns in sectoral policies and practices.

The NEPP established the ambitious objective of 'reversing environmental degradation and achieving sustainable development within one generation'. The policy discussion and public debate over this policy strategy in the Netherlands has also been enriched with the notion of 'available environmental space' which is used to indicate the limits within which resources can be abstracted from the ecosystem and the wastes absorbed such that this can continue indefinitely without harming the ecosystem. Central to this effort to achieve this dynamic equilibrium is the integration of environment into every aspect of social and economic development. The NEPP 'recognises that a high quality environment cannot be achieved through conventional pollution control measures alone'. What is also needed is a mixture of new, clean technologies and structural changes in production and consumption patterns. To this end, the plan sets out a framework strategy in which technological, social and economic change, supported by co-operative actions at the international level, offers the prospect of 'doubling Dutch GNP by 2010 while achieving emission and waste discharge reductions of 70–90 per cent (except CO_2)' (HPPE 1993:5). Although sustainable development is often defined in process terms, here the Dutch government provides a definition of its objective in measurable, substantive terms.

The overall goal of the NEPP is the achievement of a sustainable society by the year 2010. The NEPP specifies key environmental quality objectives and sets out a long-term program of actions to ensure that these are achieved. The NEPP identifies eight themes and nine target groups, and sets, for most themes, overall quality objectives. The themes are: climate change, acidification, eutrophication, dispersion (uncontrolled spread of hazardous substances), waste disposal, local nuisances, water depletion (dehydration) and resource management (sustainable use of renewable and non-renewable resources and energy). These themes are key to the policy-development process: they help to identify the physical sources of environmental degradation, the economic sectors/activities responsible and the levels at which impacts are felt (local, regional, continental and global).

The general objectives of the themes of the NEPP are further broken down into numerous reduction targets for specified substances and waste streams. For each of the environmental themes, quality objectives have been determined that are to be achieved by 2010 (or the year 2000) and the percentage reductions in key pollutants have been set that will be required to meet them. Responsibility for achieving these emission reduction targets lies with the different target groups. These target groups represent the key groups of polluting activities in Dutch society. During the last few years, continuing efforts have been underway to reach agreement, among the government and societal actors involved, on the division of responsibility for achieving specific targets between the different target groups and among the members of a given target group.

The first environmental policy plan was drawn up on the basis of evaluations of the progress that had been booked up in meeting the objectives of the measures taken up until then. At the same time that the plan presented a new vision on the nature of the environmental problem and laid out a set of ambitious objectives, there were growing doubts regarding the ability to achieve these goals with the policy instruments and management strategy then in use. In 1990 a supplement to the first plan was presented (NMP+) which was intended to make clear how the goals of the initial plan were to be achieved, taking into consideration the widespread concern regarding the limited capacity of government to influence societal developments in the desired direction. In retrospect, it is possible to see the publication of these two plans, within the context of the broader debate on the steering capacity of government, as a turning point in Dutch environmental policy (Bressers and Klok 1996:448). On the one hand, environmental policy had taken on a heavy set of tasks, while, on the other hand, confidence in the traditional policy approach had been greatly shaken. NMP+ outlined the changes considered necessary in the prevailing policy strategy. This included an extended discussion of the instruments needed, as complements to traditional forms of direct regulation, in order to realise the new direction in the environmental management.

A central element of this new strategy, in line with the proposals already advanced by Winsemius, was to bring the target groups of the policy themselves to accept greater direct responsibility for the quality of the environment. This general principle has since been worked out further in the Dutch 'target group policy', which is based on various forms of structured consultation and negotiation between representatives of government and branch organisations of the different economic sectors regarding the means for best achieving the goals of the plan. Once agreement has been reached regarding the contribution that a given sector or branch is to make to the realisation of the quality objectives, this consensus is given the form of a voluntary agreement or covenant. Such agreements are not only intended to exert direct influence on the behaviour of the firms within that particular sector or branch; they are also supposed to serve as the guidelines for the granting and enforcing of environmental permits by the competent governmental authorities.

The first voluntary environmental agreements between government and business date from the early 1980s, but the use of this instrument increased dramatically after 1990. At present there exist between 85–100 agreements, depending on which source one uses. Most covenants are based on a contract between industry representatives and one or more ministries. Most of the time, industry will be represented by its branch associations, although there have been cases where individual companies sign the agreement. In the early and mid-1980s voluntary agreements dealt with single issues (e.g., one substance-product combination). More recently covenants are either sectoral agreements covering a wide range of issues within one economic sector, or voluntary agreements focusing on a particular environmental issue that cuts across the boundaries of a number of sectors (Mol *et al* 1996:9). At present there are no legally binding procedures for drawing up such a contract; the environmental ministry has, however, issued a set of guidelines to standardise the process somewhat. It should be pointed out that these so-called 'voluntary agreements' are seldom, in fact, entirely voluntary. Quite often – and always with regard to statements of intent and implementation plans through which the 'integrated environmental task' (i.e. the branch-specific translation of national quality objectives) of a particular industrial branch are negotiated – these agreements are linked to more traditional regulatory instruments, through which these obligations can be made legally binding. Voluntary agreements are used to determine how the general quality objectives set by government can best be achieved under the conditions peculiar to a particular sector or branch of the economy. Perhaps the greatest incentive for business to conclude and carry out such agreements is the ever-present threat of formal governmental regulations should the necessary self regulation fail to materialise.

A recent evaluation of the experience with covenants, in general, has shown that despite a number of problematic aspects, environmental agreements score better than those concluded in other policy areas. Moreover, more recent covenants appear to be of better quality (Algemene Rekenkamer 1995:4–7). While environmentalist groups, and lawyers, continue to harbour doubts about the effectiveness of such agreements (due to their unclear legal status and lack of sanctions), the business community praises them for the flexibility they allow firms in pursuing environmental quality objectives. It is interesting to note that the Ministry of Economic Affairs prefers voluntary agreements in most situations, while the ME sees this instrument as a useful addition to legislation or as a temporary measure until such legislation has been passed (Mol *et al* 1996:10). Judged in terms of democratic control, voluntary environmental agreements raise a number of questions regarding who is or is not included in the negotiations, the lack of transparency of the decision making and the marginal role played by parliament.

It would not be an exaggeration to say, with Bressers and Klok (1996:446), that the target group policy has come to dominate environmental policy in the Netherlands as far as the regulation of economic activity is concerned, In this sense, a significant portion of this policy is based on one or the other form of interactive policy making, characterised by organised consultation and self regulation within the parameters set by the overarching quality objectives of government plans and legislation. This management strategy, together with the general objectives defined in the earlier plans, was confirmed in the second national environmental policy plan published in 1994. This plan stressed the need to improve the implementation of measures already taken to realise the objectives of sustainable development. It also included a number of additional measures which were considered necessary where objectives otherwise would not have been achieved with existing policy. The underlying objective continued to be the creation of the right conditions for target groups to exercise their own responsibility. As policy shifted from imposing regulations from above to mobilising the capacity for self-regulation within the normative framework set by environmental policy, it is expected that government will be able, in the long run, to limit its role to setting this framework and facilitating the contributions of the target groups to the realisation of the policy goals.

In March 1998 a third environmental policy plan was presented. Whereas the earlier plans had contained specific objectives and measures for concrete action, the present plan has more the character of a policy document which surveys progress to date and outlines a range of options to be considered for moving forward toward the quality objectives already set. The political decisions on these options have been postponed until after the pending general elections and the formation of a new government in May of that year.

Environmental administration and management

The Netherlands consists of 12 provinces, each divided into municipalities, of which there are roughly 530. Another administrative actor within the provinces are the water boards, which are responsible for surface water management. There are about 100 of these water boards. Environmental policy for the country as a whole is drawn up and implemented by the national government. Within the framework set by national policy, environmental planning and regulation is also carried out at the level of the provinces, municipalities and water boards within their respective jurisdictions.

Initially, environmental regulation was exclusively the responsibility of local authorities. For a long time the municipality remained the only government body involved in the issuing and monitoring of licenses. The environmental laws introduced in the 1970s also gave the provinces important implementation tasks. Provinces now also have license issuing powers, namely in those cases involving complex, technically complicated and potentially highly polluting companies. For its part, the national government primarily concentrates on general nation-wide legislation and regulations as well as on the planning of national environmental policy, including the setting of targets and standards. As Bressers and Plattenburg point out, this does not mean that the '...national government carries sole responsibility for determining the environmental policy which municipalities and provinces are subsequently obliged to implement' (1995:8). On the contrary, within the general framework defined by national policy decisions, these governments are authorised to conduct their own environmental policy planning. Indeed, when it comes to environmental policy, they enjoy autonomous status, for the environmental plans made at the various governmental levels do not stand in any hierarchical relation. Whatever vertical fine-tuning that is needed between the levels must, in principle, be achieved, in practice, by means of consultation, exchange of information and, ultimately, voluntary agreements. Should this fail, however, both national and provincial governments have formal instruments to 'insist' on vertical co-ordination.

At the *national level*, responsibility for environmental issues is shared among a number of ministries. The term 'environment' apparently made its first official appearance in the Netherlands in 1962, when the Minister of Social Affairs and Public Health set up a Public Health Inspectorate with responsibility for environmental protection as well. However, it was not until 1971 that environmental protection was formally institutionalised as a national government task in the form of a Directorate General (DG) for Environmental Hygiene. This unit was initially located in the then newly created Ministry of Public Health and Environmental Hygiene, after having had a more modest existence as a division within the Ministry of Social Affairs and Public Health. Initially, this DG was allocated very limited resources. But as environmental problems gained in urgency and the workload of the DG increased, the allocation of financial and human resources steadily increased in the period 1972 to 1982. At that point in time, the tasks of environmental protection were transferred to the new Ministry of Housing, Physical Planning and Environment (HPPE), that is the ME. Within this ministry the DG for environmental management is responsible for general environmental policy and co-ordinates the environmental relevant activities and policy of the other ministries. It has direct responsibility for laws concerning air, soil, water, noise, specific substances, radiation, environmental impact assessment and other matters covered by the Environmental Management Act. The DG has a staff of about 1200. There is also an Environmental Inspectorate attached to this DG. With a staff of approximately 300, both at central level and in nine regional offices, the Inspectorate assists and supports the network of local and provincial officials directly in charge of enforcing environmental laws and regulations.

The ME is not the only central actor responsible for environmental policy. The Ministry of Transport, Public Works and Water Management is responsible for general water policy and has direct responsibility for environmental policy concerning water, both sea and inland. The water management unit of the Ministry (*Rijkswaterstaat*) is also in charge of the so-called state waters: all large rivers, canals, estuaries and the coastal waters. The transport branch of the Ministry is responsible for environmental policy concerning traffic and transport. It is estimated that the equivalent of 1900 full-time staff members work on water management and environmental issues in this ministry.

Responsibility for general nature policy and for legislation concerning the protection of species and nature areas is in the hands of the Ministry of Agriculture, Fisheries and Nature, which has a full-time-equivalent staff of 1140. Within the agriculture and fisheries branches there are around 120 officials who deal full-time with environmental issues and perhaps three times as many have some environmental responsibilities as part of their jobs.

The Ministry of Economics shares responsibility with the ME for integration of environmental policy into economic activities and is directly responsible for energy policy. The Ministry of Foreign Affairs shares responsibility for international negotiations concerning environmental policy, while its Directorate

General for Development Co-operation is responsible for environmental policy within the context of development policy.

Over time that has been a number of advisory bodies and research institutions supporting the ministries in the area of environmental policy. These include: the Central Council on Environmental Protection, which has recently been incorporated into a more general advisory bottom serving the ministry as a whole; the independent Environmental Impact Assessment Commission; the National Institute of Public Health and Environmental Protection, which, in its capacity as an 'environmental planning bureau', carries out many of the scientific studies that underpin environmental protection plans and policies formulated by ME; and the Central Bureau of Statistics, which has a relatively large environmental component.

The institutionalisation of environmental management in the Netherlands has not, therefore, followed the logic of the policy concepts defining the policy strategy of the moment. Apart from having to carve out a substantive niche within already existing government policy, the environment issue also had to assert itself organisationally within the institutional order that carried and defended these competing policy interests. The organisational path of environmental management has been determined by the interaction of the substantive redefinition of the problem field, the changing tasks of environmental policy and the political and societal support it could mobilise. As we have seen, the new management strategy has led to changes in the organisational configuration of both the internal and the external relations of the ministry. Nevertheless, the policy mandate to integrate environmental concern into other policy areas has not resulted in fundamental modifications of the institutional division of powers among the existing ministries. Consequently, the amount of integration that can be achieved depends, in large measure, on the political weight of the environmental issue in a given situation. The fact that environmental policy plans, and other policy measures, are co-signed and promoted by the other environmentally relevant ministries, may be an important sign of integrated policy making. It is also a sign of the institutional and political constraints within which the environmental ministry is forced to operate. The amount of integrated policy action achieved remains a function of the balance of power among the competing societal interests and priorities embodied in the other ministries and their political representatives in the cabinet.

Environmental legislation is mainly executed at provincial and municipal levels. The 12 *provinces* are responsible for implementing national environmental legislation in their territories, for granting licences under the Environmental Management Act, and for waste-water discharges into provincial, non-state waters. In some cases, the provinces supervise the implementation of environmental tasks carried out by the municipalities and water boards within their boundaries. Every four years each province is supposed to draw up an environmental policy plan and a water management plan, based on parameters set by national-level planning documents. It is estimated that together the provinces employ between 1200 and 1500 people on environmental matters. The provinces also play an important part in physical planning by drawing up provincial plans outlining the land uses within their jurisdictions. Furthermore, they designate soil and groundwater protection areas, and select sites where wastes can be dumped. In principle, the provinces have executive tasks in water management in connection with the construction and operation of sewage treatment plants. However, most have delegated this task, together with the granting of licences for water discharges to the water boards.

Municipalities in the Netherlands also have a number of tasks with regard to the implementation of certain environmental laws. One of their tasks is the granting of licenses for smaller plants under the Environmental Management Act and the enforcing of compliance with license conditions. Municipalities are also responsible for the construction and maintenance of sewer systems (though not usually for the sewage treatment plants) and for refuse collection. In recent years, the latter task has come to include the promotion and provision of facilities for the sorting of waste products by households and small businesses. Furthermore, they play an important role in the soil clean-up operations that have been taking place since 1980.

Some municipalities are too small to execute their environmental responsibilities properly. Increasingly, in such cases, these municipalities are organising themselves into regional (sub-provincial) environmental services. Some of these bodies have been delegated all environmental responsibilities, whereas in other cases they only have an advisory role *vis-à-vis* the executive authorities in the constituent municipalities.

Together with the provinces, municipalities in the Netherlands play a key role in the actual implementation of national environmental policy. This role has grown with the decision to decentralise more tasks and responsibilities to these sub-national units. However, towards the end of the 1980s it became apparent that local authorities were not able to fulfil their tasks with regard to the granting and enforcing of environmental permits. Even more important was the realisation that environmental policy had not secured a firm position in the palette of local tasks and that, in general, the municipalities did not have the capacity to play the role which the emerging strategy of environmental management was intended to give them. In an effort to encourage these authorities to take their environmental responsibilities seriously and to help them develop the institutional capacity this required, the central government introduced a number of subsidy programs to help finance an expansion of the administrative resources available at the local level for environmental management. At first this money was to be used to improve permitting and enforcement activities within a period of 3-4 years, as well as to upgrade the status of environmental policy as such within the policy activities of local government. A second objective was to promote regional or inter-municipal co-operation. Communities of a certain size (below 70,000 residents) were given extra money if they would co-operate with other municipalities in order to reach a scale large enough to sustain effective environmental management. With the publication of the first environmental plan, the scope of local environmental responsibilities expanded beyond the traditional tasks of permitting and enforcement. This led to a new financial program to finance the execution of NMP tasks at the local level.

As of 1 January 1998, these financial subsidies have been stopped and money for environmental administration must be drawn from the general funds available to local governments. There is some concern whether or not these communities can, and are willing to, sustain the level of environmental activity achieved with the financial assistance of national government. For its part, the ME is worried that it will no longer be able to exert sufficient influence on environmental management at the local level now that it no longer can apply pressure via the subsidy programs. As far as regional co-operation among communities are concerned, this worry seems to be well founded. At present, co-operative schemes set up under the previous financial arrangement are either being reduced or phased out completely. It appears that collaboration achieved came about solely because of the extra money from the national government, and without this stimulus, there is little incentive to engage in such co-operation.

Water boards occupy a venerable and quite special place in the governmental system of the Netherlands. From the end of the Middle Ages until the middle of the twentieth century, they were primarily responsible for flood protection and water management of the polders in the low-lying regions. Since the World War II, in particular since around 1970, several water boards have been given wider powers to combat water pollution.

While money is not everything in environmental policy, the financial resources available to governmental agencies responsible for policy development and implementation does provide an indication of how serious society and the government take matters of environmental quality. In the Dutch case, expenditures of the ME increased from 289 million guilders in 1976 to 608 million in 1985. They continued to grow, rising from 1,269 million in 1990 to 1,583 million in 1993. These expenditures are primarily aimed at water pollution (37 per cent), waste management (23 per cent) and air pollution (17 per cent). The total environmental expenditures of all sectors in the Netherlands amounted to 1.1 per cent of GDP in 1980 and 1.3 per cent in 1985. After rising to 1.9 per cent in 1990, they reached 2.7 per cent in 1995 (OECD 1995:114–116). Total environmental costs were in 1991 estimated at most 10 billion guilders. The largest sources of revenue to cover these environmental expenditures are specific payments for administrative services and environmental charges, some of which have an additional regulatory effect. Municipal and provincial environmental policy used to be financed without any contributions from the national government.

Environmental policy in action

The overall quality objectives and target emission levels for the years 2000 and 2010 set in the first NEPP remain unchanged in NEPP 2, although this revised plan does propose a number of additional measures for achieving these quality objectives. In June 1993, while the plan was being formulated, an environmental status report was published reporting on the state of the environment in the Netherlands, measured against quality and emission reduction targets set out in the first NEPP. (The report also made new forecasts for environmental quality in the year 2000, based on policy measures under consideration and alternative macro-economic scenarios.) The results of this environmental survey and forecasting exercise were 'generally very encouraging' (HPPE 1993:13). Many environmental targets set for the end of the planning period (at that time 1989–93) had been met, others were on course for achievement in 2000. For example

> Emissions of the greenhouse gas methane appear likely to undercut their 2000 target and CFC consumption has been virtually eliminated. SO_2 and NH_3 emissions are dropping steadily and waste disposal targets for 2000 seem likely to be achieved. Pollution from some other key pollutants has also been substantially reduced.
> (HPPE 1993:13)

At the same time, other targets required additional efforts if they were to be met. Emissions of carbon dioxide in particular remain high. Although NO_x emissions had been reduced in terms of emissions per vehicle, these gains had been offset by rising traffic volumes. Not withstanding these 'disappointments', the report concluded that

> Overall, given full implementation and enforcement of policy measures already in place, the latest forecast indicates that total emission reductions of 50–60% will be achieved by 2000. (HPPE 1993:14)

Of course that is a big assumption. Whether 'full implementation of policy measures already in place' can be achieved will depend on how successful the Netherlands is in moving from general strategic objective to concrete measures aimed at changing behaviour of firms and other economic actors. This, in turn, will be determined, in large measure, by the effectiveness of the new environmental management strategy that is intended to bring together the different public and private actors in a joint effort to meet the country's quality objectives on the way to a sustainable future.

Developments in this strategy will have important consequences for policy and administration in this area. In particular, government will have to work closely with the targets of the policy in order to understand better their problems and possibilities for controlling environmental pollution. The emphasis on prevention also requires that these economic actors are motivated to take self-responsible initiatives to integrate systematic care for the environment into the total product cycle – from design through to disposal or reuse. This means that policy needs to be designed to encourage these groups to take responsibility for the measures necessary to prevent and control environmentally harmful impacts of their operations and activities. This strand of policy is referred to as 'internalisation' and represents an effort to ensure that the public and industry implement environmental policy themselves as far as possible, as part of the normal, daily activities. This approach not only aims at achieving more broad-based support for government policy, but also recognises that the know-how necessary to redress environmental problems can be largely found at the polluters themselves. Consequently, after national environmental objectives have been formulated, target groups and their representatives will now have a much stronger say in all further stages of the policy process.

It is clear that the effective application of the target group approach requires a number of institutional changes in order to create the organisational and management preconditions necessary. At the national level, this will entail changes in the internal organisation and procedures, as well as in the relations between the ME and other governmental actors at this level. At the level of the individual firm, it will be necessary to develop an institutional capacity to give effective operational expression to the internalisation of responsibility for environmental protection. In addition, a differentiated infrastructure for deliberation and negotiation between government and business/industry is required, through which the process is channelled by which general quality objectives are translated into sector or branch-specific 'integral environmental tasks' and then in disaggregated into implementation plans to guide the individual licensing and enforcement actions affecting individual enterprises. The national government will need to provide assistance to both the provinces and municipalities to increase their capacity for participating effectively in these activities; the same holds true for the role that will be played by branch association and industrial organisations towards the individual firms.

One visible consequence of target group consultation has been that an increasing number of companies are organising to make care for the environment an integral part of their business processes. Companies were supposed to give 'institutional expression' to their responsibility for the environmental impacts of their operations by setting up their own internal environmental management systems by 1995. Or, or at least, they were by then to have made significant progress in this direction. Such environmental management systems within individual firms are, then, seen as an important means for actively and effectively involving industry in the solution of environmental problems. Branch organisations and other industrial associations can also aid in this process by developing model environmental management systems which companies can use to create their own internal systems and procedures. A survey has shown that 90 per cent of the environmentally relevant trade associations have made an active start with EMS. Companies with high environmental emissions or with more onerous licence conditions are at the forefront in establishing EMS. The chemical sector has taken the lead in this regard (60 per cent of the companies were at an advanced stage in the introduction of environmental management systems), while small and medium-sized companies are still in the early stages. At the moment there are 24 Industrial Environmental Services (BMDs) in operation, which offer assistance in the establishment and execution of EMS, especially to smaller and medium-sized firms. Approximately 9 per cent of Dutch companies avail themselves of this assistance (HPPE 1993:10).

Bressers and Plettenburg (1995:11) note that industry organisations have generally been playing a positive role in the development of the target group approach. Still, they hasten to add, this does not mean that 'all sectors of Dutch industry are interested in reducing the pollution they cause'. On the contrary,

> many individual companies are still stand-offish or even obstructive, while employers' and employees' organisations also tend to keep their distance. (Bressers and Plettenburg 1995:11)

Still, the fact that in 1994 these organisations came out with a joint declaration affirming their commitment to a sustainable society and calling upon government to pursue a more vigorous policy does give cause for some hope.

Still, even within the new system of environmental management, the traditional permit or license remains an important instrument for dealing with the pollution from industrial, as well as other kind of economic activity. But now the effectiveness, and the efficiency, of this regulatory instrument is determined by the success achieved in combining the promotion of the general quality objectives of environmental policy with the need to tailor the granting and enforcement of such licenses to the particular conditions and possibilities of the concrete case. At present, this balance is being sought in the form of a permit that specifies the general regulatory parameters within which the particular firm is to act in an environmentally responsible manner. Such a permit defines the limits within which the enterprise, on the basis of its institutionalised capacity for environmental management, is free to seek flexible and efficient means to meet its quality objectives. Thus investment by the firm in a proven capacity to act in an environmentally responsible manner 'buys' less intensive and obtrusive interference by the permitting and enforcement authorities and more leeway for a firm-specific program for managing the environmental impacts of its activities. In this way, self-regulation based on a certified institutional capacity for environmental sound management and operations forms a core element of the Dutch system of joint regulation.

Concluding comments

Environmental policy in the Netherlands has moved from the phase of cleaning up existing damage to the management of environmental quality in order to prevent damaging from occurring. The guiding objective now, as laid out in the national environmental plans, is the achievement of sustainable development. A central element of this phase is the development and institutionalisation of new modes of environmental management, based heavily on interactive modes of policy making and the mobilisation of the assumed self-regulatory capacity of the target groups of these policy measures. At the moment, work continues on creating the legal basis for this shift from traditional vertical modes of management toward the more horizontal relations on which the new strategy is based. Again it should be stressed that – at least from the perspective of government – this self-regulation takes place within the regulatory framework marked out by public policy, defining the rules of the game and establishes the controls over the 'honesty' of play.

This system of environmental management rests on the assumption that economic actors can be brought to internalise responsibility for environmentally sound behaviour – even if they must, in some cases, be prodded by government regulations. In any case, the kinds of behavioural changes required by an integrated and comprehensive approach to pollution prevention are seen as depending on the active support and internally-driven motivation of the targets of environmental policy. The success of such policy, it is argued, requires not only a moral commitment to such objectives but also the integration of environmental values into the rationale of economic decision making. The prevailing optimism regarding the compatibility of the twin goals of economic development and environmental quality has left little room for considering the possibility that the relation between these objectives could become problematic. As long as pollution prevention does, in fact, pay, the basis for this system of environmental management remains sound. But if the pay-offs for business are not forthcoming or the need for more stringent measures to protect the quality of the environment threatens basic economic interests, what then?

Thus far, Dutch environmental policy makers have been spared such a challenge to the foundations of their management strategy. Recent evaluation studies indicate that environmental policy in the Netherlands has, in recent years, indeed achieved a marked reduction of pressures on and threats to the environment. Indications are that many of the quality objectives set in the national plans will be achieved by the deadlines stipulated (Rijksinstituut 1997:11). As far as substances such as SO_2, NO_x, NH and phosphate are concerned, present measures will lead to significant decline in emissions. But even here, without significant extra policy measures, it will not be possible to sustain this downward trend after the turn of the century. On the contrary, the development of environmental pressure will once again be coupled to economic growth. The main problem is the growing use of fossil energy and increasing mobility. In addition, intensive agriculture continues to be a source of discharges of phosphates and nitrates into the ground water and ammoniac into the air. It is becoming increasingly clear that the NO_x objectives cannot be reached with present-day technology and transportation practices. New automotive technologies, a reduction of traffic volume and the more extensive use of alternative energy sources will be required.

The situation with regard to the so-called greenhouse gases is particularly worrisome. Irrespective of the scenarios applied for estimating economic growth, the CO_2 goal for the year 2000, i.e. a reduction of 3 per cent in comparison with 1990, cannot be realised. Indeed, the recent study carried out by the National Institute for Public Health and Environmental Management concluded that even assuming the lowest rate of economic growth, CO_2 emissions will increase between 1995 and 2020 by 3 per cent; assuming higher rates of growth leads to an increase of 30 per cent (Rijksinstituut 1997:13).

In the future, solving the environmental problems of the Netherlands will require fundamental changes on the part of the industrial, agricultural and transportation sectors. However, while the need for such changes is widely recognised, in principle, faith in the ability of technological development and current policy instruments to effect them remains high. And yet, the gap between policy objectives and the effectiveness of existing policy measures grows. In particular, with regard to CO_2 and other greenhouse gases, there is talk of the 'Kyoto gap', i.e. the discrepancy between the commitment that the Dutch government has undertaken to reduce its emissions and the level of achievement likely under even the most optimistic scenarios regarding measures already on the books. Even if all the measures proposed in the new NMP3 for dealing with greenhouse gases were to be carried out, only a stabilisation of CO_2 emissions, compared with 1990, will be achieved by 2010. Without significant extra efforts, the Dutch share of the EU reduction goals accepted at Kyoto can not be achieved. There seems to be a lack of appreciation for the type of drastic measures required to meet these objectives. In any case, the financial burden on government and business, on top of what is already needed to carry out existing measures, will grow significantly if the environmental damage of economic growth is to be limited.

The new national environmental action plan makes clear that a number of hard choices will have to be made in the coming years if the Dutch are to be able to deal effectively with the environmental challenges they confront. The negotiations on the measures to be taken and the apportioning of the costs involved will test the resilience and effectiveness of the current strategy of environmental management.

Notes

1. The anti-nuclear movement was an exception to this trend, reaching its high point in the late 1970s.
2. When we here speak of the Ministry of the Environment we refer to the Ministry of Housing, Physical Planning and Environment (HPPE), established in 1982. It should be noted that there has never been a separate organisation of ministerial status responsible only for environmental policy. At the ministerial level the organisation responsible for the environment has always combined this responsibility with other functional tasks.

References

Algemene Rekenkamer (1995) *Convenanten van het Rijk met bedrijven en instellingen.* The Hague: SDU.
Arensten, M.J., Bressers, J.Th.A. and Klok, P-J. (1993): 'Van Urgentienota naar NMP en verder', in Council of Environmental Protection: *Milieu van jaar tot jaar, 1992–1993.* The Hague: Delwel, pp. 13–49.
van Ast, J. and Geerlings, H. (1993) *Milieukunde en milieubeleid. Een introductie.* Alphen aan den Rijn: Samson.
Bressers, J.T.A. and Klok, P.-J. (1996) 'Ontwikkelingen in het Nederlandse milieubeleid: Doelrationaliteit of cultuurverschuiving?' *Bestuurswetenschap* 10 (5), pp. 445–460.
Bressers, J.Th.A. and Plettenburg, L.A. (1997) 'The Netherlands', in M. Jänicke and H. Weidner (eds) *National Environmental Policies – A Comparative Study of Capacity-Building.* Berlin: Springer Verlag, pp. 109–32.
Hanf, K. (1989) 'Deregulation as regulatory reform: the case of environmental policy in the Netherlands'. *European Journal of Political Research* 17, pp. 193–207.
HPPE – The Ministry of Housing, Physical Planning and Environment – (1993) *Abridged version of the Environmental Programme, 1994–1997.* The Hague: SDU.
Leroy, P. (1994) 'De ontwikkeling van het milieubeleid en de milieubeleids-theorie', in P. Glasbergen (ed.): *Milieubeleid. Een beleidswetenschappelijke inleiding.* The Hague: Vuga, pp. 35–58.
Mattews, E. (1994) *Progress report on the National Environmental Policy Plan.* London: Environmental Resources Management.
Mol, A.P.J., Lauber, V., Enevoldsen, M. and Landman, J. (1996) 'Joint environmental policy-making in comparative perspective.' Paper prepared for presentation at the 'Greening of Industry' Conference. Heidelberg, 24–27 November 1996.
OECD (1995) *Environmental Performance Review. The Netherlands.* Paris: OECD.
Rijksinstituut voor Volksgezondheid en Milieu (1997) Nationale Milieuverkenning 4. 1997 2020. Bilthoven: RIVM.
van der Tak, T. (1993) 'Shades of Green: Political Parties and Dutch Environmental Policy.' *Dutch Crossing.* Winter, pp. 11–27.

Chapter 9

Norway: Balancing environmental quality and interest in oil

Alf-Inge Jansen and Per Kristen Mydske

Geographical and socio-economic characteristics

Mainland Norway, on the western part of the Scandinavian peninsula, covers a total of 324,000 square kilometres and extends from the same latitudes as those of Scotland and northwards as far as to the border of the Kola Peninsula in Russia. In addition to the mainland, the arctic archipelago Svalbard (63,000 square kilometres), is Norwegian territory. Nearly 37 per cent of the mainland is forested areas. Open land with vegetation covers 31 per cent, unproductive areas 16 per cent, lakes, wetlands etc. 13 per cent and only three per cent is agricultural land. Among the characteristic features of the distinct and varied Norwegian landscape are the many fjords and waterfalls. In most parts of the country there is abundant precipitation and extensive freshwater reserves in glaciers, lakes, rivers and groundwater sources. Infrastructure development, mainly associated with economic activities, has greatly reduced the extent of large areas of natural habitat, and such areas no longer exist in the lowlands of the southern half of Norway.

There are 4.3 million inhabitants in Norway, only 13 inhabitants per square kilometre. This gives the country the lowest population density in Europe with the exception of Iceland. One-fifth of the population lives in the Oslo metropolitan area. The Norwegian economy is open, with an import/export share of approximately 50 per cent of GDP. Norway's economic structure rests largely on services (35 per cent of GDP), manufacturing and oil and gas production. While the petroleum sector represents less than one per cent of total employment, it contributes about 14 per cent to GDP. Manufacturing contributes 13 per cent to GDP and 15 fifteen per cent of the labour force. Agriculture and forestry account for two per cent of GDP, but provide four per cent of total employment. Registered fishermen make up one per cent of the labour force, yet fisheries and related industries form the backbone of most coastal settlements. Norway is an increasingly important exporter of oil and gas to Europe. In 1995 Norway was the world's second biggest

181

exporter of crude oil. Norway is also a big producer of electricity based on hydropower, and hydropower also plays an important indirect role in national exports (50 per cent of exports are raw or semi-processed materials from hydropower-based aluminium smelters and ferro-alloy industries).

Development of Norway's environmental policy and organisation

Norwegian environmental policy has roots in the traditions of *public health* (e.g. public measures against cholera and typhoid fever epidemics and the Health Act of 1860), *protection of cultural heritage* (e.g. the Cultural Heritage Act of 1905, which was primarily adopted to protect Viking ships that were excavated at that time) and *classic nature conservation* (e.g. measures to preserve objects of nature, species and areas). All these traditions have contributed to and are still represented in current environmental policy in Norway. As nature conservation has been most central in influencing the development of Norwegian environmental policy, politics and organisation, we shall describe parts of this tradition in greater detail.

The earliest political entrepreneurs for protecting Norwegian natural environment were scientists and natural historians who at the start of this century raised public interest in the protection of the country's natural environment in terms of species and particular objects. They initiated environmental interest organisations, prompted already existing organisations to act, and persuaded the government to develop policies (Berntsen 1994:34–95; Jansen 1989:31–62) and even played an active role in administrating and implementing these policies (Gundersen 1988).

The first law, explicitly motivated by reasons of nature protection, was the Nature Preservation Act of 1910, which formed the legal basis for preserving species, restricted areas or specific objects of nature if these were of scientific or historic significance. The National Association for Nature Preservation was established in 1914. During the 1930s both its membership and its social basis were enhanced significantly. Concomitantly, other interest organisations developed an increasing interest in the country's nature as a social value, i.e. as a source of recreation and recuperation for their members. There was a change of motivation for conservation: nature should no longer be protected *from* the people, but *for* the people.

An emphasis on social aspects and a wider conception of conservation than mere preservation started to play a part in policy development in the 1950s. The Nature Conservation Act of 1954 laid the legal basis for establishing national parks in Norway and also for the National Council for Nature Conservation, to assist the government in implementing the law. The Open-Air Recreation Act of 1957 confirmed everyone's right of access to nature (e.g. outlying areas, seas and water courses) regardless of who owned the land. The law established a national council for open-air recreation, and also a committee for these matters in each county.

Even with this widened conception of nature conservation, environmental issues were a minor item on the public agenda during the first two decades after World War II. Post-war industrialisation was the top priority. Exploitation of waterfalls for producing electricity for the rapidly expanding industries was seen as essential to stable economic growth.

In Norway, the years after 1905 were the era of the breakthrough of process-based industry with its great consumption of hydroelectric power. In economic and industrial terms the waterfalls of Norway gave her a comparative advantage: a springboard into the industrial age (Furre 1992:17). The conflict that followed between the advocates of exploiting the waterfalls for producing hydropower on the one side and conservationists on the other side has since constituted a principal conflict in Norwegian environmental politics. In the period before World War II full exploitation of Norwegian watercourses was hampered. The main obstacles were the 'concession laws' enacted primarily to prevent foreign capitalists from gaining control of the country's natural resources, and the economic crisis and industrial stagnation that characterised economic life in Norway and in other countries during most of the period from 1920 to 1935. Still, harnessing of waterfalls continued to expand right on through to the 1960s. From 1906, when the first large man-made dam for regulating fresh-water for production of hydropower was built, until 1962 between a third and a half of the fresh-water area was regulated. Planning was, moreover, proceeding to regulate a number of large and well-known watercourses.

During the 1960s, a new mobilisation of conservationist interests took place. The national organisation was reorganised. Public discussion of nature conservation, land use policy and pollution increased. Conservationists developed a more ecologically oriented ideology and promoted policies consistent with this orientation. More action-oriented activities and organisations were established. The development in public perception of environmental problems and in the saliency of environmental politics is reflected in an ensuing development of environmental policies, as manifested in various laws and governmental programmes.[1]

In 1972, the Ministry of Environment (ME) was established. The ME was to co-ordinate policies for pollution control, physical-economic planning, nature conservation and open-air recreation, and international environmental co-operation within the framework of a general policy of *growth with conservation* (Jansen 1989). The ME took many initiatives, not only in making new laws and revising existing ones, but it was particularly active in the field of pollution abatement in which it mainly pursued *the Polluter Pays Principle*. Notably the Ten Year Action Plan for Cleaning Up of Pollution from Old Industry of 1974 was an important precondition for major improvements. Financial incentives like subsidised loans, state guarantees for private loans and state grants were employed to implement the new regulations. For instance, in the case of paper and pulp industry, these regulations contributed to otherwise necessary structural changes (Reitan

1995:5). The action plan was initiated at a time of economic growth. Its implementation was continued even during the first years of the recession from 1975 onwards, as it was consistent with the Keynesian economic policy pursued by the government until the late 1970s (Asdal 1995:106–109). During the late 1970s the programme to clean up Norway's largest lake, Mjøsa, was undertaken. State grants to the municipalities and to the agricultural sector were seen as part of the conditions for this successful environmental programme (Hallén 1981).

More than 90 per cent of the acid precipitation over Norway originates from abroad and significant marine pollution is caused by ocean currents carrying pollutants to the Norwegian coastline. It is, therefore, not surprising that Norway was one of the first countries to identify environmental damage caused by transfrontier air pollution. With the ME as the leading actor, Norway from the early 1970s was a vigorous actor for international environmental co-operation, e.g. in the OECD, the UN and regional international co-operation as well as in bilateral co-operation. Norway pushed to establish and implement measures against sulphur emissions internationally as well as nationally. She is party to the Nordic Environmental Protection Convention (1974), which introduced the principle of non-discrimination in positive international law. As to marine pollution, Norway and other Scandinavian countries were front-runners in establishing the Oslo Convention (1972), which regulated the dumping of industrial waste, sewage sludge and other wastes in the north-eastern-Atlantic Ocean. Norway also played an active role in establishing other conventions like the Paris Convention (1974), which regulated pollution from land-based sources to adjacent marine areas, and the Hague Declaration (1990), which is directed primarily at solving pollution problems in the North Sea.

By contrast, the ME did not become a principal political actor in the most contested environmental issue areas during the first years of its existence, i.e. conservation or exploitation of watercourses for hydropower. During the 1970s several large watercourses were regulated, but against stiffening and growing opposition. The Labour government, the Labour Party and the Confederation of Trade Unions together with the Conservative Party and the Confederation of Industries were main political actors in the vigorous promotion of hydropower and these parties also constituted core elements of the alliance that pressed ahead for laying down the basis for production of nuclear power in Norway. Most other political parties rejected atomic power in Norway, and there was growing opposition against atomic power in all political parties. In this situation, the reservoirs of petroleum on the Norwegian continental shelf provided the dominating coalition an opportunity to pursue its policies for economic growth without having to launch a programme of building atomic power plants in Norway. The main cleavages in environmental politics, i.e. in areas such as the exploitation of water courses, the level of energy consumption, drilling for oil and gas north of the 62° latitude, production volumes for oil and gas on the continental shelf as well as support or opposition to the Labour government's recommendation that a principal watercourse in northern Norway, the Alta river, should be exploited for hydropower production[2], cut across the left-right dimension among the political parties as well as in the electorate (Petersson and Valen 1979; Aardal 1993; Knutsen 1996:299).

The crystallisation of a green cleavage in Norwegian politics can partly be explained by the fact that the growing concern about environmental issues coincided with the political struggle over Norwegian membership of the European Community. The issue of the EC, later the EU, activated politics on a number of significant dimensions in the Norwegian political system. The green movement and the anti-EU-movement did not only coincide; they mutually influenced each other and to some degree coalesced.

Turning from environmental policy and politics to the organisation of the environmental policy process, the establishment of the ME was a significant event. The new ministry set about to organise a more integrated environmental administration. Notably in the area of pollution control, the ministry took the lead. Existing regulating bodies were merged and reorganised as (central agencies) under its authority. The ME also made efforts to develop administrative structures for environmental matters at county level, and in 1982 a separate environmental department was established at the County Governor's Office in the 18 counties. In 1986 the ME initiated a trial programme, elevated to a reform in 1992, establishing municipal responsibility for environmental considerations.

From the mid-1980s there was a marked increase in public and political attention given to environmental matters. The field of environmental policy was widened and its content and structure changed, i.e. new problems were included, new notions and principles were added, reorganisation mobilised new actors and generated new patterns of conflict and co-operation in environmental politics. The World Commission on Environment and Development (named the Brundtland Commission after its leader, Norwegian Prime Minister Gro Harlem Brundtland) played a significant role in this development. As expected, the Brundtland government put quite an effort into making the Commission's proposals Norwegian policies; for instance, the government in numerous statements, strongly supported the Commission's concept and policy of *sustainable development*.

Current politics: actors, interests and issues

The current pattern of environmental politics is to a significant degree dependent on the development during the formative period of the environmental policy sector. Recent changes in the definition of environmental problems, in the interests involved as well as in the pattern of interaction between actors, are also important for current environmental politics.

During the 1980s much of the debate on conservation of watercourses and hydropower production centered around the making of a Master Plan for

Watercourses, a national co-ordinated plan for the management of undeveloped hydropower sources in Norway.[3] This process led to a more analytically oriented discussion on the exploitation of water resources. Due to the political opportunities offered by the increasing volumes of petroleum produced on the continental shelf and of the work on the masterplan for the watercourses and the two protection plans that followed,[4] the issue of development or conservation of watercourses, contested for decades and constitutive in Norwegian environmental politics, was partly removed from the political agenda. The greater consensus on the need to protect a significant number of watercourses, notwithstanding a green cleavage cross-cutting the left-right dimension, was repeatedly demonstrated in issue areas like the level of energy consumption and the drilling for oil in the Skagerak and in the northern waters.[5]

In spite of the existing green cleavage between the political parties, most of them have, however, promoted themselves as environmentally oriented. Because of the strong support for environmental values in the population, political parties find it difficult to argue directly against environmentalism, in other words, environmentalism has a strong symbolic value. Interestingly, in 1989, when environmental issues ranked as the second most important issue for voters, the Green Party received only 0.4 per cent of the votes in the national election. The Socialist Left Party, however, increased its number of seats in the parliament from six to 17. The major explanation for the failure of the Green Party is that there is no 'vacant' position in the party system, waiting for a new environmental, left of centre party (Aardal 1990:150). Both the Socialist Left Party and the Centre Party, as well as the Liberal Party, have to a great extent promoted themselves as green parties, leaving little room for a new green party in the electorate.

A distinct feature of environmental politics and policy of the 1990s is the increased degree of internationalisation. As Norway actively participates in efforts of international environmental co-operation, Norwegian actors of various types have interacted both multilaterally and bilaterally with numerous international actors (e.g. representatives of international organisations, foreign states as well as foreign environmental interest organisations).[6] Such international co-operative efforts as well as the agreements they lead to, are increasingly used as political resources by actors in Norwegian environmental politics. Ministries and politicians, as well as environmental and other types of interest organisations, use them in their arguments for or against specific policies or in order to enlist foreign and international allies to carry heavier weight in shaping Norwegian policy at a later time.

The green cleavage in environmental politics seems to be continuously confirmed in most policy areas. Of particular significance has been the coalition between the Labour Party (which formed the minority government during the periods 1986–89 and 1990–97) and the Conservative Party in climate policy issue areas.

The Labour government led by G.H. Brundtland agreed with the Brundtland Commission's identification of the greenhouse effect as one of the most important global challenges. In order to act as a driving force in achieving an international climate agreement for specific reductions of greenhouse gases (GHGs) as soon as possible, the government declared unilaterally that its goal was to stabilize CO_2 emissions during the 1990s and at the latest by the year 2000, and later to reduce the emissions. The parliament, in its debate on the follow-up programme, concluded that the national goal regarding future CO_2 emissions should be even stricter: stabilisation by the year 2000 with 1989 as the base year.

The implementation of the national target for reduction of CO_2 emissions meant that environmental policy instruments and measures had to be applied to influence and regulate numerous social activities that traditionally had been seen as part of policy fields other than that of environmental policy (e.g. agricultural production, industrial processes, oil and gas production and transportation, road traffic, off-shore transportation). In other words: climate policy meant a significant expansion of the environmental policy field as well as the involvement of actors that represent the core characteristics of the Norwegian political order.

In June 1995 the government declared that '... it is not possible to find a practical policy that will ensure stabilisation of Norway's total CO_2 emissions by the year 2000' (*Report to the Storting*, No. 41 (1994–95:29).[7] The Labour Party, the Conservative Party and the Progressive Party (extreme right) as well as leaders of trade unions and of organised interests of big industry have given their consent to this position as well as to building two natural gas-powered electricity plants that have been estimated to increase Norwegian CO_2 emissions by six per cent. The new minority government that was formed by the centrist parties (the three parties that are regarded to be at the centre of the political spectrum) after the election in September 1997 has so far sought positions of consensus and compromise. This government has, however, publicly stated its intention to propose new general green taxes to reduce CO_2 emissions, and Labour and the Conservatives have voiced strong reservations.

As for national environmental interest organisations, there are more than 20, and they vary both in strength as well as in type of issues they address.[8] To a great extent, the development of environmental interest organisations takes place in interaction with the environmental administration. In general, characteristics of the relations of environmental NGOs to the public authorities now resemble those of other types of NGOs (Rommetvedt and Melberg 1997: 42–56). A paradox of this interaction is that as environmental organisations succeed in making environmental authorities accept their policy positions and points of view, their platform for mobilisation is eroded. This paradox is probably particularly relevant to organisations with a platform and profile built on a single or few issues. Another consequence of successful campaigning and lobbying by these interest organisations has also been suggested. When stronger environmental policy positions are taken on board by the central administration, the logic of action is changed: the dynamics of issues and politics fizzle out and the logic of routines and standardisation takes over (Jansen and Osland 1996:223).

The general tendency in recent years has been that participation of environmental organisations in most phases of the routine policy-making process has been formalised. In an institutionalised manner, they submit proposals to the environmental authorities, they are represented on boards and in committees, and through the 'hearing institution' they comment on recommendations made by investigating commissions as well as on draft bills and policy programmes. In this context, the environmental interest organisations participate together with other types of national interest organisations, in particular the Confederation of Norwegian Business and Industry (CNBI) and the Norwegian Confederation of Trade Unions. In addition, representatives of the interests of 'big industry', e.g. Norsk Hydro and Statoil, are also influential actors in direct interaction with the authorities. In this way the government has expanded Norway's traditional consensus-corporatist style of policy making into the environmental policy field.[9]

During the last 10 to 15 years, some environmental interest organisations have been modernised. Their secretariats have been 'streamlined' and they have set up special units and programmes that often are staffed with specialists who prepare statements on environmental policy matters and deal with the media. From the beginning of the 1980s there has also been a tendency among these organisations to adopt the vocabulary and analytical models that are applied by the environmental administration (Gundersen 1991:27–28). In sum, a process of professionalisation and bureaucratisation of leading environmental interest organisations has taken place. This process is interpreted as a tendency on the part of these organisations to model themselves on characteristics of the core environmental administration with which they interact (Jansen and Osland 1996:222).

The Norwegian environmental policy organisation

Norway's environmental policy organisation grew out of, and functions as part of, her general governmental structure. In this section we refer to how environmental policy is dealt with in the parliament (*Stortinget*) and point out recently developed features of the environmental organisation at the central administrative level. In addition we describe the core environmental administration at the central, county and municipal level, and finally we discuss the environmental policy system in practice.

Environmental policy in the parliament

Most of the work of the parliament takes place in its standing committees. In environmental matters, the normal procedure is to send propositions and reports from the ME to the Committee on Energy and the Environment. A more informal and effective method of co-ordination takes place as every party's members on the committee report to their party's parliamentary group.

At the core of the parliamentary government of Norway is the principle of ministerial responsibility. This functions as the linchpin in the pattern of actions according to which ministers undertake their tasks. This is demonstrated by the role of ministers in parliamentary debates as well as by the application of parliamentary control mechanisms. Of particular significance for the Minister of the Environment is the debate on her/his annual Environmental Policy Statement, first introduced in 1987. Its purpose is to provide the parliament with information on aspects of the current environmental situation and on conditions of environmental quality, both nationally and internationally. This statement by the minister is later followed by a general parliamentary debate. The system of ministerial questions and interpellations is one of the parliament's most important means of both defining problems and obtaining information from government ministers about matters which have not yet been submitted for deliberation.

Environmental policy organisation at the central administrative level

The government also complied with the Brundtland Commission's recommendation as how to deal with the trans-sectoral dimensions of environmental administration. In order to develop environmental integration cross-sectorally the government adopted the sector responsibility approach, according to which each ministry was explicitly given responsibility for environmental policy goals within its sector.

Combined with an expansion of the environmental policy field from the late 1980s (such as the inclusion of climate policy) the government's adoption of the sector responsibility approach has changed the pattern of environmental politics and policy-making at the central level of administration. Environmental issues are now, both because of the broader implications of environmental policy and because of the new procedures adopted for handling such matters, of central interest to other ministries. This has also led to an increased importance of these ministries in environmental policy making. Of great consequence to this development has been that Norway's positions in international environmental politics in general, and in climate policy in particular, have generated new patterns of politics between the ministries. While the ME was, until the end of the 1980s, the most important Norwegian actor in international environmental negotiations, in recent years both the Ministry of Finance and the Ministry of Foreign Affairs have played increasingly important roles in deciding Norwegian positions in international environmental negotiations. In environmental matters in which Norway's petroleum resources are involved, the Ministry of Industry and Energy has turned out to be a formidable political actor. Although their relative importance may vary in different periods and with regard to different issues the most important ministries in addition to the ME and the three ministries referred to above are the Ministry of Transport and Communications, the Ministry of Agriculture, the Ministry of Fisheries and the Ministry of Local Government and Labour.

A characteristic consequence of this widened scope of environmental policy and the increased importance of a number of ministries in environmental policy making has been the increasing processes of co-ordination between ministries on environmental issues. As mechanisms of co-ordination, budgets and plans are crucial. The Governmental Memorandum No. 1 for the preparation of the budget of 1991 stated that each ministry was to draft a memorandum of the environmental profile of its budget (i.e. what later has been referred to as the progress report on the ministry's environmental protection and improvement activities), and to translate these into corresponding budgetary items for the coming budgetary year. The ME and the Ministry of Finance maintain close contact with the other ministries on the preparation of these reports. This part of the budget is also published separately as the Green Book.

Of great importance are a number of temporary and more or less permanent interministerial committees and working groups that have been established to address environmental policy issues. Some of these working groups have members only from the ME and one other ministry, while others have members from the ME and several other ministries, and they also differ as to whether the members are high-status or lower-status civil servants.[10] Undoubtedly commissions and committees like these function as arenas in which the various ministries can try out their policy proposals and present them for discussions and negotiation; i.e. as mechanisms for co-ordinating and integrating environmental concerns and information into legislation, plans and programmes in the various policy sectors.

The core environmental policy organisation of Norway

Notwithstanding the important role of the sector ministries in environmental policy, the ME and its agencies, together with environmental departments at the County Governor's Office and environmental officers of the municipalities, constitute the core environmental organisation of environmental policy and administration. We will describe this core environmental policy organisation, starting with the organisation at the national level.

The question whether competence to make binding decisions should be delegated from ministries to central agencies/directorates, and if so the question of how the relation between ministries and directorates should be organised, has regularly been the object of public debate since the 1820s. Pursuant to the doctrine followed since the 1950s, professional and technical as well as routine matters should be transferred to central agencies, which on the basis of general instructions and regulations issued by the ministry, were empowered to make independent and binding decisions. The expectation was that by delegating competence in such matters to directorates, the ministries would be allowed more time to work on matters of policy and principles. The ministries would be transformed into more political secretariat-like organisations, better qualified to function as political secretariats for their ministers. By setting up central agencies, each headed by a director and subordinate to a ministry, effectiveness in meeting requirements for professional and technical differentiation could be combined with capability in meeting an increasing demand for political planning and co-ordination as well as for strategic policy making.

Relationships between the ME and its various agencies differ, but the general relations are described by the following characteristics: the agencies are administered (and organised) by the ministry. The tasks and the legal authority of the agencies are delegated from the ministry. The agencies can make professionally independent decisions as they are authorised by the ministry and the government to do so. On the basis of their instructions their task is to administer and implement laws and budgets under the ministry, Reports to the *Storting* and national policy guidelines as well as other policy programmes and provisions, various types of circulars and regulations issued by the ministry. The agencies make regulatory decisions, but it is typical that a significant part of the implementation takes place through advice, guidance and information. The knowledgeability and technical competence of the agencies give most of them great professional authority. Because of their combination of legal and professional authority, their decisions not only carry much power but great legitimacy as well. The decisions of the agencies are, however, subject to appeal and the ME is the appeal agency for their decisions. In 1994 the ME and its five agencies had approximately 1500 employees, implying that from 1973 to 1994 the total number of employees had risen by almost 150 per cent.[11]

The ME was reorganised in 1989 and has now five departments: (1) Organisational and Economic Affairs (2) Nature Conservation and Cultural Heritage (3) Water, Waste Management and Industry (4) International Co-operation, Air Management and Polar Affairs and (5) Regional Planning and Resource Management. When the ME was established in 1972, lawyers were the dominant profession, occupying over one-third of positions in the Ministry, and an even larger share of higher positions. All other professions held only small shares. Today the ME is heterogeneous with regard to the personnel's educational background. As of 1 June 1995 the educational background of the staff in the ME was as following: lawyers, 20 per cent, scientists, 11 per cent, economists, 13 per cent, social scientists, nine per cent, candidates educated at the Agricultural University, 12 per cent, architects, six per cent, engineers, six per cent and others, 23 per cent (N=235). At present no profession occupies a dominating share of higher-level positions. Furthermore, less than 35 per cent may be regarded as 'specialists' (natural scientists, engineers, architects, nature managers etc.) Although lawyers still are the largest group in the ME, the percentage of lawyers has decreased during the last ten years.

Among the five central agencies subordinate to the ME, the State Pollution Control Authority (SPCA), the Directorate for Nature Management (DNM), the Directorate for Cultural Heritage, the Norwegian Mapping Authority, and the Norwegian Polar Research Institute, we will here point specifically to the SPCA and the DNM.[12]

The SPCA's main responsibility is related to the enforcement of the Pollution Control Act and the Product Control Act. This enforcement has in particular been related to major industrial pollutants (offshore activities included). In addition to its activities enforcing the Pollution Control Act in direct interaction with major industrial pollutants, its activities are currently more directed at providing guidance to other authorities for preventing pollution in the oil and energy sector, the municipal and agricultural sector, and the transport sector. The SPCA is responsible for preparing surveys on the state of the environment, assisting the ME in drawing up environmental goals, and for monitoring the extent to which these goals are achieved. In 1994 its staff numbered 238, almost the size of that of the ME, with engineers and scientists as the dominant professions.

The legal authority of the DNM is derived from acts and regulations, e.g. the Nature Conservation Act, the Outdoor Recreation Act, the Wildlife Act and the Act Relating to Salmonids and Fresh-Water Fish. The DNM has also been given responsibility associated with the implementation of the Planning and Building Act. In addition to its statutory tasks, the DNM is responsible for identifying, preventing and resolving environmental problems through co-operation with other authorities and user groups, and for providing advice and information. The DNM is also responsible for participating in international negotiations on agreements concerning nature conservation and biological diversity. In 1994 its staff numbered about 140 employees with over one-third educated in the sciences, nine per cent lawyers and 14 per cent educated at the Agricultural University.

The County Environmental Departments (CEDs) at the County Governor's Office and the County Municipalities are today significant actors in environmental policy and administration. The CEDs are regional administrative entities subordinate to the ME with its agencies at the central level.[13] To understand the activities of the CEDs one should keep in mind that they are part of a hierarchical structure of state authority. They are organisational instruments by which the central authorities in Oslo seek to implement national environmental policies throughout the territory. Traditionally they have had responsibilities for the enforcement of the Pollution Control Act with regard to municipal sewage treatment, municipal waste and agriculture, and since 1993 also for the regulation of smaller industrial pollutants. Furthermore, an important aspect of the role of the CEDs is that of guiding and giving advice in the processes of municipal planning and in the preparation of nature conservation plans as well as in ensuring that wildlife and natural stocks of salmonids, fresh-water fish and their habitats are managed to maintain natural diversity and productivity.

Without doubt the CEDs are crucial actors in the implementation of national environmental policies at subnational levels. However, it should be noted that CEDs formally are part of the administrative staff of the County Governor and are subordinate to the County Governor. Formal authority rests with the County Governor, and one of the County Governor's main responsibilities is to co-ordinate environmental policies with other types of policies.

According to the Planning and Building Act, it is *the county municipality* that organises county planning work. The planning department of the county municipality is the principal executor of the planning activities, and the CED co-operates with the planning department in the making of environmental sections of such plans (Vabo 1995:136–140, 161–166). It is also the county municipality (mainly the planning department), that is responsible for making guidelines on land use that municipal plans are to comply with (ibid.:145–161). Furthermore, the county municipality is crucial to the administration of the Cultural Heritage Act. The CED has no legal authority in the process of municipal planning, but municipal plans must be submitted to the CED, and inter-changes take place. Informally, but also by making written statements, and in particular, by proposing that the County Governor objects to plans, the CED can make a significant impact on municipal plans. Again it is the professional authority of the CED that has the effect, as the County Governor accepts the CED's proposal and applies his right to object to municipal plans.

The municipalities have for decades been assigned tasks which today are categorised as environmental matters (e.g. tasks concerning pollution abatement, the planning of water supply and water management, land use and other aspects of physical planning). One important characteristic of environmental policy at the local level, at least until 1992, was that it had been fragmented and that the degree of fragmentation had been increased by the fact that the various parts had been entrenched in different sectors of the municipal organisation (Jansen and Osland 1996:215).

In 1992 the ME launched a nationwide programme, Environmental Protection in the Municipalities (EPM), designed to transfer responsibility for solving environmental problems of local significance to the municipal authorities and to increase the municipalities' capabilities in protecting environmental quality. To strengthen environmental expertise in environmental administration at the municipal level, the ME provided economic incentives for the municipalities that participated in the programme by giving them financial support, roughly equivalent to the salary of an environmental protection officer (EPO), during a trial period.

Although this programme contributed significantly in making municipalities crucial actors in the politics of environmental policy implementation, the results of the EPM programme are somewhat ambiguous. On one hand it can be pointed out that with the exception of the administration of the Wildlife Act, few initiatives have been taken to transfer responsibility for environmental tasks to the municipal level. On the other hand, it is a fact that by April 1994 only 19 of Norway's 439 municipalities had not created a position for an

EPO (Jansen and Osland 1996:216). Consequently, the EPM programme led to the entrance of a generation of environmental specialists into the municipal administration. If these two results are seen as related, the establishment of all the EPOs at the municipal level may not necessarily mean real decentralisation of environmental policy making and administration, but could equally well be interpreted as an attempt by the central environmental administration to penetrate new territories at the municipal level. The future of the EPO in municipal administration is, however, uncertain as the government has stopped the subsidies ear-marked for EPO salaries. From 1 January 1997 money for EPOs must be drawn from the general state funding for local governments. As of now, a number of local governments has abolished the position of municipal EPO.

As for the response of municipal councillors to the EPM programme, it has been reported that these are concerned about local environmental problems, and less interested in taking responsibility for global environmental problems. Environmental attitudes vary among councillors, and this variation does to some extent follow party cleavages. Left-wing and green councillors have a tendency to give higher priority to environmentally oriented positions than councillors of right-wing parties (Hovik and Harsheim 1996:4).

Environmental policy content: notions and principles

In the 1980s a more comprehensive and coherent environmental legislation was adopted. In 1981, the parliament passed a new Pollution Control Act which encompassed the protection of all types of recipients (water, air, ground) against pollution-related emissions, and also established the legal basis for an integrated licensing system.[14] A new more comprehensive Planning and Building Act was passed in 1985. Both the Pollution Control Act and an amendment in 1989 of the Planning and Building Act establish the legal basis for the authorities to require an environmental impact assessment (EIA). The Planning and Building Act is the principal law for regulating the use of land and water areas as well as natural resources. Besides these core acts the legislative offensive included amending the Open-Air Recreation Act (1990), the Nature Conservation Act (1990), the Cultural Heritage Act (1992), the Wildlife Act (1993) and the passing of the Gene Technology Act (1993). In addition the government's general policy for land use was stated in *Report No. 31* to the *Storting* (1992–93).

This legislation and the more specific plans and programmes for nature conservation, biological diversity and conservation of the cultural heritage that were adopted reflect the significance of the application of a number of new concepts and policy principles that were taken on board by the government during the latter half of the 1980s. Paramount in governmental policy documents since the end of the 1980s has been the policy principle of *sustainable development*. Even taken in the ambiguous way that the notion of sustainability is applied by the Brundtland Commission, one cannot help but notice a principal weakness of sustainable development as a policy principle: because it emphasises both sustainability and development this principle gives no clear directive for what to choose when there is a conflict between these two objectives.

On one hand, a number of notions and principles, derived from or related to *sustainability,* have informed and shaped the development of a strategy of anticipation and prevention of environmental degradation that has been converted into programmes for various areas of Norwegian environmental policy. A principle, which in recent years has informed and shaped a number of laws, programmes and plans is *the precautionary principle.* In consonance with this principle is the requirement for impact assessment in the case of activities that have significant environmental effects. Associated with the precautionary principle is the application of *the notion of critical loads on nature's carrying capacity.* This notion implies a focus on the effects on the recipient, and individual permits for major pollutants are based on the cleaning capacity of different recipients. According to governmental documents, another principle has been to move away from an emphasis on repair and end-of-pipe solutions towards pollution control at the source and cleaner technology. Characteristically, such notions and principles are derived from the sciences, in particular from biological and technical sciences.

On the other hand, notions and principles primarily related to *development,* such as *cost effectiveness* and *cost-benefit analysis,* have been given increasing priority in Norwegian environmental policy. Logically, these principles are applied to pursue the objective of both

> (1) designing the policy instruments in such a way that each given level of environmental quality is achieved at lowest cost to society (*cost-effectiveness*), and (2) reach(ing) the 'right' level on the basis of a cost-benefit analysis. If both these criteria are fulfilled at the same time, then the goal of a *socio-economically efficient* environmental policy has been achieved. (Report of the Environmental Policy Instruments Committee, NOU 1995:4, pp. 13–14)

By applying concepts and principles like these, issues of environmental degradation are made calculable in monetary terms. In this way environmental issues can be treated as *analytical* problems, and by means of the 'tool kit' of macro- and microeconomics, environmental problems and their solutions can be evaluated and chosen on the basis of economic calculations.

The applicability of both sustainability-related and development-related concepts is dependent on choosing appropriate policy instruments and collecting data on the empirical impacts of applying these instruments. Along with the emphasis on the above principles, one has been witnessing an increasing effort of developing information systems from nationwide environmental monitoring and surveillance, the collection of statistics, research and management. The above concepts and principles, applied to the relevant large quantities of data, are seen to provide the authorities and the policy implementing agencies with a common, co-ordinated basis for continuously evaluating developments and adjusting the instruments and measures used. By means of these data, organised according to the above concepts and principles, environmental issues can be defined in terms of problems that can be measured, discussed with the concerned parties, and solved analytically. In this manner, seemingly, they can be made into questions of problem-solving and management; they are no longer primarily conflicts of interest.

Policy content: the national environmental policy strategy

It was the reservoirs of petroleum and the production of oil and gas on the Norwegian continental shelf that provided the dominant coalition an opportunity to pursue its policies for economic growth without building nuclear plants and also making what in many quarters was considered a reasonable compromise in restricting the number of rivers to be exploited for hydropower production. The general policy that was pursued was *growth with conservation*. As a policy strategy, it recognised that both economic and conservationist interests were legitimate and that a balancing of these interests had to be achieved. The logical structure of this balancing process was that economic growth interests functioned as generating premises and conservationist, later environmental, interests as testing premises.[15] However, in striking the balance, economic-growth interests had top priority. Although conservationist, later environmental, interests commanded less weight, they were not to be ignored (Jansen 1989:248–272). This core feature of the general environmental policy strategy of the dominating coalition was from the mid-1970s regularly neglected, and even partly overshadowed, by the many initiatives and increasing activities of the ME, in particular during the period of active promotion of the report of the Brundtland Commission and the follow-up activities of its recommendations.

Norway's role in the area of international environmental co-operation is illustrative. Norway has repeatedly declared herself to be a 'driving force' both in regional and global environmental negotiations. She contributed effectively to a series of bilateral and multilateral co-operative agreements in several areas of pollution and was among the first states to unilaterally decide to reduce emissions by certain percentages by specific years. As Norway is one of the big 'net importers' of pollution, this has to a significant degree been decided by self-interest. Her self-declared role as a driving force has, however, been widely accepted and recognised because she, at least partly, has been leading from the front and has been a creative actor in the game of symbolic politics.[16]

As a national policy strategy, *sustainable development* was quite in harmony with the general policy of *growth with conservation*. In an equally ambiguous manner, the policy recognised both economic growth interests and environmental interests as legitimate. Logically it made it possible for the government to both pursue actively its policies of pollution control and management of natural resources at home and be a driving force for such policies in international co-operation, while it actively pursued economic growth policies and even financed the cost of environmental policies with the enormous income from oil and gas export. It was the emergence, and later what was seen as the imperatives, of climate policy that demonstrated the conflict between sustainability and economic interests by highlighting that Norway could not unilaterally pursue national CO_2 emission goals without significant consequences for the value of her petroleum resources and for her economic competitiveness as well.

When the Labour government argued that it is not possible to find a practical policy that will ensure stabilisation of Norway's CO_2 emissions by the year 2000, it posed the problem as follows:

> An international climate agreement that reduces the consumption of fossil fuels will also cause a drop in the producer price of crude oil, thus reducing the value of Norway's petroleum reserves. The resulting loss of revenue may be substantial, and is the largest expected cost to Norway of an international climate agreement.
> (*Report to the Storting*, No. 41, 1994–95:7)

Moreover, it has been pointed out that, for instance, the same CO_2 tax in all industrialised countries will have a far greater impact on emissions in industrial countries other than Norway. An internationally harmonised CO_2 tax imposed at a rate high enough to stabilise global CO_2 emission will not stabilise Norwegian emission by the end of the century. If Norway is to stabilise national CO_2 emissions in addition to taking part in an international agreement designed to stabilise global emissions, the costs of the adjustment process (e.g. in terms of loss of competitive strength and drop in employment levels) will be significantly higher.

In order to understand Norwegian climate policy, we must take note of the contrast in developments of CO_2 emissions and emissions of other greenhouse gases (GHGs). CO_2 emissions were in 1996 15 per cent above 1989 levels, but this increase was balanced by the reduction in emissions of other GHGs. Moreover, according to a 'business as usual' alternative, it is assumed that CO_2 emissions will increase by 33 per cent while emissions of other GHGs are assumed to be reduced by about 23 per cent (*Report to the Storting* No. 58, 1996-97:194 and 197–98).

It is emissions from the petroleum sector that are the main cause of the Norwegian CO_2 problem. It has been expected that at least 70 per cent of the rise in Norwegian CO_2 emissions from 1989 to 2000 will be generated by this sector. The core of the problem is related to the fact that about 70 per cent of the projected rise in emissions from the Norwegian petroleum sector is expected to be a result of an increase in gas production and transport (*Report to the Storting* No. 41, 1994-95:9). Therefore, if efforts of developing new technologies that significantly reduce CO_2 emissions and is economically competitive as well, for instance in the case of gas-powered electricity plants, are not successful, the increasing gas exports will have a very negative effect on Norway's emissions of GHGs. At the same time it should be noted that gas used as a fuel generates smaller emissions of CO_2 per energy unit than does coal or heavy fuel oil, and it does not generate any SO_2. The government argues that Norwegian exports of gas thus improve *other* countries' emissions inventories and actually reduce overall CO_2 emissions in Europe. Consequently, instead of establishing national goals based on the same per cent reduction in emissions for the various countries, the government's strategy now is to pursue binding international agreements with flexible mechanisms of implementation which allow Norwegian emissions to be evaluated in terms of the impact they have in replacing, or preventing further expansion of existing energy sources such as coal and heavy fuel oil in other countries. In sum, Norway's strategy in climate policy is (1) internationally to pursue the most cost-effective solutions for *all* countries, sectors and greenhouse gases considered *jointly* and (2) to strengthen climate policy instruments gradually, economic, legal (both traditional ones and voluntary agreements with industry), informational as well as physical instruments, and to ensure that in Norway these instruments, in particular economic and legal ones, are applied not to a greater, but to an equal extent compared with the EU and most other OECD countries.[17] The change in rhetoric has been dramatic, but what is significant is that when it comes to striking the balance between environmental quality and interests in oil the national policy strategy remains stable: *environmental quality counts, but national economic interests decide.*

The Kyoto agreement of December 1997 allows Norway to increase her emissions of GHGs by one per cent above 1990 levels in the commitment period (while most other OECD countries will have to reduce their GHG emissions during that period). It also states that the commitment covers for the aggregate anthropogenic CO_2 equivalent emissions of six GHGs (i.e. CO_2 is only one of these gases). Among the provisions of the agreement are also that it opens up the possibility for joint implementation and emissions trading between industrialised countries as well as the establishment of a clean development mechanism which makes it possible for OECD countries to finance emission reducing projects in developing countries and be credited for this in terms of their own emission budgets. Both the new centrist government and the major parties of the dominant coalition in Norwegian politics welcomed this agreement.

The environmental policy system in practice: interplay of parts and relations of authority

In this section, the distribution of authority between the ME and other core environmental organisations on one side and other ministries and parts of the administrative apparatus and important societal interests on the other, will be addressed.

A striking feature of the Norwegian environmental policy system is that it has been conventionalised and incorporated as an integral part of routine politics, standard processes and familiar structures of authority that are characteristic of the country's institutionalised political system. The authority the ME has in the policy process can be based on (1) jurisdictional competence (the *legal authority*) of the ministry or on (2) its technical competence (its *professional authority*).

The ME's legal authority is based on its responsibility for administering certain laws, primarily the Planning and Building Act and the Pollution Control Act. Pursuant to the Planning and Building Act, the ME has the main administrative responsibility for the national authorities' functions in physical planning and in the making of authoritative decisions on land use. Both the Planning and Building Act and the Pollution Control Act state that the ME has the authority to issue regulations for the purpose of implementing and supplementing the provisions of the act[18] as well as deciding on complaints against decisions made by a directorate or by decisions made at the county level. It is the sector ministries, however, that within their own sector have been given the legal authority to decide whether or not to require a developer to carry out an environmental impact assessment (EIA) and to accept such EIAs. Such decisions are not to be made, however, until after the cases have been submitted to the ME (Backer 1995:101). If the ME disagrees with a sector ministry, the matter has to be discussed and decided upon in the government.

Because they administer legislation and programmes that are crucial for the use and management of natural resources and other environmental matters, and in consonance with the 'sector responsible' principle, some of the sector ministries exert great legal and professional authority in matters of core significance to environmental quality. For instance, the Ministry of Transport administers the Road Traffic Act as well as specific programmes and policies for developing the communication and transportation sector. The Ministry of Industry and Energy administers acts like the Watercourse Act, the Watercourse Regulation Act, the Energy Act and the Petroleum Act, as well as numerous programmes for stimulating industrial growth and development of energy resources. The Ministry of Local Government administers acts and policies for economic development and employment in municipalities and regions, while the Ministry of Fisheries administers the act concerning fish farming and also implements programmes for the management of fish stocks as well as of fish farms. The Ministry of Agriculture pursues policies for agriculture and forestry. Since the ME has no legal authority over these ministries, it is dependent on its professional authority to co-ordinate the integration of environmental considerations and values into the policies of the above ministries. There are scant empirical data on the professional authority of the ME in these relations, but we would point out that this distribution depends on the area of environmental policy.

In an area like agricultural pollution, a co-operative relationship has been established between the ME, on the one side, and the Ministry of Agriculture and the organised interests of the farmers on the other. In other areas, environmentally motivated considerations and values are in conflict with well-established societal interests. Environmentally motivated policy in such areas, therefore, regularly collides with institutionalised solutions in other spheres or sectors of public policy. In these cases sector ministries and the organised representatives of the societal interests constituting the core political order of these sectors pursue other values and logics of action from the ME and its agencies. Thus as in climate policy, when the Ministry of Industry and Energy and its dominating constituency of the organised interests of industry and labour mobilise for economic growth policies, the ME has indeed limited professional authority.

In the area of pollution control, the ME and the SPCA have dominating legal and professional authority.[19] There is, however, a paramount exception to this; i.e. pollution control from transport. The SPCA is responsible for monitoring pollution, while the Ministry of Transport and Communication and its agencies – responsible for transport policy – also are responsible for the environmental impact of such policies. One distinct feature of this policy sector is the great authority of its agencies, both at the central level and at the county level. In addition, the county municipalities are responsible for important areas of public transport and, furthermore, municipalities and county municipalities are the responsible planning authorities for the transport system, including county and national roads. Consequently, the manner in which authority is divided between the core environmental administration, the transport administration, counties and municipalities is striking. In practice, the transport administration carries the greatest authority.

With regard to nature management, certain aspects of the relations between the central agency, DNM, and the CEDs and the municipalities should be noted. The DNM and the CEDs have consistently pursued national standards and policies. In this area, however, not all municipalities are acting in unison with the CEDs' and their masters' wishes. We thus refer to the EPM programme. This programme made municipalities more capable of implementing national environmental goals. However, since the EPOs are employed by the municipalities, neither the ME nor the CEDs have any authority over these officers. It was the municipalities that recruited their EPOs, and more than one-fifth of them were recruited from the municipality's local organisation; they were, by implication, already socialised to the definition of problems and general administrative culture of their municipality. This, together with other indications, lead to the conclusion that a number of municipalities made use of the financial support from the ME in order to strengthen their administrative and professional resources. However, research project reports indicate that authorities in municipalities, using these additional administrative resources, have partly taken the opportunity to develop counter-strategies to the nationally standardised policies in the area of nature management. In other words: the authority of the core environmental organisation is dependent not only on the area of environmental policy but also on the tier of government.

One characteristic of recent Norwegian environmental policy is the increased use of economic instruments and models. A significant step in this direction was the report from the abovementioned investigative Environmental Tax Commission. The commission, chaired by a leading official of the Ministry of Finance, cogently presented an economic approach emphasising the cost-effectiveness criterion, and strongly recommended an increased and more effective use of economic instruments. This commission's report and recommendations significantly raised the profile and professional authority of the Ministry of Finance in environmental policy, and doubtlessly also had implications with regard to the distribution of competence between ministries, between the ME and the Ministry of Finance, in particular. Correspondingly, this development demonstrates the authority of economists on environmental policy in Norway.

It is also significant that the Ministry of Finance during the 1990s has specifically exerted greater authority in the preparing of Norwegian positions in international environmental affairs. During the 1970s and 1980s the ME was the initiator and authoritative actor in international environmental affairs. The Ministry of Foreign Affairs has also gained greater authority in this area.

As to the authority of external actors in Norwegian environmental policy one should keep in mind that international commitments have been authoritative premises for adopting quantitative targets and for applying instruments and measures for implementing them as well. We see the OECD as of great consequence. It should be noted that Norwegian authorities, and in particular the Ministry of Finance, have paid a great deal of attention to the OECD, to its activities and recommendations. Similarly Norwegian high-quality expertise, in particular top-level economists, have been highly regarded as well as employed in OECD activities. Mediated by the interaction of leading economists, co-alignment of interests, philosophy and analytical models has to a great extent taken place between the OECD and the dominating coalition in Norwegian environmental policy-making. With regard to the impact of the EU, it should be noted that Norway's environmental legislation is usually consistent with relevant EU directives and in some areas (e.g. chemicals, dangerous substances and pesticides) is even stricter. The extent to which an EU-fication of Norwegian environmental policy is taking place, and the consequences of such a process, has not yet been the object of any systematic empirical research.

Concluding remarks

Almost a quarter of a century after the ME was set up, the SPCA and the DNM make the observation that the development is positive in environmental policy areas where it is the core environmental organisation that, on the basis of updated information of the state of the environment, administers laws and regulations to ensure that necessary measures are taken.[20] In contrast, the development is clearly negative where the ME and its agencies have no effective means to counterbalance environmental damage. In important areas, such as energy and climate policy, Norway is departing from the original intentions of sustainable development (Lafferty, Langhelle, Mugaas and Holmboe Ruge 1997:19). Also, biological diversity is being reduced because of the damage to, and changes in, various types of natural habitat areas for plants and animal species (e.g. wilderness areas, river deltas, old forests). In these policy areas it is up to the responsible sector ministries to ensure that effective measures are implemented in order to achieve national targets (*Direktoratet for naturforvaltning/Statens Forurensingstilsyn* 1996:4). We interpret this statement both as part of a strategic action in environmental politics and as a lucid observation of the institutionalised distribution of authority in the Norwegian environmental policy system.

The practice of the Norwegian environmental policy system is the outcome of the interaction between the strategies of the various actors and the institutional structure of the system. The environmental policy system grew out of and has since been part of the institutionalised political system of the country. Its development over the last twenty-five years demonstrates that indeed prior institutional choices limit future options (Krasner 1988:71), but also that the strategies of actors do count.

We have pointed out how the defining of environmental problems and the policies and instruments chosen to cope with these problems, vary historically as well as between collective actors anchored in different sectors of Norwegian public policy (e.g. agriculture, energy and oil, health, industry, national economy, transport) and according to different areas of environmental policy.

Norwegian environmental policy in the 1990s has been characterised by the processes of choosing policy objectives and quantitative targets and of the operationalisation and implementation of different policy principles including the choice of relevant instruments for achieving objectives and targets chosen. The institutionalised interests and logics of action that are typical for the various policy sectors have shaped the goals ministries and agencies have pursued as well as influenced the choice of notions and principles they suggest should be applied to achieve the goals, have given them their 'political weight', and left their imprint on the political outcomes.

Norway's political system has grown out of and is based on rather fixed or even rigid lines of representation. The dominant political parties and interest organisations aggregate and represent interests mainly in accordance with socio-economic criteria. Decision-making about collective arrangements are made between political and organisational leaders who are acting on behalf of the socio-economic interests of their voters and membership. In general policies are made through adjustments of socio-economic interests and bargaining. This consensual and co-operative policy style is typical for the corporatist character of the Nordic welfare state, and the government has deliberately tried to bring it into the environmental policy field. In some environmental policy areas it has worked, but in areas where basic economic interests of the dominating coalition in Norwegian politics have been at risk, *green corporatism* has not worked, if seen from an environmental point of view. Environmental interests have not been able to mobilise enough support to win the upper hand in the kind of bargaining that is typical for corporatist arrangement. As of now the prospects for a significant change are not bright[21].

The above characteristics notwithstanding, it should be pointed out that the existence of the ME and the other parts of the core environmental policy organisation constitute corporate actors of the state that routinely and persistently make efforts to ensure that environmental considerations are integrated into, and given authority in, public policy. As part of these efforts they have increasingly acted to establish and make use of national and international data-gathering systems that continuously provide up-dated information on the state of the environment, aiming at institutionalising the use of such information as authoritative premises in environmental policy. In sum: During the 1990s strategic action by core environmental actors seems to have generated a current in favour of institutionalising environmental values in new and widening areas of public policy. This apparently general development has, however, been broken when significant economic interests have been at stake, and the government has had to make a choice between economic growth values and environmental quality. On most such occasions the ME has lost the battle. The current has indeed not turned the tide.

Notes

1. The Building Act of 1965 established a framework for land-use planning. From 1970 onwards a string of new laws followed which aimed at regulating specific forms of environmental effects and certain social activities that have significant effects on natural resources or environmental qualities. These laws include new laws against pollution, e.g. the Water Pollution Control Act of 1970 and the Oil Pollution Control Act of 1970, and a new Nature Conservation Act of 1970, the Product Control Act of 1976, the Act Relating to Motorised Traffic in Marginal Land and Watercourses of 1977, the Cultural Heritage Act of 1978 and the Wildlife Act of 1981.

2. The Labour government's position and handling of the matter led to extraordinary events in Norwegian politics. Large demonstrations took place in Oslo as well as at Alta. In an unprecedented demonstration at Alta, hundreds of people chained themselves together in front of the bulldozers. A special police force of 600 was brought in from other parts of Norway. During January and February 1980, hundreds of demonstrators were cut out of the chains, carried away and fined. About a thousand 'river saviours' were removed by the police (Berntsen 1994:217).

3. From the conservationists' point of view, the political point with such a plan was national co-ordination. Since 1906 the Norwegian authorities, after considering each case individually, had granted a licence permitting regulation or exploitation of almost 500 large or small watercourses.

4. A majority in the parliament in 1986 passed the Protection Plan III, and in 1993 the Protection Plan IV, an implication of which was that watercourses representing 20 per cent of the potential hydropower were protected.

5. There is, however, no disagreement on the need for co-operative efforts in the Barents region, including the increasing emphasis on bilateral co-operation with Russia to curb pollution on the Kola peninsula and to assess the risk of radioactive contamination in the Barents and Kara seas (Scrivener 1995).

6. Here we point to the long-established interaction within the framework of the OECD and to the co-operation with the European Union through the European Economic Area Agreement (EEAA). Increasingly, Norway has also emphasised co-operative environmental efforts in the Barents region, and in particular bilateral co-operation with Russia.

7. Doubt as to the government's willingness to uphold this national goal had been growing during the 1990s. The decision to build a methanol factory at Tjeldbergodden, which was calculated to increase Norwegian CO_2 emissions by 1.5 per cent, was regarded as a test case by the environmental movement (Sydnes 1996:286).

8. Among the most important ones are the Norwegian Society for Conservation of Nature (NSCN) (28,500 members), and its youth organisation Nature and Youth (6000 members), The Future in Our Hands (about 17,000 members), Bellona (3000 members) and Greenpeace (700 members). Although the organisations do differ in terms of support in the population, the development of the NSCN is quite illustrative of a general pattern of development in membership. From the mid-1980s (when membership was about 24,000) the NSCN experienced a stable growth until 1991/1992 (about 40,000), and has since had a significant decrease in membership.

9. For instance, all these different types of interest organisations contributed to Norway's national report to the Rio Conference (1992), and representatives of these organisations were included in the official Norwegian delegation to the conference.

10. Of particular significance within the environmental policy field have been two recent investigative commissions which both had as its members high-status officials from a number of ministries. The first of these commissions was the Environmental Tax Commission (*Miljøavgiftsutvalget*), which also included two top economists at the University of Oslo/Statistics Norway as its members and the other was the Commission on Instruments of Environmental Policy (*Virkemiddelutvalget*).

11. This increase reflects that both organisational changes in which administrative units have been transferred from other ministries and a substantial increase within already established administrative entities have taken place. The total budget of the ME quadrupled during the period from 1973 to 1991, from NOK581 million (adjusted to 1991 values) to NOK2,336 million (Norway's National Report to UNCED, 1992:58). If we take the increase of the total state budget into consideration, this picture is less dramatic. From 1972 to 1994 the ME's share of the total budget varied between 0.4 per cent and 0.7 per cent (Jansen and Osland 1996:209).

12. Presently discussions are going on about ways to reorganise and merge these two agencies.

13. In 1994 the employees of the eighteen CEDs numbered 301. As for educational background of their personnel, about 33 per cent were educated at the Agricultural University (14 per cent in nature management and 19 per cent in agronomy). About 28 per cent are engineers and 23 per cent were educated in science (Jansen and Osland 1996:214).

14. This act also provides the legal basis for issuing regulations that are pollution preventive; e.g. regulations that define specific requirements to physical installations as well as activities in agriculture, industry (in combination with the Petroleum Act the petroleum sector is included). Included are also requirements as to the quality of equipment and requirements as to the qualifications of persons assigned with certain tasks. The Product Control Act provides the legal basis for issuing similar regulations, and these two acts also provided the legal basis towards obligating industrial companies to develop systems of internal control and monitoring of the environmental impact of their activity.

15. On generating and testing premises see Herbert Simon, 'On the Concept of Organisational Goal', *Administrative Science Quarterly*, vol. 9, 2 (1964), pp. 1–22.

16. It may appear as the Norwegian government, on one issue, has neglected the game of 'symbolic politics'; i.e. whaling. As to the national follow-up of the UN Convention on Biological Diversity we note in the case of whaling that on the one hand Norway resumed whaling on the basis of recommendations both from Norwegian scientists and from the scientists under the International Whaling Committee (IWC). On the other hand we note that Norway, which in many areas in environmental policy emphasises the importance of complying with majority decisions and honouring international obligations, abstained from the majority decisions on whaling in the IWC (Skjærseth and Rosendal 1995:177–178).

17. In a speech at a big oil conference in September 1996, the then prime minister, Gro Harlem Brundtland, stated that Norway will not reduce her emissions if the great industrial countries do not demonstrate willingness to act. It was reported that she indignantly referred to specific countries and that she also pointed out that a reduction by the same percentage for all countries is not a reasonable solution; 'Those who pollute twice as much as we do, will still do so after such a reduction', she said. (Bergens Tidende, 28.09.96)

18. Similar authority to issue regulations is given to the ME under acts like the Cultural Heritage Act, the Nature Conservation Act and the Wildlife Act.

19. In their implementation of the Pollution Control Act with regard to the municipalities' obligations in areas like sewage treatment and collection and treatment of waste as well as with regard to pollution from small manufacturing industry, the CEDs are loyal and effective instruments for the ME and the SPCA.

20. For instance, national targets for the emissions of ozone depleting pollutants, SO_2, phosphorus to the North Sea as well as some types of hazardous substances.

21. After having lost a number of bouts in the government, the Minister of Environment, in an interview with a Labour Party journal, stated that 'The Labour Party is still anchored in the economic growth "philosophy"' (*Aktuelt Perspektiv*, 14.09.96:3). Most commentators interpreted the statement as being addressed to the Labour government.

References

Asdal, K. (1995) 'I gode og onde dager... miljøvern i en skiftende norsk økonomi', *Sosiologi* 25: 101–118.

Backer, I.L. (1995) *Innføring i naturressurs- og miljørett*. Oslo: Ad Notam.

Berntsen, B. (1994) *Grønne linjer. Natur- og miljøvernets historie i Norge*. Oslo: Grøndahl Dreyer.

Direktoratet for naturforvaltning/Statens forurensningstilsyn (1996) *Miljøtilstanden i Norge, 1996*. Oslo-Trondheim.

Furre, B. (1992) *Norsk historie 1905-1990. Vårt Århundre*. Oslo: Det Norske Samlaget.

Gundersen, F. (1988) *Vitenskap, naturforvaltning og politikk. En studie av relasjonene mellom naturvitenskapelige forskningsmiljøer, offentlig naturforvaltning og frivillige naturvernorganisasjoner i Norge fram til 1965*. Magisteroppgave i sosiologi, Universitetet i Oslo.

Gundersen, F. (1991) 'Utviklingstrekk ved miljøbevegelsen i Norge', *Sosiologi i dag* 21 (2): 12–35.

Hallén, A. (1981) *Mjøsaksjonen 1977–80. Studiar av iverksettingsprosessane*. Oslo: NIBR-rapport 1981:12.

Hovik, S. and J. Harsheim (1996) *Miljøvernets plass i kommunepolitikken*. Oslo: NIBR-rapport 1996:5.

Jansen, A.I. (1989) *Makt og Miljø. Om utformingen av natur- og miljøvernpolitikken i Norge*. Oslo: Universitetsforlaget.

Jansen, A.I. and O. Osland (1996) 'Norway', in Christiansen, P.T. (ed.) *Governing the Environment*. Nord 1996:5. Copenhagen, pp. 179–251.

Knutsen, O. (1997) 'From Old Politics to New Politics: Environmentalism as a Party Cleavage', in Strøm, K. and L. Svåsand (eds) *Challenges to Political Parties*. Ann Arbor: University of Michigan Press, pp. 229–62.

Lafferty, W.M., Langhelle, O.S., Mugaas, P. and Holmboe Ruge, M. (eds) (1997) *Rio + 5. Norges oppfølging av FN-konferansen om miljø og utvikling*. Oslo: Tano Aschehoug.

Norway's National Report to the United Nations Conference on Environment and Development, Brazil (1992). Oslo: Ministry of Environment.

NOU (1992:3) *Towards a More Cost-Efficient Environmental Policy in the 90's* (Mot en mer kostnadseffektiv miljøpolitikk i 1990-årene). Oslo: Ministry of Finance.

NOU (1995:4) *On Instruments in Environmental Policy* (Virkemidler i miljøpolitikken). Oslo: Ministry of Environment.

Petersson, O. and H. Valen (1979) 'Political Cleavages in Sweden and Norway', *Scandinavian Political Studies* 2 (4): 313–31.

Reitan, M. (1995) 'The Deviant(?) Case of Norway'. Paper prepared for presentation at 'Makt- og Miljøforum', Sundvollen, 23 and 24 August 1995.

Report to the Storting, No. 31 (1992-93): *Regional planning and land-use policy*. Oslo: Ministry of Environment.

Report to the Storting, No. 41 (1994-95): *On Norwegian Policy against Climate changes and NO$_x$ emissions*.
Report to the Storting, No. 58 (1996-97): *Miljøvernpolitikk for en bærekraftig utvikling*. Oslo: Ministry of Environment.
Rommetvedt, H. and Melberg, K. (1997): 'Miljø- og næringsorganisasjonenes politiske påvirkning' in J.E. Klausen and H. Rommetvedt (eds) *Miljøpolitikk. Organisasjonene, Stortinget og forvaltningen*. Oslo: Tano Aschehoug, pp. 37–57.
Scrivener, D. (1995) 'Environmental Co-operation in the Euro-Arctic', *Environmental Politics* 4 (2): 320–27.
Simon, H.A. (1964) 'On the Concept of Organizational Goal', *Administrative Science Quarterly* 9: 1–22.
Skjærseth, J.B. and Rosendal, K.G. (1995) 'Norges miljø-utenrikspolitikk', in T.L. Knutsen, G.M. Sørbø and S. Gjerdåker (eds) *Norges utenrikspolitikk*. Oslo: CMI/Cappelen Akademisk Forlag a s
Sydnes, A.K. (1996) 'Norwegian Climate Policy: Environmental Idealism and Economic Realism' in T. O'Riordan and J. Jäger (eds) *Politics of Climate Change*. London: Routledge, pp. 268–297.
Vabo, S.I. (1995) *Forskning om fylkeskommunen – en kunnskapsstatus*. Oslo: NIBR-rapport 1995:137.
Weber, M. (1966): *The Theory of Social and Economic Organization*. New York: The Free Press, pp. 58–60.
Aardal, B. (1990) 'Green politics: A Norwegian Experience', *Scandinavian Political Studies* 13 (2): 147–64.
Aardal, B. (1993) *Energi og miljø. Nye stridsspørsmål i møte med gamle strukturer*. Rapport 93:15. Oslo: Institutt for samfunnsforskning.

Chapter 10

Spain: Environmental policy and public administration. A marriage of convenience officiated by the EU?

Francesc Morata and Nuria Font

Introduction

With an area of 503,478 square kilometres Spain is, after France, the second largest country in Western Europe. It is also the second highest country after Switzerland. Historically, its rugged topography has been a political and socio-economic obstacle, as it has separated the Iberian Peninsula from the rest of Europe (Tamames 1992). It has also encouraged a progressive depopulation of the mountain areas in favour of the more fertile and hospitable plains and the coast.

Along with its geographical position, the physical characteristics of Spain have given the country two distinct types of climate: the relatively wetter Atlantic climate, affecting only the north-west, and the dry climate characteristic of the rest of the country, where in some areas rainfall does not reach 300 mm per year. This explains why erosion and desertification are the main environmental problems in Spain (MOPU 1990). According to official data, in 1990, medium and high erosion affected 44 per cent of the Spanish soil (Martinez 1993). The country loses approximately 1000 million tons of fertile soil each year, and is the only European country included in the United Nations Program for the Environment risk map. Moreover, this problem is intensified by the frequent forest fires, which affected almost 400,000 hectares in 1989 and five per cent of the total forest area from 1988 to 1992, and the uncontrolled commercial lumbering activities of the pulp and paper industries.

With regard to the pollution of ground water, the main problem is one of controlling the discharging of waste from industrial activities in urban areas (more than ten million tons of waste per year) and agricultural activities (about a million tons of nitrates and huge quantities of pesticides). The discharge of these waste materials has serious effects on water quality. As a result, the purification of water (for drinking and industrial use) requires increasing investments at the local level. Therefore, the conservation and the efficient use of water continues to be a top political priority.

According to official data (MOPU 1990), Spanish flora and fauna represent one half of the whole European Community in terms of quality and quantity. There are 1300 species of flora found only in Spain (Greece, 750; Italy, 250; France, 100; UK, 12; and Germany, 5) and 50 species of animals endemic to the country. However, both public authorities and conservationist groups have pointed out that Spain faces a great risk of losing these biological resources: 350 vegetal species and 169 animal species are considered in 'danger of extinction'.

In 1916, Spain was the first country in the world to pass legislation establishing natural parks. At present, with a forest area covering 31 per cent of the country, there are nine national parks, whose management depends on central authorities, and many regional parks and nature-protected areas managed by regional governments. However, this network remains incomplete and the protection measures are considered inadequate, especially in the regions of Cantabria, Catalonia, Castille, Canaries and Andalusia (El País 9/6/90). In some cases, under the pressure of tourism and forest-products interests, regional and local public authorities are trying to reduce the protected areas. In reaction to these threats, the former national agency for nature conservation ICONA (now a General Direction has replaced and restructured ICONA) is promoting an ambitious program seeking to increase to 15 the number of national parks which will then cover 27 per cent of the national surface by 1997 (El País 8/1/94).

The socio-economic structure

The economic growth of the 1960s and early 1970s was due to the combination of different developments (Recio 1992): massive emigration of rural population to the industrial areas; uncontrolled industrialisation and the tourist boom, supported by foreign investments; and the public management of basic industries.

Spain has a population of nearly 39 million inhabitants with a population density of 77.2 inhabitants per square kilometre (INE 1992), which represents almost one half of the Community average. The population is distributed unevenly over the different regions of the country. Only 37.5 per cent of the population lives in the interior regions of Spain, which represent 70 per cent of the total territory of the country. Therefore, 62.5 per cent of the total population is concentrated along the coast and on the islands, which constitute roughly one-third of the area of the country. This pattern of irregular population distribution is also found at the local level: 60 per cent of the 8000 municipalities have less than 1000 inhabitants, while 100 municipalities contain more than 20 million inhabitants. The same tendency is also found at the regional level where the metropolitan areas of Barcelona and Madrid represent about two-thirds of the respective regional population. On the whole, the Mediterranean coast is subject to the highest demographic pressure, even without taking the summer tourist rush into consideration.

The demographic imbalance is a result of the industrialisation process of the 1960s. During this decade more than three million Spaniards emigrated from the Southern and Eastern rural regions to the industrial areas of Barcelona, Madrid and Bilbao. This had dramatic effects for housing, urban congestion and air pollution. The lack of democratic controls encouraged the rapid expansion of urban areas without public planning and services.

Modern industrialisation took place with little or no attention being paid to the impacts of these economic developments on the environment. The automotive, petro-chemical, paper, and iron and steel industries were the industrial sectors that carried the development of Spanish economy. The whole process was supported by an expansion of the energy industry, mainly based on domestic low-quality coal and lignites. With a population and territory representing respectively 11.9 per cent and 22.3 per cent of the European Union, Spain only emits 7.7 per cent of the EU global carbon dioxide (CO_2) (MICYT 1991). Nonetheless, Spain is the third highest European sulphur dioxide (SO_2) polluter, after the UK and Italy. The reorientation of Spain's energy policy towards the consumption of domestic coal after the energy crisis of the 1970s together with the low quality of this coal account for that score. However, the dispersion of power plants across the country, and their location far from the major urban centres, prevents the concentration of emissions. In addition, because of the alkaline nature of Spanish soils, acid depositions are less destructive than in other countries. The main threat faced by Spanish soils is erosion, not corrosion.

In addition, around 60 per cent of small and medium cities discharge untreated water directly into the nearest river or into the sea. Most of the toxic waste generated by the industry is also discharged without any treatment.

In many rural areas, tourism has become the dominant economic activity to the detriment of both the traditional agricultural sector and the quality of the environment. Wetlands of ecological value have been drained to provide land for intensive agriculture, tourism and industrial plants. The growing prevalence of industrial farming has created serious problems regarding the treatment of animal waste.

Representing nearly 25 per cent of national income, tourism has played a crucial role in Spanish economic growth. Nevertheless, its negative impact on the environment has also been quite evident, especially in terms of land destruction, sea pollution, unregulated construction activity and inadequate public investments in necessary infrastructure. Since the natural resources along the Mediterranean coast have been almost exhausted, private promoters, supported by many local and regional governments, are now investing in mountain leisure activities in order to develop new markets for recreation and tourism.

The development of public action on environmental problems

In Spain it is generally accepted that public action on environmental problems started in 1838, with some health regulations adopted by central authorities to cope with the negative effects of industrialisation on human health at the local level (Martin Mateo 1977). During the first quarter of the present century, in addition to the already mentioned national parks legislation, some other measures were adopted aiming to protect the local environment against harmful activities. For instance, according to a central government regulation of 1910, municipalities were supposed to set up sophisticated laboratories to carry out pollution analysis. In the middle of the 1930s, the Spanish Republic paid special attention to the improvement of both worker safety and health conditions, as well as urban health.

During the Franco period the environment was never an issue of political concern. Its preservation was mainly considered a local issue. The first relevant measure was a national regulation of 1961 on 'Unhealthy, Harmful and Dangerous Activities', which established a complex set of environmental rules apparently giving municipalities a prominent role in the area of industrial pollution and other disturbances. However, only a few cities were able to implement these regulations: bureaucratic complexities and administrative controls made it ineffective (Martin Mateo 1977). The first piece of national legislation, the Air Pollution Act of 1972, was in many regards conceived as a response to the Stockholm Conference which was held the same year. It established the basic framework of air pollution monitoring and control in Spain (Alvárez 1980) which is still run by a national agency under the jurisdiction of the Ministry of Health. However, as it lacked the necessary financial, technical and human means, its practical effects were quite limited (Martin Mateo 1977). Ironically, during the Franco period, many nature areas were preserved, not as a result of a specific policy protecting the habitat, but as a consequence of the multiplication of game preserves in the rural regions.

As regards the organisational structure, the establishment (in 1972) of a General Directorate of Territorial Action and the Environment, lodged in the Ministry for the Economic Development Plan, reflected the political influence of technocratism in the late Franco period. However, the environment as a whole was not treated as a separate issue in its own right but was rather dealt with in terms of the different sectoral bureaucratic concerns. Surface waters became the exclusive domain of civil engineers; industrial pollution was handled by industrial and chemical engineers; nature conservation by agricultural and forestry experts; environmental health by doctors; and so on. Consequently, environmental management functions and responsibilities were dispersed over 11 ministries. Co-ordination of these myriad activities was supposed to be provided by two interministerial bodies created in 1972:

the Delegate Commission for the Environment and the Interministerial Committee for the Environment (CIMA). Both proved to be totally ineffective.

At the local level, the lack of democratic representation and public accountability, the pressures of housing lobbies and central controls left little room for effective environmental action (Bassols 1984). Moreover, the regulations governing local planning were easily circumvented. According to the new Local Government Act of 1975, municipalities were supposed to pay special attention to the environment. However, the lack of democratic institutions and pressure from influential private interests reduced the law to little more than a well-intentioned declaration.

The public demand for environmental measures was also quite weak during the 1960s. The environmentalist movement did not start until 1973 with strong protests against nuclear power station projects, especially in the Basque Country and Catalonia (Costa 1981). These demonstrations were even supported by the Basque terrorist organisation, ETA, whose attack against a nuclear power station under construction in Lemoniz (Basque Country) brought the project to a complete standstill. This action became a rallying point for the anti-nuclear-power movement. These actions were then followed by many local protests against regional and urban plans (highways, yachting harbours, tourist towns), polluting industries and in favour of green belts and nature areas. The usual pattern of development was that green groups arose at the local level without an overall view of the environmental issue. Given the absence of democratic channels for popular participation and the lack of public information, environmental problems were articulated by university experts and professionals, usually linked to the clandestine parties and unions and local democratic movements. At that time, any form of protest was automatically viewed by government authorities as a problem of public order and a potential threat to the stability of the Franco regime.

The strong influence of the anti-nuclear-power movement, localistic activism and ideological diversity characterised the beginning of the Green movement in Spain (Piulats 1992). This set of factors has led to two basic organisational models for environmentalist groups. First, there is an assembly-type anti-party movement, mainly based on unstable local groups which are related to each other through co-ordinating bodies at the regional level. The second model includes stable and less radical organisations that are financed through contributions from its members and by public grants. These revenues allow the organisations to carry out public campaigns and to publish their own journals. The most representative organisation of this kind at the national level is probably the confederate *Coordinadora de Organizaciones de Defensa Ambiental* (CODA). However, some organisations have tried, with some degree of success, to combine stability with activism, such as the *Asociación Española de Defensa de la Naturaleza* (ADENAT), Friends of the Earth and Greenpeace Spain have done. There is also evidence of a trend whereby environmental organisations utilise contacts with EU institutions to circumvent domestic authorities while applying more effective pressure on them (Font 1996).

In the 1970s, despite the growing damage to the environment in Spain, the issue of environmental protection was not high on the political agenda. At a time when the environment was drawing increasing political attention in other Western European countries, domestic political actors in Spain devoted little time and energy to the problem of environmental protection policy. They were more concerned with re-establishing democracy. In the late seventies, during the initial period of democratic transition, the priorities of political parties and public opinion were defined by the need to consolidate the new democratic institutions and overcome the economic crisis in which Spain was embroiled. The Green movement, which in that period had already peaked in some European countries, had little room in Spain for effective action. In the 1980s, despite the growing attention paid by the media to environmental protests, the Green movement repeatedly failed in its attempts to set up a single political party. Green parties were established in 1983–84 with a strong ideological and organisational influence of the German *Grünen* (Baras 1992).

The first Green list was *El Partido Verde* (the Green Party) which competed in the Catalan elections of 1984 and scored 0.30 per cent of the vote. A year later, the Greens organised a Unity Congress at the national level. However, this attempt failed since many local and regional groups refused to accept any kind of centralisation within the party. In the general elections of 1986 and 1989, which, with 3 per cent of the total vote, represented the high point of their electoral success, and in the European elections of 1987, the Green presented four separate candidates. Finally, both a favourable electoral outlook and intense conciliatory efforts undertaken by some leaders led to the presentation of a single Green list in the general elections of 1993. Nevertheless, the outcome of the vote was once again disappointing. As a result, the Greens continue to be an extra-parliamentary force at the national and regional level. Moreover, they are hardly represented in the city councils.

Many factors account for the political failure of the Greens in Spain, including the generally low level of public environmental awareness. In 1989, only 3.9 per cent of population gave first priority to environmental protection (Eurobarómetro 1989); among the EC member states, only Greece and Portugal were less sensitive in this respect. Moreover, a survey conducted in 1990 pointed out that 62 per cent of Spaniards still were not sufficiently informed by the media about environmental problems (El País 4/10/1990). However, in 1992, 82 per cent of the Spanish public declared that 'environmental protection and pollution damages' were 'immediate problems needing prompt actions' (Eurobarómetro 1992). In the last few years, opinion polls have detected an increase in the ecological sensitivity of the public (Eurobarómetro Magnum 1996). Whether or not this will be incorporated into everyday behaviour is another matter, since the predominance of consumerist values in the Spanish society operates as a powerful obstacle to environment values (Pridham 1993).

The 1978 Spanish Constitution guarantees the citizen's right to enjoy a high level of environmental quality, and commits public authorities to the defence and protection of the environment. Environmental legislation fixes penal and administrative sanctions on those infringing this right. In addition, most of the constitutional laws defining regional autonomy recognise the protection of the environment as a public responsibility. Nevertheless, the environment continued to be relatively neglected during the economic expansion of the 1980s. The modernisation process of Spanish society followed the same path as other European countries during the 1970s. The aim of the socialist government, backed by regional and local authorities, was to invest large sums in urban and regional infrastructure in order to overcome the historical deficit in this regard with respect to the rest of the EU.

Only the accession of Spain to the EC in 1986 effectively led to the inclusion of the environment on the political agenda. However, this new behaviour emerges only after 1988, due to the initially permissive attitude of the European Commission with regard to the new Iberian member states. According to the Commission's report on the enforcement of Community legislation in 1991 (CCE 1992), Spain was one of the member states least adapted to the task of implementing the Community's environmental directives. Measures with regard to industrial waste water, designation of bathing areas, pollution in certain urban areas, treatment of solid waste and sewage were particularly troublesome for Spanish public authorities.

In short, Community pressures are an important factor to keep in mind when trying to understand the present situation in Spain with regard to environmental problems and their solution. As we will see further on, another set of developments that has had a significant impact in this connection is the general process of decentralisation.

Spanish environmental policy and administration

According to the constitution and the statutes regulating regional autonomy, responsibilities for environmental functions are shared among all levels of government, although regional authorities play a leading administrative role.

The central level

Central government is empowered to negotiate international agreements, including EU legislation, establish basic norms and general plans on environmental policy, and co-ordinate central-regional programs, including state grants. In addition, it has specific functions in the area of environmental management regarding the protection of state natural heritage and inter-regional river basins, whose agencies (*Confederaciones Hidrográficas*) also include representatives of the regions.

Like many other countries, until recently Spain did not have a single ministry responsible for the environment. Environmental policies were fragmented among eight ministries (El País 11/10/1995). The lack of a single ministry for the environment was long viewed as an indicator of the political and administrative weakness of Spanish environmental policy. This situation was officially justified by pointing to the degree in which the different responsibilities for environmental issues had been concentrated at the regional level. Nevertheless, another explanation could be the prevailing view that environmental protection policies were not compatible with socio-economic policies. It has also been argued that the creation of a ministry for the environment would provoke political and bureaucratic conflicts with the remaining departments. Pressure from environmental groups and certain political parties was not able to change these views during the 1980s.

In the early 1990s, the central government's approach to environmental policies began to shift. In 1990 a Secretary General for the Environment was created within the Ministry of Public Works and Housing. This new Secretary General, which consisted of a Directorate General for Environmental Policy and a Directorate General for Environmental Co-ordination, occupied an intermediate position in the Spanish administrative hierarchy. It replaced the former General Office for the Environment, which had been established in 1978 to carry out research and advisory tasks but had no institutional capacity for effectively collecting information required from the other ministries (Ortega 1991).

It can be argued, however, that this organisational change was little more than a formal move, since the Secretary General was neither given adequate functional authority nor guaranteed sufficient political clout. Growing recognition of these shortcomings, together with renewed pressures from the European Commission to provide it with a competent interlocutor, led to a new revision of the organisational framework in 1991. This change involved moving responsibility for the environment to the level of Secretary of State for Water and Environmental Policies. The former Secretary General was abolished, and the two Directorates General were merged into one in charge of Environmental Policy. After the 1993 general election, the new socialist government decided to provide the Environment with (shared) ministerial status by combining it with Public Works and Transport.

The Ministry of Public Works, Transport and the Environment (MOPTMA) formally acted as the co-ordinator of all national environmental policy and was responsible for contacts with the EU. In addition to the Secretary of State, it included an Under-Secretary of Public Works, Transport and Environment and a number of directorates general. It was also given supervision and management responsibilities for a number of related organisations. The Directorates General were: hydraulic works, water quality, coasts, environmental policy, territorial and urban planning, the Geographical Institute and the Meteorological Institute. A further reorganisation created a Directorate General for Environmental Impact Assessment and Information.

The DG of environmental policy played a central role in the formulation of national policy, since it was responsible for data collection and analysis, environmental planning and programming, the preparation of environmental legislation as well as for the negotiation and national transposition of international and Community legislation. It had sectoral responsibilities in the fields of waste and air pollution and co-ordinates common or joint actions with the regions, through the Sectoral Environmental Conference.

As has already been pointed out, the MOPTMA shared responsibility for the environment with seven other ministries. The Ministry of Agriculture was responsible for the implementation of several areas of basic legislation of environmental importance, including forestry and general aspects of nature conservation. The former Institute for Nature Conservation (ICONA), an autonomous agency reporting to the same Ministry, had a delegation in every region in order to ensure the management of national parks and the maintenance of the national register of sites of ecological importance as part of the national government's responsibility for natural resource inventories.

The Ministries of Industry and Energy and Public Health and Consumer Protection also had important power with respect to the control of industrial pollution and of environmental health, including international negotiations on these matters. The former was also responsible for the rational use of mineral resources, for the reduction of the environmental impacts of industry, and for industrial innovation and technology. It included three autonomous institutes of research: the Research Centre of Energy, Environment and Technologies (CIEMAT), the Centre for Technological Industrial Development (CDTI) and the Institute for Energy Diversification and Conservation (IDAE). The Secretary of State of Energy was in charge of the network of monitoring stations which measure air pollution emission levels.

The Ministry of Public Health was responsible for health monitoring and for the integration of environmental aspects in health issues. It maintained the National Centre for Environmental Health.

Since it negotiated regional development programs at the Community level, the Ministry of Economic and Financial Affairs was also involved in environmental matters.

Finally, the Ministry of the Interior had the duty of ensuring compliance with the laws concerning nature protection and hydraulic resources. Together with the regional governments, its Civil Protection DG was responsible for the prevention of major industrial accidents. It also had a unit responsible for the surveillance of nature protection.

The problem of inadequate co-ordination among central authorities was not solved with the creation of the Secretary of State for Environmental Policies and Housing. In particular, this agency did not have the power to intervene in matters concerning nature areas (the responsibility of the Ministry of Agriculture); energy policy, technological innovation, air emissions and mines (Ministry of Industry and Energy); and the negotiation of environmental aspects of Community regional policy (Ministry of Economic Affairs). These other ministries were supposed to notify the Secretary of State about actions having environmental impacts through the Delegate Commission for Economic Affairs. But such notification procedures in themselves were no guarantee that the central government would develop an integrated environmental policy. Rather, the contrary seemed to be the case: some of the departments formally involved in environmental policies were and remain scarcely concerned about this issue.

Such an overview of organisational changes shows that it was very difficult to find an institutionally secure, let alone prominent position in the administrative structure of central government for the environmental issues. As a consequence of ministerial rivalry, the public management of the environment remained fragmented, to the benefit of the different bureaucratic sectoral interests. Central authorities even delayed the adoption of a general law for the environment which would have brought together all those ministries which were involved in the environmental policy-making process.

In May of 1996, this situation changed dramatically, at least on paper. At that time, the new conservative government created the first ministry in Spain responsible for the environment. This organisation, which brings together the environmental functions previously spread among eight ministries and 21 directorates-general, includes one State Secretary for Water and Environment. The development of the National Water Plan and reforestation policy will be the main priorities of the ministry. In order to assist in these activities, the minister intends to strengthen the Advisory Council on the Environment (*Consejo Asesor del Medio Ambiente*). This body was established by the former socialist government and has, most recently, been heavily criticised by the green groups as an ineffective instrument for public participation. In the coming period, the ministry is supposed to work on implementing a number of laws developed by the socialist government. These include a general law on waste management, which finally reached parliament after a lengthy and complicated process of consultation with the affected parties; a new law on packaging waste (which was finally passed in April 1997 after four years of negotiations); and laws on atmospheric pollution and noise.

The new government of the Popular Party has also agreed to transfer to the Autonomous Communities a part of the jurisdiction contained in the law on coastal protection. This partial decentralisation of environmental responsibilities has been seen by many as a threat to the effective preservation of the environment. A further questionable development with consequences for the country's capacity for managing the quality of its environment, is the reduction in public investment for the natural environment as compared with previous years.

Pollution control policy in Spain has traditionally relied primarily on instruments of direct regulation. Moreover, Spanish policy style tends to be closed, bureaucratic and intangible. The administrative process has not traditionally been open to significant participation by societal actors and lacks institutionalised channels for consultation with interests involved (Pridham 1993). Consequently, there has been very little use of voluntary agreements and without real participation by industrial interests. These groups, in turn, have not put any significant pressure on central authorities, both because of the low priority that they themselves have attached to this issue and because of their fear of the consequences that such an institutionalised relationship with public authorities could entail.

Although there has been little convincing evidence of a shift with regard to both process and substance of policy, towards a new paradigm in which sustainable development or prevention are taken into account, there have been some first signs of movement in this regard. Since 1991, the need to meet the requirements of the EU Environmental Directives and the growing environmental awareness of the public, large industries are becoming active in their efforts to comply with EU environmental legislation. In addition, the Secretary of State has developed a double strategy consisting of both the imposing of hard economic sanctions on polluting industries and the funding of the retrofit measures required to meet Community standards financed through the Community Cohesion Fund, the money from which can be used, in part, for environmental projects. It is ironic that since it offers technical assistance and financial incentives, the Ministry of Industry and Energy has become the central governmental actor most sensitive to the demands coming from the modern industrial sectors (El País 13/12/92). Breaking with a long-standing environmental policy style, which often ignored private demands, under the Socialist government the MOPTMA increasingly promoted contacts with large firms for the formulation of environmental adaptation plans through an industrial, technological and environmental programme which offers technical assistance and financial incentives (Lopez-Novo and Morata 1998).

Since 1994, in connection with efforts to implement EU directives on air protection, without resorting to intensified use of legislation and regulations, the previous socialist government began promoting voluntary agreements with the main refineries in order to comply with EU standards. In a similar vein, the program Commitment to Progress (where progress is defined in terms of responsible care by industry and business) was established on a voluntary basis between the central government and the chemical industry. The implementation of this program involves roughly 65 per cent of this sector. The Commitment is divided into six sets of rules with regard to several practices. One of these is intended to improve environmental management within the firms, including research on new products linked to sustainable development. According to one public official involved in the negotiations,

> the new culture support for the Commitment is not limited exclusively to large companies. The principles of the programme are applicable to the whole sector and

the small and medium-sized companies will get the support of the most advanced companies, according to the international cohesion aims of Responsible Care.

The autonomous communities

The Spanish Constitution establishes a hybrid system which is neither federal nor regional, but based on the principles of the 'unity of the Spanish nation' and the 'autonomy of the nationalities and regions which constitute it'. The constitution establishes that each Autonomous Community shall have a legislative assembly elected by universal suffrage, a government headed by a president and a high court of justice. The distribution of functions is based on two lists. Under the first one, section 148.1, all the regions can take responsibilities for urban planning, housing, public works, agriculture and fisheries, communications and transport, environmental protection management, regional development, tourism and others. Those with full autonomy are also responsible for education, health, regional police and local administration. However, section 149.1 gives the central state the power to set basic legislation or principles in a range of these fields, including the environment, economic policy and planning, agriculture, health, education, local administration, general transport and communications. In some areas, R&D and culture, both levels are fully competent, producing a lack of clarity and the danger of one encroaching upon the jurisdiction of the other.

Autonomous Communities maintain their own administration to carry out their functions, but central government also retains field services to deal with supraregional or 'general interest' matters. However, central control over regional administration is limited to just those functions delegated from the centre.

The general decentralisation process in Spain has, then, generated great complexity in the area of environmental management. Due to the pattern of allocation of responsibilities and competencies to the different levels of regional authorities, the development of environmental administration has not taken place in a uniform way throughout the country. The lack of adequate technical and human resources makes it difficult for some regional governments to reach a level of effective management.

The regions are the most relevant actors in the environmental policy process. Regions enjoy some financial autonomy. Their responsibilities include the implementation of central and EU legislation; the setting of supplemental standards (some regions have established higher standards than those of the Community); and the performance of a number of crucial implementation and management functions. In terms of substantive policy, these activities deal with a multiplicity of issues such as regional policy, housing, public works, air pollution control, solid waste and sewage treatment, control of hazardous activities, mountain and nature areas and water quality.

The range of variation in organisational structure among the regions is quite wide. Catalonia and Valencia have created a single ministry responsible for the environment, though this does not mean, as we will see below, that these ministries actually exercise all the corresponding functions. Other regions merge the environment with another function such as economy (Basque Country) or culture (Andalusia), in a single ministry. The latter and four other regions (Aragón, Extremadura, Madrid, Murcia) have created special agencies. The remaining regions continue to combine environmental responsibilities with other governmental units such as the presidency, agriculture, regional planning, health or economy. It is clear that there is no single approach when it comes to organising environmental administration. As a result, the organisational choices made have led in some cases to redundancy, organisational difficulties and bureaucratic conflicts.

This organisational setting tends to reproduce some of the traditional characteristics of central administration: segmentation of functions and decision-making centres; lack of co-ordination; and an absence of adequate bureaucratic, technical and financial resources. As far as regulative functions are concerned, even though regional standards, mostly copied from one region to another, cover almost all aspects of environmental management, they are more often well-meant declarations of intentions without sufficient means to guarantee the effective implementation and evaluation of regional programmes. On the whole, regions have tended to develop a centralised model of environmental policy management at the expense of local authorities.

Much like the central pattern, the regional mode of environmental decision-making is rather closed, with limited opportunities for lobbying. The situation, in this respect, is quite different with regard to the implementation process. For instance, representatives of unions, business associations and Chambers of Trade participate in the advisory board of the Catalan industrial waste agency. In some regions such as the Basque Country, Catalonia and Andalusia, regional authorities negotiate with industrial groups over the implementation of the urban and industrial refuse treatment directive. Moreover, the regional councils for the management of nature areas generally include members of environmental groups with advisory tasks. Lastly, some regional authorities have recently concluded environmental agreements with industries located in highly polluted areas.

Local government

According to the Local Government Act of 1985, local authorities (municipalities, provinces and other supralocal agencies) are responsible for many of the activities which have consequences for the environment. One characteristic of the system is that the obligations of municipalities with respect to infrastructure provision and the management of local environmental services differs according to their population. Besides the traditional functions common to all municipalities (housing, public works, street cleaning, rubbish collection, sewage systems, drinking water supply), those with more than 5000 inhabitants must carry out sewage treatment, and those with more than 20,000 must also prevent and extinguish fires. Only municipalities above 50,000 inhabitants are responsible for the general protection of the environment, including waste-water treatment, monitoring and analysis of water quality, installation of water purification plants and monitoring air pollution. The province takes care of some essential environmental services and infrastructures for the smaller communities. Therefore, there is often an overlapping of several regional and local (municipal or provincial) competencies.

In addition, municipalities play an important role in the protection of the environment as implementors of national legislation on 'Bothersome, Unhealthy, Harmful and Dangerous Activities' and on some aspects related to air pollution protection. Industries are legally obliged to ask for municipal permits before starting their activities. Local authorities can also establish specific regulations in order to protect the environment. They participate in the national and regional air pollution monitoring networks, prepare reports on the local environmental situation and can take the initiative on cases of environmental emergency. Finally, they can impose (low) fines under the heading of environmental damages.

The main problem that local authorities confront in meeting their responsibilities is the lack of sufficient technical expertise and financial resources for effectively implementing regional and local environmental policies in their communities. The limited capacities of individual municipalities has led to a growing tendency to set up various forms of local co-operation, either by delegating functions to the province or by establishing inter-municipal bodies to perform these functions. In the case of the province of Barcelona, 84 per cent of the municipalities participate in such inter-jurisdictional agencies, mostly dealing with waste collection and treatment.

Intergovernmental relations

The governmental actors involved in the implementation of environmental policy in Spain are highly interdependent. Furthermore, environmental policy processes are characterised by a good deal of complex overlap of responsibilities among different levels of government. This is typical of the Spanish constitutional system of shared responsibilities. However, the different political and administrative actors tend to guard jealously their own competencies and jurisdictions from possible encroachments by others. With regard to Community decision-making, instruments allowing for regional participation in the internal formulation of Spanish positions or in the negotiations at the EU level have this far been limited.

Spain has delayed or blocked negotiations on some issues such as the protection of habitats, waste-water treatment or reduction of carbon dioxide emissions, by arguing that its particular circumstances would make it difficult to meet EU standards. For example, with respect to carbon dioxide emissions, Spain has demanded that it be allowed to increase its emissions, arguing that the 1989 Spanish per capita level was below the Community average and that since Spain had opted for a nuclear moratoria, it would have to rely more on coal-fired power plants. With this argument Spain sought to obtain a higher level of economic development and be allowed to maintain current levels of pollution.

On the other hand, Spanish negotiators have put forward claims for financial compensation for the costs associated with the adjustments that will have to be made in order to comply with EU standards. The greatest Spanish victory during the Maastricht summit was the decision to establish the Cohesion Fund to finance infrastructure and environmental programs. Spain threatened to block the agreement on Political Union if the Fund was not included in the EU treaty. In exchange for the agreement of the other member states on the Cohesion Fund, Spain accepted majority voting for a number of environmental decisions. Spain will receive 7.9 billion Ecus from the Fund during the period 1993–99 of which only one-third is specifically intended for projects to improve the environment (water management, forests, nature areas, urban pollution reduction and technology programmes) (El País 30/4/93). In line with the Spanish proposal approved in the European Council summit in 1992, money from the Fund is to be allocated in national and not regional shares. Here is a new source of conflict between the national government and the regions.

The fact that regional authorities have the primary responsibility for implementing environmental policies has led to the establishment of formal and informal mechanisms for promoting co-operation between the regions and central government. These include the sectoral conference on the environment as well as bilateral informal contacts. However, the sectoral intergovernmental conference on the environment, that was established 1990 to channel contacts between both authorities, has not been effective so far. There are three main reasons for this lack of success: first, there is the great diversity in the political-administrative status of the regional actors, including ministries, general secretaries, and general directors; second, Catalan and Basque authorities have given greater priority to bilateral contacts with central authorities; and finally, participants in the conference have not been able to reach agreement among the different interests represented with regard to overall priorities.

Despite such formal arrangements, intergovernmental relations often have an informal nature. According to regional officials, in the past informal co-operation with central ministries used to be easier between experts than between politicians. Some regional governments are also active in establishing contacts with public agencies or institutions other than the central administration. For instance, the Catalan government is active in the follow-up of recent developments at the EU level, while the Andalusian Environmental

Agency has been entrusted by the Assembly of the European Regions with the formulation of the European Network of Natural Areas.

Moreover, Community directives can be complemented and adapted by the regional governments. Central authorities are not in a position to prescribe specific measures or to control the way in which these authorities carry out their tasks. On the other hand, the formulation of general plans by central authorities depends on information provided by the regions which rely largely on information made available by the local level or by social groups. Such information is, however, often incomplete and in a form that makes it nearly impossible to use directly in deciding on matters at hand. The implementation and monitoring of both Community programmes and actions by central government with regard to industrial waste, sewage, air pollution control and maritime pollution ultimately depends on the strength of the commitment of regional politicians, and on the attractiveness of the financial incentives offered them by central authorities. Consequently, regions may reject certain programmes, as recently happened with the General Waste Plan, formulated by central authorities to comply with EU directives. Some regions refused to accept the incineration installations planned for their territories because of strong popular opposition and the lack of adequate financial compensations (El País, 10/9/93). As a result, the original plan has been replaced by a new one, effective since 1994 until 1998. Of course, intergovernmental conflicts discredit the central government, which then has to answer to the Commission for the infringements of the EU directives.

This situation may explain why the EU appears to be concerning itself more and more with the allocation of environmental competencies in Spain between the central and regional governments (TC 1992). This 'interference with the internal affairs' of a member state can be seen in the growing pressures being put on regional authorities directly by the Commission, for example, on the Catalan government because of the industrial discharges into two rivers and due to insufficient protection to a Pyrenean park (Aigües-Tortes); on the Valencian authorities because of the pollution of the coastal area in that region.

There are not only problems with respect to the division of labour and collaboration between central and regional authorities. The lack of horizontal co-ordination between the regions themselves gives rise to contradictory standards and criteria for the protection of common resources. For instance, management of the Pyrenees parks is shared by four regions and by the national forest agency without any permanent co-ordination. Moreover, in many cases, local, supralocal, and regional authorities are simultaneously, but singly, concerned with the treatment of sewage discharged into rivers.

In order to improve intergovernmental co-ordination, in 1994 the central government, through the MOPTMA, proposed to the autonomous communities the formulation of a *National Plan on the Environment* which was to represent the most important environmental initiative undertaken by central authorities since the entry into the EC. The plan was justified as the only means able to reduce the fragmentation of competencies and to deal with the lack of intergovernmental co-ordination. In addition, the central government assumed the responsibility of 'reviewing both sectoral policies and economic policy from the environmental perspective'. The proposal establishes four main priorities: the struggle against desertification, water quality and rational use of water, industrial, urban and agricultural waste management and stricter control of urban environmental quality. A national fund for the environment, estimated at 700 million Ecus and integrating national and Community allocations (Structural and Cohesion Funds), has been set up to enable the effective implementation of the plan which has been in effect since 1995. Finally, the proposal provided for the creation of an advisory committee including environmental organisations, trade unions and industrial associations.

In recent years, some regions and local governments have taken the initiative to promote programmes of co-operation at the European level in order, among other things, to obtain additional funds from the EC (Molins and Morata 1993). The *Four Motors of Europe* is a co-operation agreement signed in 1988 between Catalonia, Baden-Württemberg, Lombardia and Rhône-Alpes. The agreement also contains programmes for co-operation in the area of environmental protection. In addition, the Working Community for the Pyrenees, which is made up of seven Spanish-French regions, aims at developing environmental co-operation programmes to be funded by the EC. At the local level, Barcelona has been very active in creating a transnational network made of five other cities (Montpellier, Toulouse, Palma de Mallorca, Valencia and Zaragoza). Two of its objectives are to analyse the administrative impact of the EU directive on waste water, and to develop a computerised monitoring system on industrial waste water data in the six urban areas.

Conclusions

Spanish environmental policy is still emerging. Its recent development is closely linked to exogenous factors and, particularly to the Spanish access into the European Community in 1986. Before this date, environmental protection took a very secondary place on the political agenda. In Spain, protection of the environment became a political issue much later and with less intensity than in other Western European countries. In trying to account for this relative lack of social and institutional environmental concern some facts regarding the Spanish context should be kept in mind: the uncontrolled economic development of the 1960s, the late democratisation of the political system, the context of socio-economic crisis in which it was consolidated, the neglect of the environment during the economic expansion of the 1980s, the predominance of consumerist attitudes among the population and, lastly, the weakness of the environmental movement.

In Spain, public opinion and pressure groups supporting environmental protection have been traditionally weak. In spite of the environmental degradation inherited from the Francoist period, affecting particularly the Mediterranean coast and urban areas, the new democratic political elites have given priority to economic prosperity through industrialisation and infrastructure development.

Moreover, the decentralisation process which started in the early 1980s has introduced higher complexity in environmental management. As a result, the policy process has been fragmented in a multiplicity of interdependent policy-making and implementation decision centres.

Although a law on air pollution control had already been enacted in 1972, Spain lacked any consistent environmental administration until 1991, the subsequent establishment of a more comprehensive Ministry for the Environment took place only in 1996. The resistance on the part of the various ministries involved in environmental tasks to any attempt to deprive them of their environmental functions has so far been an obstacle to the creation of a single Ministry for the Environment. The environmental bureaucracy gradually has become more prominent, but the long tradition of administrative segmentation has left its trace.

The environmental policy style has until now been closed and bureaucratic. Social participation has been restricted to some aspects of the implementation process. However, there is some evidence that, since 1991, environmental problems are becoming a more relevant issue on the public agenda, and this should increase the system's capacity for change and performance. Two key elements may account for the growing political prominence of the environment: the entry of Spain into the European Community and the decentralisation process.

The devolution of power to the regions has introduced a great deal of complexity in the environmental management. The regions now have the primary responsibility for effectively implementing both national and Community regulations. However, due to both bureaucratic rivalries and clientelistic pressures from private interest groups opposed to higher environmental standards, regional governments have not taken advantage of their formal position to develop more effective organisational arrangements for carrying out their management responsibilities. In most cases, they merely reproduced the institutional arrangements characteristic of environmental management at the central level. In addition, the Secretariat of State for the Environment lacks the necessary means to co-ordinate Community-central-regional policies and to negotiate in Brussels. All decisions on these matters must be agreed to by the other national ministries involved. At the local level, environmental strategies, when they exist, depend on the municipalities' size and on the political priorities of the parties in power. The large cities, which have been given sufficient competencies but no financial resources, tend to develop their own policies, often in association with other local partners. These policies are, however, sometimes in conflict with those of the regional administrations, which the local authorities criticise for both their centralist tendencies and their failure to provide financial resources to implement environmental legislation. On the other hand, small municipalities are more dependent on provincial and regional assistance and on private lobbies that are interested in exploiting natural resources in one way or another.

It was Spain's entrance into the EC in 1986 that was a determining factor for the development of the environmental policy in Spain. Ironically, in light of the relatively low salience of this issue at the time of accession to the EC, Spain, unlike Portugal, did not attempt to negotiate special provisions in the accession treaty granting a delay or other derogations on the execution of Community environmental policy. Consequently, the relevant directives came immediately into effect upon Spain's admission to the Community. Therefore, the effect of EC policy was more sudden and disruptive than in other member states. This has created serious problems for the different levels of Spanish government, as can be seen by the increasing number of non-compliance proceedings initiated by the Commission against Spain (CCE, 1992).

In spite of difficulties in enforcing Community rules, Spain has developed a rather successful bargaining strategy at the European level that combines both the claim for special treatment, according to its specific conditions and its relatively low industrial development, and the demand for additional funds to enable it to comply more quickly and effectively with Community requirements, which are considered rather Northern in their emphasis on industrial pollution. Worried by the enormous costs involved in adapting to the requirements of the Community's legislation, which far exceed available budgetary resources, both public and private sectors have exerted great pressure to get financial compensations from the European Community. The main result of these efforts can be seen in the newly created Cohesion Fund, which is partly designed to provide Spanish administrations with more financial assistance to fill the environmental gap. The effective expenditure of its allocations from the Cohesion Fund will be a real challenge for Spanish authorities at all levels in dealing with environmental problems. Nevertheless, political priorities linked to the fund express a difficult compromise between the emphasis given to territorial infrastructures development and the need to deal with environmental issues.

To sum up, the EU has been crucial in the emergence of environmental policy in Spain, as it is the origin of most domestic environmental legislation and is co-financing essential infrastructures. However, the dominance of development values, the weakness of environmental structures and political will in addition to the lack of resources prevent Spain from fully implementing EU environmental policies. Much improvement is still needed.

One area in which such improvement will be needed is that of climate change policy. Extreme shifts in weather patterns over the last few years, combined with action by the environmental movement and increasing media attention have raised the visibility of the issue of human-induced climate change in Spain. Should global warming continue, higher temperatures and lower levels of percipitation would likely exacerbate the country's already serious environmental problems of desertification, soil erosion and water scarcity. In reaction to growing domestic pressure as well as to increased activity at the international level, the former Socialist government declared that climate change constituted one of its primary concerns and represented the most serious challenge to Spanish environmental policy (Labandeira Villot 1997:147).

However, even though the present government shares the general commitment to combating global warming, popular concern and the potential negative effects for the country have not led to any real climate change policy in Spain. Of course, there are a number of public policies that may have an impact on climate change, but these have been developed to serve other purposes. During the EU debate on measures to be taken to meet the goal of overall emissions of CO_2 at 1990 levels by the year 2000, Spain argued that it needed 'room' to increase its emissions instead of stabilising them. Increased emissions were justified, according to the Spanish government, because of the country's lower per capita emissions and its below average economic development compared to other member states. Spain only signed the Framework Convention on Climate Change (FCCC) because it assumed that it could get the leeway it needed within reduction target of the EU as a whole – where growth in the emissions in some countries would be balanced by reductions in other countries. While the EU internal agreement seeks to achieve a global reduction of 15 per cent by 2010, Spain was to be allowed an increase in its CO_2 emissions of 17 per cent above the levels reached in 1990. In 1990, Spanish emissions represented 5,8 tons per capita, lower than the EU average (8,4) and far below the German (11,1) and British emissions (9,8). However, given the efforts of government and business in Spain to 'catch up' with other, economically advanced members of the EU, this target will require strenuous efforts on the part of Spanish industry, including the use of renewable energies and the introduction of cleaner and more efficient technologies. Without such measures, the present rate of CO_2 emissions would mean that in 2010, Spain would reach a level 40 per cent higher than that of 1990.

Spain's exemption from the FCCC stabilisation commitment through the informal EU target-sharing arrangement informs and determines most of Spain's actions in this area. It determined the position taken by the Spanish government during the Kyoto Conference on Climate Change, a stance strongly criticised by Spanish environmental groups monitoring the meeting. And although the Spanish environmental movement does not have the same weight as many of its counterparts in Northern Europe, its influence and audience are growing. With this in mind, Labandeira Villot concludes, in his brief overview of the climate change strategy of Spain, that it is likely that 'the Spanish authorities will face a growing pressure from the voting public and from the environmentalist movement', which could, in the coming years, encourage the government to take a more active stance on the issue of climate change (Labandeira Villot 1997:163). Still, he concludes, there will be few changes in the current trends. Spain will continue to seek its national advantage within the requirements set at the EU level. The government will also attempt to limit action to measures outside the areas of transport and energy supply. Consequently, in the medium term, the levels of greenhouse gas emissions will depend on the way in which the Spanish economy develops (ibid., 163–164).

References

Aguilar, S. (1993) 'Corporatist and Statist Regimes in Environmental Policy: Germany and Spain', *Environmental Politics*, 2: 223–47.
Alvárez C. (1980) *Derecho Ambiental: Manual Práctico*. Madrid: Penthalon Ed.
Baras M. (1992) 'Los Partidos Políticos Verdes en Cataluña'. Barcelona: Institut de Ciències Polítiques i Socials, WP/47.
Bassols, M. (1984) 'El Medio Ambiente y la Ordenación del Territorio', *Documentación Administrativa*, 103.
Casademunt, A. (1992) *Las Organizaciones Empresariales del Sector Químico en Cataluña: su influencia en la política de medio ambiente*, paper presented in the session 'Asociacionismo y Estatutos' (group 22), IV Congreso Español de Sociología, Madrid, Sept. 1992.
Commission des Communautes Europeennes (CCE) (1991) 'Contrôle de l'Application du Droit Communautaire, 1990', COM (91), 321 final.
Comision de Las Comunidades Europeas (1992) (CCE) 'Control de la Aplicación del Derecho Comunitario, 1991', COM (92).
Commission of the European Communities (1993) (CEC) *Administrative Structures for Environmental Management in the European Community*, Directorate General XI, Environment, Nuclear Safety and Civil Protection. Luxembourg: Office for Official Publications of the European Communities.
Costa, P. (1981) *El Medio Ambiente en España*. Madrid: Fundación IESA, Informes y Documentos.
Costa, P. (1985) *Hacia la Destrucción Ecológica de España*. Barcelona: Grijalbo.
Diputacio de Barcelona, Sistema Territorial i Medi Ambient. Barcelona: March 1992.
Eurobarómetro (1989, 1992). Brussels: CEC.
Eurobarómetro Magnum (1996). Brussels: CEC.
Font, N. (1996) 'La Europeización de la política ambiental en España. Un estudio de implementación de la directiva de evaluación de impacto ambiental.' Ph.D. thesis. Barcelona: Universidad Autónoma de Barcelona.
GP (1993) Green Peace España (personal interview)
INE (1992) *Boletín Mensual de Estadística*, December. Madrid: INE.
Labandeira Villot, X. (1997) 'Spain: Fast Growth in CO_2 Emissions' in U. Collier and R.E. Löfstedt (eds) *Cases in Climate Change Policy*. London: Earthscan, pp. 147–64.

Lopez Bustos, F.L. (1992) *La organización administrativa del medio ambiente*, Universidad de Granada. Madrid: Civitas.

Lopez-Novo, J. and Morata, F. (1998) 'The implementation of IEAs in Spain: from national energy considerations to EU environmental constraints', in K. Hanf and A. Underdal (eds) *International Environmental Agreements and Domestic Politics*. London: Ashgate.

LV, La Vanguardia (newspaper).

Martin Mateo, R. (1977) *Derecho Ambiental*. Madrid: Instituto de Estudios de Administración Local.

Martinez, L. (1993) 'La situación del medio ambiente en España', *Economistas*, no. 55 extra.

MICYT (1991) *Plan Energético Nacional, 1991–2000*. Madrid: Ministerio de Industria, Comercio y Turismo

Molins, J. & F. Morata (1993) 'Spanish Lobbying in the EC', in R. van Schendelen (ed.) *National Public and Private EC Lobbying*. Aldershof: Dartmouth.

MOPU (1990) *Medio Ambiente en España*. Madrid: Ministerio de Obras Públicas y Transportes.

Ortega, L (1991) 'La Gestión del Medio Ambiente', *El Medio Ambiente en España*. Madrid: Congreso de los Diputados.

Recio, A. (1992) 'Los Problemas del Movimiento Ecologista en el Estado Español', *Ecología Política*, 3, 79–85.

Piulats, O. (1992) 'Teoría y Praxis de la Política Verde en el Estado Español', *Ecología Política*, 3, 65–78.

Pridham, G. (1993) 'National Environmental Policy-making in the European Framework: Spain, Greece and Italy in Comparison', Paper for the Workshop Environmental Policy and Peripheral Regions in the EC, ECPR Joint Session, University of Leiden, 2–8 April 1993.

Sierra, V. (1990) 'Reflexiones en torno al Problema Medioambiental Español', *Boletín de Información sobre las Comunidades Europeas*, 28–29/1990.

Tamames, R. (1992) *Estructura Económica de España*. Madrid: Alianza Editorial, Chap. 2.

TC (1992) Tribunal de Cuentas de las Comunidades Europeas: 'Informe especial n. 3/92 sobre el medio ambiente al que se adjuntan las respuestas de la Comisión', DOCE C 245, 23 Sept.

Valero, V. (1992) 'Los ríos no ríen: Políticas de saneamiento a través del análisis de los instrumentos preventivos y sancionadores', Working Paper on Intergovernmental Relations, Bellaterra: UAB, Departament de Ciència Política i de Dret Públic (unpublished).

Chapter 11

Sweden: From environmental restoration to ecological modernisation

Lennart J. Lundqvist

Sweden: country, society, environment

Sweden is the third largest country in what was before 1989 called Western Europe. From its southern Baltic shores to its northernmost region way above the Polar Circle, the distance is longer than that between Copenhagen and Rome. If the country's area were equally divided among all inhabitants each Swede would have about 50,000 square metres at his or her disposal. Lakes and streams cover 9 per cent of the country. Eight per cent is used for agricultural purposes, while 3 per cent is urban areas. The other 80 per cent comprises forests, marshlands and mountains.

This should not be taken to mean that Sweden is untouched by environmental problems. Rapid urbanisation and industrialisation have concentrated more than four-fifths of Sweden's 8.9 million inhabitants in cities and suburbs. Unprecedented economic growth in the post-war period, and the consumption patterns associated with an expanding welfare state had some astounding impacts. Water pollution forced the closing of beaches in the 1950's and 1960's. Agricultural use of pesticides led to prohibition against fishing in lakes around the country. Industry and transportation together account for 25 per cent of Sweden's GDP. These sectors have had, and still have, very strong impacts on the quality of Sweden's natural environment. Pulp and paper mill emissions of BOD_7 and COD_{Cr} in 1992 were several times larger than those from municipal sewage. Domestic SO_2 emissions have been brought down, but almost 90 per cent of the sulphuric downfall in Sweden originates in other countries. Prevailing south and south-westerly winds cause acidification of water and soil over and above the critical level, causing permanent damage in important agricultural and forestry areas of southern Sweden. Nitrogen emissions from Sweden's more than 3.5 million cars, and from intense use of fertilisers in agriculture, cause trouble in populated areas and along the country's southern shores (Lundqvist 1996:294ff.).

Since the adoption of 'modern' environmental policy in the late 1960s, much has been achieved to improve environmental quality. Emissions to water of phosphorous and BOD_7 from municipal sewers have come down to pre-World War II levels (Lundqvist 1995a:44). BOD_7 emissions from the Swedish pulp and paper industry are down by two-thirds since the 1960s. Emissions of heavy metals to air and water are in many cases down to a trickle compared to the situation in the early 1970s. The total consumption of CFCs decreased by 75 percent from 1986 to 1994. VOC emissions have continued to go down since 1988, and will have decreased by 45 per cent by the end of the 1990s. While CO_2 emissions from fuel combustion have been cut in half, they have increased from transportation (Lundqvist 1997). The Swedish Environmental Protection Agency's (SEPA) 1994 report on the implementation of nine environmental policy objectives stated that most of the goals for the year 2000 will be achieved. However, some thorny problems will remain. The decrease in NO_x emissions will not be on target without draconian measures, such as prohibiting lorries in city centres, forcing more long-range transport to railways, and drastically raising the price of gasoline. The program for preserving natural and cultural values in agricultural areas has come only halfway because of insufficient funding. Nature conservation in forested areas still has a long way to go before the present objectives are reached (SEPA 1994:7f.) Emissions of CO_2 will *not* become stable at the 1990 levels till the year 2000, and then decrease. This has become particularly clear in the wake of the 1997 Energy Policy Deal. The deal included a shutdown of the Barsebäck nuclear power station beginning in 1998; it is projected that despite massive energy savings measures, much of the *Ersatz* energy needed has to be imported from old oil combustion plants in Denmark and Poland (Cabinet Bill 1996/97:84, p. 20; Ds 1997:26, p.116).

Towards a 'modern' environmental policy: actors, institutions and strategies

The earliest advocates for the protection of Sweden's natural environment were natural scientists. Their key word was *conservation*; society must protect what was still left of original, 'untouched' nature. The criteria for selection of areas to be protected were a matter for science; no claims for economic exploitation within these areas should be acknowledged. Right after the turn of the century, the Parliament set aside a string of national parks in accordance with principles developed by scientists. From the mid-1930s and onwards, there was a fundamental shift in outlook; nature should no longer be protected *from* but *for* the people. A 1951 proposal stated that decisions on nature conservation should secure a comprehensive weighing of all interests concerned with the use of natural resources. Two NGOs, the Swedish Society for Nature Conservation (*Svenska Naturskyddsföreningen*) and the Society for Homestead Preservation (*Samfundet för Hembygdsvård*) were given semi-official status as 'nature interests' in decisions on, for example, the taming of rivers for hydroelectric power production (Lundqvist 1971).

In other fields of environmental protection, pragmatism dominated over science. The *balancing of interests* became the leading principle. Proposals for a Water Inspectorate in the late 1910s tried to secure the co-operation of industrialists benefiting from unregulated pollution. The 1950s saw a gradual worsening of water quality, not the least as a result of existing policy. The National Highway and Water Construction Agency provided state grants for building sewage pipelines, but not for the construction of sewage treatment plants! The Water Inspectorate set up in the late 1950s was charged with examining *all* plans for extension of water and sewage systems in Sweden, and was to monitor and supervise *all* industrial pollution, an impossibility given its minuscule resources.

Up to the early 1960s, Sweden's environmental policy and administration was given low political priority. The quality of the natural environment was one among many legitimate societal interests, to be weighed against others in the competition for political attention. The policy response was to deal with environmental issues one by one. The result was institutional *fragmentation*. There were small national administrations for nature conservation, and for water and air pollution control, respectively. Politically, these issues were handled by the Ministry of Agriculture.

The mid-1960s brought environmental issues to the top of the political agenda. Several events and developments, mercury in fish leading to the blacklisting of many waters and streams, mercury in birds threatening many predatory species, blatant water pollution causing bans on bathing, air pollution from burgeoning traffic, and the threat to all still unexploited major rivers in the north, caused a dramatic upswing in the public debate over environmental issues. The number of newspaper editorials on environmental topics increased more than sixfold between 1963 and 1968. 80 per cent of people in a 1969 poll said they were for higher local taxes if that money were spent on water pollution abatement, and supported environmental charges on polluting industries. There were signs of an increasingly radical environmental opinion, manifested in the growth of 'alternative movements' from the late 1960s and onwards. The politicians' response was quick. The number of environmental bills tabled in the Swedish parliament (*Riksdag*) doubled between 1965 and 1968. Several investigatory commissions looked into legislative, administrative, and information-gathering problems.

The first policy decisions in the parliament concerned policy organisation and main strategy. They were accompanied by discussions on how to treat the environmental issue. One emerging, albeit numerically small, line of reasoning reflected the old scientific approach. The objective was *ecological balance* (Lundqvist 1971:108ff.). Environmental quality should supersede other societal concerns, and the boundaries of environmental administration should be drawn in accordance with 'natural' areas; biotopes, water drainage basins etc. The other, dominating perception of the problem was pragmatic; the keyword was *interest balance*. Environmental quality is one 'interest' among others, and should be considered together with other concerns in policy decisions on how to use the nation's resources. Environmental administration should be established in geographical correspondence with other sectoral administrations.

Both the policy organisation and the major strategy reflected the second line of thought. National policy-making remained within the Ministry of Agriculture. An Environmental Advisory Council *(Miljövårdsberedningen)* was set up in 1968 to provide counsel and information to the Prime Minister and the Cabinet on long-term environmental issues involving several sectors in society. The Council held many thematic meetings on such issues as climatic change and acidification (Lundqvist 1971:148f.; Loftsson *et al* 1993:77). More recently, the Council has been given important functions as an information and inspiration centre for the Local Agenda 21 process (SOU 1994:128).

The new Swedish Environmental Protection Agency (SEPA; *Naturvårdsverket*), set up in 1967, consolidated existing agencies in the field. The agency was given wide responsibilities, ranging from protection of rare species and management of national parks to promulgation of pollution-control guidelines and payment of state subsidies to sewage treatment plants. An Environmental Research Board was established in 1968 within the SEPA. The 1969 Environment Protection Act (see below) gave SEPA responsibility for developing the general principles of the Act into more precise emission guidelines and other criteria for implementation measures against polluters. This work was carried out with vigour during the first few years, providing SEPA with important links to both environmental researchers and industrial branches (Lundqvist 1971).

A special National Environmental Licensing Board (*Koncessionsnämnden för miljöskydd*) was set up in 1969 to handle applications from polluters for permits required under the Environment Protection Act. Appointed by the Cabinet, the Licensing Board has an experienced judge as Chairman, and three ordinary members with environmental, technological, and industrial experience. When municipal permits are handled, there is a fourth member with municipal background. The Board has about 30 employees. The permit cases handled by the Licensing Board are those which may cause serious environmental disturbances (Loftsson *et al* 1993:76).

At the local level, the municipal Health Protection Committees (*hälsovårdsnämnderna,* politically appointed by the Municipal Councils*)* already had supervisory and control responsibilities for such issues as local water and air quality and waste management. However, the regional administrative arm of the central government, i.e., the County Administrations (*länsstyrelserna)*, were given the power to supervise and control all local polluters requiring permits under environmental law.

The policy strategy from the late 1960s onwards had a strong regulatory emphasis. The principal idea of the 1969 Environment Protection Act was to promote *end-of-pipe* solutions by requiring all potentially harmful activities conducted on a property or some permanent facility to obtain a permit spelling out the conditions under which they may be carried out. A *balancing of interests* was built into the law; the polluter must take only such precautions and tolerate such restrictions 'as may *be reasonably* demanded' to protect environmental quality. Assessments of what is 'reasonable' should be based on what is 'technically feasible' and 'economically reasonable'; attention should be paid to the balance between disturbing effects on the one hand, and the usefulness of the activity and costs of protective measures on the other. The policy organisation just described was to determine the permissibility of the activities, set the conditions for polluters and supervise the compliance with these conditions (Lundqvist 1971).

This balancing of environmental and other interests also contained *economic* measures. The *polluter pays principle* was temporarily set aside by a program of subsidies to industries for investment in pollution control measures in already existing facilities. Such state subsidies could cover up to 25 per cent of the costs of investment. Throughout the 1970s, subsidies were also paid to local government investments in sewage treatment plants (Lundqvist 1995a). Since then, however, there have been only some small programs of subsidies to technologically advanced, experimental pollution control facilities.

The politics of environmental policy: actors, arenas and issues

It is fair to say that the formation of the new environmental policy in force from the late 1960's was very much an affair for the politicians and the national agencies. The venerated Swedish Society for Nature Conservation (*Svenska Naturskyddsföreningen*, founded in 1909) lost its status as the 'recognised' representative of the environmental interest. The short outburst of environmentalism in the 1970s, symbolised by the new 'alternative' movements, had no common organisation and a limited number of followers. Their quite unorthodox tactics precluded easy access to the then tightly knit policy networks of Swedish central government.

After the 1973 oil crisis, environmentalists were active in the 'Peoples Campaign Say No To Nuclear Power'. The Centre-Right coalition broke up in 1978 over the nuclear issue, and the Harrisburg incident in the US led the politicians to bring about the nuclear referendum of March 1980. Then the environmentalists splintered; some went to the Green Party, others remained at the margins, and still others went into the established environmental organisations. The Swedish environmental movement has been 'subjected to a range of incorporation pressures that have made it difficult for an autonomous social movement to develop'. The very strong and early response to environmental issues from the political parties led to an 'ideological incorporation of environmentalism as part of the established parties'. Because of these incorporation pressures and the 'instrumental orientation of the national culture, environmentalism has had difficulty carving out a public space of its own' (Jamison et al 1990:13ff.).

The emergence and formation of the Green Party (_Miljöpartiet – de gröna_) in 1981 was very much a result of the 1980 referendum on nuclear power. Many of the party founders were active in the _People's Campaign Say No To Nuclear Power_. The 1988 election campaign was totally dominated by environmental issues (Asp 1990), which no doubt helped the Green Party to become the first really new party in 70 years to enter the Swedish parliament (Vedung 1989; Gilljam & Holmberg 1990). The Green Party has been very active on environmental issues in the parliament; 23 per cent of all Green Party MP bills presented to Parliament in 1991, and 35 per cent of all Green Party MP questions and interpellations to Ministers in that year concerned the environment. The other greenish parties, the Left and the Centre Parties, have much lower environmental shares (Lundqvist 1996:266). Environment was no issue in the 1991 fall elections, and the Green Party did not reach the four per cent needed to gain continued representation in the parliament (Gilljam & Holmberg 1993). However, the Greens got over five per cent of the votes in the 1994 elections, thus becoming the first Swedish party ever to return to the parliament after being ousted in an earlier election. The Greens ran an energetic 'No to EU' campaign in the 1995 elections to the European Parliament; gaining 17 per cent of the votes, they became the third largest of the 11 contending parties (_Från Riksdag & Departement_ 29/1995:13). Green Party membership is presently about 5500 (_Vem gör vad i Miljösverige_ 1994:114).

There are now more than 50 pro-environmental interest organisations in the arena of environmental policy making. About half of them are active in ecotechnology, ecoinformation, and ecoproduction. One-quarter consists of associations of environmental professionals. Seventeen may be classified as environmental 'movement' organisations. The three largest, the Society for Nature Conservation, Greenpeace and the WWF, together have 460,000 members, or five per cent of the Swedish population. A majority of their followers are only 'chequebook members'. The smaller 'movement' organisations with a more activist profile gather about 20,000 members (Lundqvist 1996:287). Nearly 70 special periodicals with some 'green' label were available in 1994. All leading newspapers, as well as the news programs on radio and television, have special reporters working exclusively on environmental issues.

Most citizen groups give strong support to generally stated proposals to save the environment, but some are more 'environmentally oriented' than others. Women and elderly people are more active in recycling, composting and using collective transportation systems (Bennulf & Gilljam 1991; Bennulf & Holmberg 1991). Young and middle-aged people in urban areas tend to score high on support to the Green Party and other environmental organisations. But it is notable that proposals to restrict private car use are unpopular across large segments of the Swedish population (Holmberg &

Gilljam 1987; Bennulf & Holmberg 1992; Gilljam & Holmberg 1993; Bennulf & Lundin 1997).

Polluting industries worked closely with SEPA officers and environmental researchers in special committees set up by SEPA to develop emission guidelines under the 1969 legislation. Since then the environment has become an important business. Environmental investments in Swedish industry tripled in ten years, reaching six per cent of all investments in 1991 (SCB 1994). There are now 35 business and industrial environmental organisations, and all individual enterprises of some size have environmental divisions. These divisions serve as information and lobbying vehicles on environmental issues, and represent industry in state commissions which investigate environmental problems and propose solutions (*Vem gör vad i Miljösverige* 1994, see below).

The development of policy content: concepts, instruments and implementation strategies

The policy principles promulgated in the late 1960s reflected the *interest balance* principle. The environmental policy should promote a 'fair adjustment between interests with competing demands on natural resources', while 'the preservation of natural resources for coming generations' was only secondary. Throughout the 1970s, the policy of *clean up* rested on *regulatory* instruments, with rather general laws clarified by specific guidelines for different polluting activities. Specialised policy-implementing agencies determined the permissibility of the activities, set the conditions for polluters and supervised the compliance with these conditions. This they did not do in 'a command and control'-centred way, but rather through co-operation with the polluters in order to reach consensus on 'reasonable' protective measures. The co-operation between SEPA, researchers and industrial interests aimed at specifying branch guidelines, and the seeming reluctance to bring polluters to court both indicate the emphasis put on *co-operation and consensus* rather than command as the main implementation strategy.

The major 1987/88 Environmental Bill provides evidence of a shift in emphasis. Now, environmental policy should be *preventive* and *cross-sectoral*; 'A successful environmental management presupposes that care for the environment is integrated into the development plans for different sectors of society'. All sectors would now have 'a responsibility to prevent new environmental damage, and to solve existing environmental problems', and the costs should be borne by those causing the problems (Cabinet Bill 1987/88:85, p. 35f.).

The 1990/91 Environmental Bill reveals a further shift in emphasis; '...responsibility and care for the environment should permeate all walks of life and society. The mission of the 1990's is to readjust all societal activities in an ecological direction'. Based on the *polluter pays* and *precautionary* principles, future environmental policy should 'protect human health, preserve biological diversity, manage natural resources for sustainable use, and protect the natural and cultural landscape' (Cabinet Bill 1990/91:90, p. 11). Intensified international co-operation to limit transnational and global environmental problems, improved production of knowledge and monitoring of the state of the environment, and increased sectoral responsibilities and decentralisation were seen as important strategies for implementing this approach.

The Right-Centre coalition government in power between 1991 and 1994 saw the objective of environmental policy as 'a long-term sustainable development ... where future generations can take over a national wealth, including environmental and natural resources, of at least the same size as today.' (Cabinet Bill 1991/92:150, Annex I:12, p. 1):

> Concern for the environment ... means giving prior consideration to the environmental impact of every decision capable of exerting a major effect on the environment. It also means that the 'precautionary principle' must be more widely applied; measures to prevent environmental destruction must not be postponed owing to lack of scientific certainty ... Economic growth must not be allowed to threaten the conditions of human, plant and animal life ... The objective is to preserve 'biodiversity', i.e. the complete range of animals and plant species found in our country.

At the level of principles, Sweden's policy for the environment has thus come a long way from the 1965 objective of 'trying to take care of the environment in the hardening competition with other interests' (Lundqvist 1971:183). The policy change can also be detected in instrument choice and implementation strategies. As point-source pollution gradually came under control, and with increasing insight into the problems of diffuse and long-range pollution, the traditional regulatory approach was supplemented by *economic* and *information* measures.

The 1990 tax reform included a quite spectacular 'green tax exchange' (*skatteväxling*). To cover the losses in state income following the radical cuts in direct income taxation, the government began to tax fuels and products posing potential hazards to environmental quality. Environmental taxes and charges to induce behavioural changes generate around SEK10.5 billion in state income. A 'greening' of the whole tax system was proposed by a special investigatory commission in 1997. However, this comprehensive 'green tax exchange' is expected to need an implementation period of at last 15 years, and will depend on Sweden's success in getting a common European and Nordic strategy adopted (SOU 1997:11).

Much of the recent policy debate has focused on such concepts as *recycling* and *sustainability*. The Eco-cycle Bill of February 1993 and subsequent decisions amount to what could be labelled a system of *ecological corporativism*; the state establishes demands and targets for reuse and recycling of different products and materials, but leaves the actual implementation to the producers. The branch organisations do this by providing the organisation and running the processes which would guarantee that the eco-cycle is closed for the products and materials in question (Cabinet Bill 1992/93:180). Such use of 'corporatist' agreements between state and market actors has spilled over to more and more branches (Cabinet Bill 1995/96:120, pp. 48 ff.).

Information has come to play an ever-increasing role as both a detecting and an effecting environmental policy instrument. There are nationwide programs for monitoring, and a host of regional and local measurement programs to aid decision-makers and planners. Local governments run environmental information campaigns on such things as recycling, composting, separating of garbage, etc. Information is a crucial, consensus-generating activity in the Local Agenda 21 process. All local governments are engaged in a dialogue with affected local interests to develop and anchor local plans for sustainable development. Both the government and responsible environmental agencies at the central level envisage that the information obtained in such local strategic plans will provide the input to a national plan for sustainable development (SOU 1994:128). The quest for information in policy implementation is seen also in the introduction of environmental impact assessment procedures in decision-making on the use of natural resources. Such assessments are increasingly required for a wide variety of undertakings, involving a widening circle of actors (Cabinet Bill 1994/95:230).

The 1990/91 Environment Bill clearly shows how the *scope* of environmental policy has grown since the early 1970s. Environmental policy comprises measures not only in relation to atmosphere and climate, trans-national pollution, the urban environment, nature conservation, sea and water, but is also directed towards numerous other sectors and activities in Swedish society (Cabinet Bill 1990/91:90, Table of Contents, p. 578 ff.). The changed content of the annual Cabinet Bill on *the State of the Swedish Environment* is another sign of the efforts to lend environmental policy an integrated sustainability outlook. Unlike earlier reports, the one from January 1997 does not deal with individual environmental problems. Departing instead from the argument that the political objective of an 'ecologically sustainable society' requires integration of "environmental care and resource management ... into all processes in society", the report then analyses the interplay between 19 different societal sectors and environmental/natural resources. As a means to integrate eco-sustainable objectives and practices into these policy areas, the report in effect calls for environment- and sustainability-adapted national accounts (Cabinet Report 1996/97:50, p. 5 ff.).

The organisation of environmental policy

The design of Sweden's environmental policy organisation did not emerge in an institutional vacuum. The existing organisation of the political system, and prevailing trends within it, provided a (limiting) framework. Figure 1 outlines, in a general form, the main elements of the Swedish governmental structure.

The most characteristic feature of Swedish central government is the division between 'political' ministries and 'administrative' central agencies (the following builds on Petersson & Söderlind 1992). The Cabinet and the ministries in Sweden almost exclusively deal with the formulation of policy. The Cabinet as a collective, not the individual minister, issues directives to the central agencies. At the same time, the Cabinet and the Ministry of Environmental Affairs have access to the national agencies and boards and their expertise when formulating new policies or revising old ones. Policy clarification and implementation in individual cases is a matter for the about 70 central national agencies and the regional and local bureaucracies.

Governmental branch Geographic level	*Stat*	*Landsting*	*Kommun*
National	Parliament* Ministries Central agencies		
Regional	24 county administrations	24 county councils* County committees	
Local	Local agency branches (e.g. police, tax authorities, health insurance)	Local health services	288 municipal councils* Special boards and committees

* popularly elected

Figure 11.1. Sweden's governmental structure: a general outline

The regional level in Sweden is divided into one state-directed and one popularly elected branch. In the 24 'counties' (*län*), the County Governor (appointed by Cabinet) and the County Administration (*länsstyrelsen*) function as the regional arm of central government, implementing national policies and controlling regional developments. The popularly elected County Councils (*landsting*) may be best described as a special governmental unit for providing health care to the inhabitants of the county, as they have very few other areas of legal competence. Councils have the right to tax the income earners in the county in order to finance the provision of health services.

The presently 288 local governments have a constitutional right to act independently of central government on issues of 'common concern' to their inhabitants and an equally guaranteed right to tax the inhabitants of the municipality.

This institutional structure of Swedish government is reflected in the organisation of environmental policy. Politically, environmental policy is a matter for the central and local levels. The regional level comes in only administratively (see below, Figure 11.2).

Governmental branch Geographic level	Stat	Landsting	Kommun
National	Ministry of Environment (110)*		
	Swedish environmental protection agency (470)*		
	Licensing board for environmental protection (30)*		
	Chemical inspectorate (90)*		
Regional	24 county administrations		
	Environmental units (700)*		
Local			288 municipal councils
			231 have environmental & health protection committees
			57 handle environmental issues in other committees (2100)*

* number of staff

Figure 11.2. Sweden's environmental policy organisation

It took 20 years before 'modern' Swedish environmental policy got a ministry of its own. The 1987 Ministry for Energy and Environment was to have 'an offensive and co-ordinating role within the Cabinet'; it should instil environmental aspects into other policy sectors, and thus also into other Cabinet Ministries (Cabinet Bill 1987/88:85, p. 28). Energy issues were brought back into the Ministry for Industry in 1990, when the Ministry of Environment took over issues of national physical planning from the Ministry of Housing and Physical Planning. There were even aspirations to make this 'specialised' Ministry of Environmental Affairs a 'super ministry' as co-ordinator of environment-related issues at the Cabinet level (Loftsson *et al* 1993:71).

The Ministry of Environment and Natural Resources (ME) handles such policy issues as regulation for environmental protection, nature conservation, control of chemical products and waste management, and the development of economic instruments for environmental policy. The management of Sweden's natural resources is within the ME's jurisdiction, as is protection against radiation and control and supervision of nuclear technology with regard to its potential effects on man and the environment. Under the Natural Resources Act, permit cases which concern particularly large projects with extensive environmental implications can be decided only by the ministry (Cabinet Bill 1991/92: 100, part 15, p. 30 f.). Still, the ministry is a small body with only about 150 employees, three-fifths of whom have an academic background. Governmental expenditures for environmental purposes are far less than one per cent of the total state budget (Lundqvist 1996:272).

During the 1980s and 1990s, the Swedish Environmental Protection Agency's area of responsibility became increasingly dominated by research planning and funding, information gathering, counselling to implementing units and other national agencies, and advice to government on long-term policy issues. Since 1978, SEPA has been responsible for the Program on Measurement and Control of Environmental Quality (PMK). Through systematic monitoring, data gathering and inventories, the Agency tries to create a firm basis for its policy advice and priority recommendations. The SEPA presents itself as working with 'knowledge, actions, and evaluation/monitoring'. Action involves development and implementation of environmental programs and operative goals, as well as new policy instruments. The traditional activity of supervision is mentioned only in passing (SEPA 1996:132). The SEPA thus became an environmental policy 'think tank'; its emphasis was on research-based development of programs and action plans as crucial inputs in the Cabinet's environmental policy. However, the 1996/97 Research Bill brought profound changes. The Agency's research budget was annihilated; funding for environmental research would now come through the Foundation for Strategic Environmental Research, MISTRA. In a longer perspective, SEPA has thus lost its almost total access to research results, which will have profound repercussions on its future standing in the policy formulation process.

The 1990/91 Cabinet Bill on Environment strengthened the environmental responsibilities of the national agencies for road, air and railway traffic, agriculture, fisheries, and forestry. This role is further emphasised in the 1996/97 Bill on the 'State of the Swedish Environment' (Cabinet Bill 1996/97:150). The charters of these agencies now require them to elaborate sectoral plans and programs with precise environmental objectives in order to 'implement nationally determined goals in the most cost efficient way'. The central environmental agencies, SEPA and the Chemical Inspectorate (*Kemikalieinspektionen*), are to provide information and knowledge to the sectoral agencies, and to co-operate actively with them to formulate, follow up and evaluate sectoral agency plans (Cabinet Bill 1990/91:90, p. 66).

The Environment Protection Act of 1969 empowered the regional County Administrations to issue permits for a variety of polluting activities potentially harmful to the environment and to supervise and control such activities within the county borders. The 1988/89 decentralisation of environmental administration set out to create a coherent administrative system for both permit decisions and supervisions. The polluting activities listed in the Environment Protection Act and its Ordinance were regrouped according to their duration, scope, and effects on man and the environment. This cut in half the number of permit cases handled by the National Licensing Board. County Administrations would now take over permit decisions for a total of 4200 plants and industrial activities, and continue to be responsible for supervising their performance.

A 1992 analysis of the County Administrations found that their Environmental Units are the second largest among the specialised CA sub-units, accounting for 11 per cent of total man power and monetary resources within the County Administrations (Cabinet Bill 1992/93:100, Part 14:34 f.). The decentralisation of the late 1980s is successively changing the regional role, from supervision and control to development of new environmental policy instruments and cross-sectoral Regional Environmental Action Programs (Cabinet Bill 1992/93:100, Part 14:55 ff.). A 1997 report from the National Audit Office states that the gradual decrease of grants to CA environmental activities have so undermined the Environmental Units' capacity that they are getting unable to fulfil the environmental policy objectives set by the national government (RRV 1997:10ff.).

The local environmental organisation has been affected by both environmental and constitutional reforms. A 1981 reform enabled municipalities to take over, by agreement, some or all supervision of polluting activities within the municipality. Eight municipalities had taken over supervision across-the-board of polluting activities by 1987, while about 40 had limited their participation to minor polluting activities (SOU 1987:32, p. 200). The 1989 environmental reform transferred the regional permit power concerning 6000 medium-sized and small plants and facilities over to the municipal Environmental and Health Protection Committees (EHPCs). This was in addition to the more than 4000 activities requiring prior notification to the EHPCs (SOU 1987:32, p. 309 f.). Local environmental inspectors are also responsible for supervising the environmental performance of some 93,000 farming units, 115,000 units handling chemicals, 44,000 places where food is processed or marketed, as well as some 3000 water works.

Since this decentralisation of responsibility was not accompanied by any economic compensation from central government, municipalities felt obliged to fire supporting staff in order to afford more inspectors to handle this supervisory workload. Total manpower in Sweden's municipal environmental inspectorates is now about 2100, of whom 1500 are directly engaged in the field. Since they have become loaded with administrative work, the inspectors spend just about 25 per cent of their time on direct field supervision of environmentally harmful activities (SOU 1993:19, pp. 73ff.). This makes for some discrepancy between what is legally expected and what is actually achieved. Many of the environmentally harmful industries are only visited once every three years. Two-thirds of 305 randomly surveyed firms had not implemented required measurements since previous supervisions, or were actively obstructing action through administrative or court procedures (SOU 1993:19, p. 85f.). The environmental policy's share of local government expenditures oscillated around 0.5 per cent in the early 1990s. Local governments thus spent an average of SEK90 (USD12) per capita on the implementation of environmental and health policies (SOU 1993:19, p. 73).

The 1991 Municipal Act allows Swedish local governments to organise their activities as they see fit. Previously mandatory boards and committees can be substituted by a political board and committee organisation tailored to the municipalities' needs. Within one year of the Act, 57 local governments abandoned their Environmental and Health Protection Committees in favour of amalgamating them with the Building and/or Fire and Rescue Committees. As many as 73 municipalities integrated their environmental protection officers in other municipal administrations, such as the Real Estate and/or Building Administrations, the Technical and Infrastructure Offices, or the Municipal Central Chancellery (SOU 1993:19, pp. 43 ff.). A late 1997 report from the Environmental Officers' Union indicates that just about half of Sweden's municipalities now have a specialised EHPC, the rest mixing environmental issues with infrastructure, rescue services, and technical works. The Union's survey reveals that half of the municipalities have decreased funding to environmental protection, while officers in *all* the municipalities say they suffer from increased work loads (MHTF 1997).

To understand and appreciate how this organisation works, and what types of actor relationships are developed, let us take a look at the environmental policy process.

Specially appointed investigative commissions perform a very important role in the formulation of new policy. The frequent combination of parliamentarians, interest organisation representatives, and administrative experts from different administrative levels indeed provides a vehicle for producing consensual proposals. Somewhat less than five per cent of the commission reports (*utredningsbetänkanden*) of the last 20 years have dealt with environmental policy problems, while the share of environmental ministerial committees (*departementsutredningar*) was four per cent (Lundqvist 1996:289).

All Commission Reports do not result in Cabinet Bills to the parliament, and Cabinet Bills are not solely based on such reports. The variety of inputs to environmental policy formulation can be clearly seen in the 1991 Omnibus Environment Bill. Of the 39 different reports, action plans, memoranda etc. used as the basis for the bill, one item was produced within the ME itself. Seven were Commission Reports, and two emanated from ministerial committees. The SEPA alone provided ten 'action plans' and/or memoranda, and was involved in another four inter-agency documents. The Licensing Board for Environmental Protection and the Chemical Inspectorate provided three items. Agencies with 'sectoral' environmental responsibilities produced 12 input reports (Cabinet Bill 1990/91:90, p. 6ff.).

As for policy *implementation*, pollution control legislation spells out an elaborate formal permit procedure, whose characteristic features are intensive *counselling* between polluters and supervisory authorities at the early stages, and *judicial* or *administrative* regulation at the end. For applicants and affected parties the elaborate process of issuing permits provides several possibilities of participation. There is extensive consultation provided for among public authorities at different governmental levels. A very elaborate division of responsibility is established among licensing or scrutinising authorities, based on the assumed importance or potential effects of the polluting facility. Reforms have streamlined the process, by relieving the National Licensing Board of all but the most heavily polluting cases, and by giving the County Administrations a wider circle of industries, facilities and activities for which permits are mandatory. Furthermore, the local-level authorities have become much more important in the early stages of pollution control; their advice has to be sought before small polluting facilities or activities are started or changed. Given the local supervisory agencies' lack of resources to cope effectively with industrial polluters (see above) one may ask whether this elaborate scheme of permits actually works as intended.

To provide adequate information and counsel to polluters applying for permits, the general principles of 'technical and economic feasibility' are clarified through guidelines. The guidelines are subject to review and reformulation in order to reflect scientific, technical, and economic development. The intense co-operation on emission technologies and guidelines between controllers and polluters in the early years established an important feature in the implementation of Swedish environmental policy. Relations between the controllers and the polluters are based on *co-operation*, aiming at *consensus* about pollution control measures. This approach was firmly established by Mr. Valfrid Paulsson, SEPA's legendary Director General during the Agency's first 25 years. A climate was to be created where controllers and polluters would reason and take each others' arguments seriously. One key was *information*; if only each side knew exactly what the other wanted and why, the relations between controllers and polluters would lead to rational and balanced decisions. Another was *trust;* polluters should be relied upon to execute agreed pollution control programs and prescribed control measurements, without day-to-day interference from regional or local environmental officers (Lundqvist 1971:127ff.). The prevailing strength of this approach is revealed by the fact that although 'corporatist' interest representation on agency boards and councils was officially abandoned in the early 1990s in favour of individual 'competence', central administrators, industrial environmental experts, and environmental researchers still make up most board and council members. Environmentalist organisations have almost no access (Uhrwing 1995).

The answers to questions in a 1991 survey of the Chief Environmental Inspectors (CEIs) in Sweden's municipalities suggest two important features of the local implementation process; lively *intermunicipal co-operation*, and the emergence of *local environmental administrative networks*. Three out of five municipalities are members of a water drainage basin association or board. More than half are engaged in formally organised co-operation on air quality management, while about one-third reported joint company co-operation in waste management. More than half of the registered co-operative contacts concern neighbouring municipalities. Almost nine out of ten answers concern municipalities in the respondents' 'own' county. We seem to find an institutionally determined pattern; municipalities co-operate more with non-neighbours inside the same county than with neighbours across the county border, even if they may have environmentally more in common with the latter. There are, however, also patterns which could be predicted from the character of the environmental problem. Water-courses seem to unite neighbouring and non-neighbouring municipalities across county borders. Air quality management, although organised mostly within counties, also involves municipalities situated far from each other. The intermunicipal co-operation is to a large extent informal. Only one out of six CEIs said that the co-operative arrangements actually make decisions which are binding on the participating municipalities. Very few say that the co-operative institutions have the competence to impose taxes or levy charges on the municipalities.

The survey gave a rather clear picture of the relationships among actors and interests at the local governmental level. The CEI *is contacted* by local citizens, by local media, by local environmentalist groups, and by local interest associations. On the other hand, contacts go *from* the CEI to local enterprises, as well as to environmental administrators in other municipalities and at the county level. These intermunicipal contacts are even more intense than those with politicians within the CEIs 'own' municipality. These lively contacts among local CEIs, and between local and county-level Environmental Officers usually concern advice and counselling before decisions are taken. We may thus speak of *a local* and *regional network in environmental affairs*. The strong *professional* emphasis of this network is revealed by the fact that both local media and local environmental groups seem to have difficulties in getting access to local environmental administrators (Lundqvist 1994). On the other hand, reports from the earlier phases of the Local Agenda 21 process indicate a very lively interaction between citizen groups and environmental administrations (Norén 1996; Boije af Gennäs 1994).

The comprehensive challenge: sustainability and climate change

Two general observations can be made from this overview of Sweden's environmental policy. First of all, policy *content* is being pushed from an earlier focus on pollution clean-up and restoration to a more comprehensive and integrative view, where precautionary concern for the sustainability of eco-systems is expected to influence decision-making at all levels and in all sectors of society. Second, the policy *organisation* is still very much built upon the sectorally organised administrative structure in Sweden. Sweden thus has a very small policy-making Environmental Ministry, a specialised central environmental agency being reduced mainly to advice and counsel in its efforts to push pro-environmental views *vis-à-vis* other actors in the policy process, and regional and local environmental administrations with some legal implementation powers but with extremely thinned-out resources at their disposal.

The question, then, is whether this organisational pattern is indeed adequate for the challenges posed by the drive for a comprehensive and integrated policy content. Does the present organisation really 'allow for prior consideration of the environmental impact of every decision capable of exerting a major effect on the environment'? Is the principle of 'sectoral responsibility' really enough to tackle the issues of sustainable development and climate change?

The present environmental policy organisation is evidently *not* structured in accordance with such 'natural' or 'ecological' boundaries – catchment areas, air sheds, connected landscape types – that may be necessary for achieving integrated, sustainable resource management. There are, however, signs of organisational changes along such lines. Over 50 intermunicipal water catchment associations are presently active (Gustafsson 1995), and a proposal for organising ten to fourteen integrated 'water district authorities' was presented in late 1997 (SOU 1997:99). Regional associations for coastal water management cover the entire Swedish west coast from the Norwegian border to the Öresund (Engen 1995). Air quality management associations are at work in more than half of Sweden's counties (Lundqvist 1995b:120). But there is one crucial setback; municipal and county members have not given away decision-making competence which would enable these associations to make binding decisions for these ecologically defined areas.

Another way to achieve integrated and sustainable resource management is through mandatory environmental impact assessment. By way of both EU action and national legislation, the Swedish government has expressed its intent to make EIAs a key factor all the way from strategic long-term issues at the international level to decision-making in the local planning process (Cabinet Bill 1995/96:120, p.67 f.).

Still, there are signs that these organisational measures do not suffice to enforce the principles of precaution and sustainability, particularly when up against formidable economic interests. During the 1990s, Sweden has experienced infrastructure investments on an unprecedented scale. The huge highway ring around Stockholm, the Öresund Bridge between Sweden and Denmark, and the construction of a railway tunnel through Hallandsås in southern Sweden (the latter causing one of the greatest environmental scandals ever in Sweden) are conspicuous examples.

These projects have been launched despite intensive and comprehensive processes of environmental impact assessment all the way up to the Cabinet level. There has been no conclusive evidence as to the environmental acceptability of these projects. Evidently, the 'ecological capacity' within the Swedish environmental policy organisation (Lundqvist 1997) does not suffice to enforce the precautionary principle or promote sustainability.

Admittedly, the question may be wrongly put: It is not 'ecological capacity', manifested as 'impact knowledge', but rather the political power position of economic interests, that ultimately determines the fate of environmental policy.

This strain on the traditional environmental policy organisation is further evident in the development of Swedish climate policy. As indicated earlier, there will be an increase in CO_2 emissions up to 2010. The main contributors are domestic traffic and fossil-fuelled energy production; they will produce CO_2 emissions in 2010 which exceed those of 1995 by fourteen and thirty per cent, respectively (Ds 1997:26, pp. 116 ff.).

This development contradicts the objective formulated in the climate policy decision of 1993; CO_2 emissions were to be stabilised at their 1990 levels by the year 2000 and then decrease (Cabinet Bill 1992/93:179). A variety of administrative, economic and other policy instruments have been used to achieve these objectives. Established in 1993, and reorganised in 1995, the Climate Delegation was charged with co-ordinating research results and information on climate change, and with producing reports and briefs to aid the Swedish Government in its national and international climate policy strategy (SOU 1994:138, p. 1; SOU 1995:96). There have been subsidies to production of electricity and heating from biofuels, as well as to the development of energy saving measures. An environmental classification system for cars has been developed; it now covers sixty per cent of the Swedish car fleet. Energy and environmental taxes on fossil fuels used for heating and transports have been gradually increased. It is, however, particularly notable that these taxes have been levied at much lower rates on Sweden's industries (Ds 1997:26, pp. 59 ff.). A 1996 Cabinet Bill suggested a doubling of the CO_2 tax for 'electricity intensive' industries to make it half of that for other CO_2 emitters. There would, however, still be a maximum limit in relation to the industries' sales value. Furthermore, the tax increase would only take effect if the EU Commission gave its blessing; this came in 1997 (Cabinet Bill 1996/97:29, *Från Riksdag & Departement* 14/1997). The above-mentioned investments in railroads have been presented as part of the overall strategy to reduce climate-affecting emissions.

The single toughest blow to the prospects of achieving the climate policy objectives came with the February 1997 Energy Policy Deal, struck between the Social Democratic Government and the Centre and Left Parties. The shutdown of the Barsebäck nuclear power station between 1998 and 2001 creates a need for electricity in spite of the massive support to programs for energy saving (Cabinet Bill 1996/97:84). Much of this electricity will emanate from fossil-fuelled combustion plants in Denmark and Poland. Sweden's radical stance on nuclear energy will thus contribute a further five million tons of CO_2 up to 2010 (Ds 1997:26, p. 121).

It is evident that the Energy Deal has presented problems to Swedish environmental policymakers, both in relation to the EU and the environmental opinion. For while the 1993 climate policy decision leaves open *how much* the post-2000 Swedish decrease of greenhouse gas emissions should be in per cent, the Swedish Government expresses satisfaction with the eight per cent EU reduction by 2008-2012 agreed upon in Kyoto (Cabinet Office 1997:3). However, even with the 1996 arsenal of climate policy measures applied in full, Sweden's total emissions are expected to increase by sixteen per cent up to 2010 (Ds 1997:26, p. 23). This may lead to some tough negotiations with the EU Council on exactly how large a burden of the EU reductions should actually be borne by Sweden. But even if that burden is higher than the average for the Union, it may still not suffice to achieve the national 1993 policy objective. If so, the originally 'environment-oriented' nuclear shutdown's ecological consequences may politically discredit the 'eco-sustainable' image so eagerly promoted by the Social Democratic Government (see below).

The political answer: Towards centralised 'ecological modernisation'

Economic growth, employment, and the state budget deficit have been the core political issues in Sweden during the entire 1990s. This has had repercussions for the political standing of the environmental issue. In fact, the centrality of the environmental issue 'mirrors' that of the Swedish economy. When the economy presents 'no problems', the environment ranks high in citizens' minds. But during the downturn and crisis of the Swedish economy in the first half of the 1990s, the economy rose and the environment plunged in the citizens' ranking (Bennulf 1994). The Greens have not been able to influence policy making since their comeback in the parliament; after the 1994 elections, the parliamentary situation enabled the Social-Democratic minority government to find enough support from other parties, even across earlier 'enemy' lines.

Up until the economic downturn, Sweden's environmental policy record was quite impressive. The country managed to set up an organisation and make use of instruments which contributed to a comparatively high capacity to understand different environmental problems and quite successful action to solve the. Long-term intersectoral planning by national and regional agencies, policies based on problem-oriented environmental research, policy implementation by way of a widening variety of policy instruments and involving every level of government all contributed to this record.

In its attack on the first-generation problems of large-scale industrial pollution, the government and responsible authorities could build on a long-standing Swedish tradition of close co-operation with well-organised, responsible target-group interests. A consensual and co-operative policy style developed in the late 1960s, and has continued ever since. The target groups have had a well entrenched system of 'opposite numbers' dealing with their 'peer' environmental administrators at different levels of government on an equal professional basis. More recently, this consensus approach has manifested itself in the administration of programmes for recycling. Targets are set by the government, while the affected branches are trusted with implementing the policy in such organisational ways as they see fit (Lundqvist 1996:300 f.).

As pointed out above, however, the demands posed by such second-generation problems as climate change and ecological sustainability – in combination with the extremely stringent measures to overcome the Swedish budget deficit – have put strains on the environmental policy organisation. The political answer has been centralisation; while responsible professional agencies and boards at all levels have suffered from budget cuts, the central government has taken a firm political grip on policy developments.

A case in point is the Social Democratic drive to make Sweden 'an internationally driving force and a forerunner in the endeavours to create an ecologically sustainable development'. This was introduced as a new 'noble mission' for the Swedish Social Democratic Party by its new leader Göran Persson in his acceptance speech to the extra Party Congress in March 1996, and reiterated in the presentation of his new Cabinet's platform (Parliamentary Record, March 22, 1996).

Words were swiftly followed by action. At the September opening of the 1996/97 parliamentary session, the new Prime Minister announced (1) the beginning of the phase-out of Swedish nuclear power, (2) a comprehensive environmental and natural resource legislation, and (3) a strong investment program for an ecologically sustainable infrastructure (Parliamentary Record, September 17, 1996).

An important step towards integrated political leadership came in January 1997 with the creation of a Delegation for Ecologically Sustainable Development (DESD). Situated within the Cabinet, it consists of the Ministers of Environment, Agriculture, Taxation, Basic Education and the Junior Minister of Labour. Already in March 1997 this Ministerial Delegation presented a report called *A sustainable Sweden*, calling for (a) an annual Sustainable Development Report in connection to the State Budget, (b) a set of environmental objectives for all relevant sectors in society, (c) a 'Greening of National Agencies' Program and (d) a Sustainability Investment Program (SIP). To run from 1998 to 2004, the SIP allocates one billion SEK to eco-cycle adjustment of built environments and infrastructure, nine billion SEK to eco-cycle transformation of the Swedish energy system, and six billion SEK to sustainability investments by municipal governments (DESD 1997).

Again, swift action followed. In the April 1997 Economic Bill to the Parliament, the Delegation's proposals were fully accepted; as much as 12.6 billion SEK were allocated for the period 1998-2000 to local investment programs and infrastructure and energy conversion projects (Cabinet Bill 1996/97:150, pp. 87 ff.).

The drive for political centralisation is quite obvious. In June 1997, the Delegation sent a proposal to the Cabinet on the specific demands to be met by investment programs receiving support (Ministry of Environment 1997).

It is the Cabinet, *not* the Ministry of Environment or the SEPA, that sets the criteria for and makes the decisions on applications for eco-sustainability subsidies (DESD 1997). Another important driving force in this process of centralisation is Sweden's membership in the European Union. The Social Democratic Government stated in November 1994 that its environmental policy towards the EU would be one of *activism* (Ds 1994:126). However, the recent cumbersome work on the new Environmental Code, as well as the preparation of Swedish stands on common EU policy formulations, indicate that *adaptation* is a more adequate description. Out of political necessity, the centre of gravity thus moves towards the upper levels of government.

But political centralisation and international adaptation are not exactly 'winners' in the Swedish opinion, half split over the EU, with a massive distrust towards politicians, and marred by the all-time-high unemployment figures. It thus becomes crucial how the massive measures to achieve an 'ecologically sustainable development' are legitimised. Prime Minister Persson's vision puts a strong emphasis on ecological sustainability as the fourth step in building the *folkhemmet*, the 'people's home', which is one of the most revered symbols in Swedish politics ever since the 1930s. In his April 1997 speech to the Social Democratic Youth and the Construction Workers' Union, the Social Democratic Prime Minister uses what Maarten Hajer has labelled 'ecological modernisation' arguments (Hajer 1995:25 ff.). He presents sustainable development as a 'win-win' issue; it creates demands for investments in new infrastructure and technology, which in turn creates new growth and job opportunities (Göran Persson 1997).

The issue, then, for Sweden's environmental policy, is whether and if so, how far, such symbols will suffice to meet the environmental challenge.

References

Asp, K. (1990) 'Medierna och valrörelsen', in M. Gilljam, & S.Holmberg *Rött blått grönt. En bok om 1988 års Riksdagsval*. Stockholm: Bonniers.
Bennulf, M. (1994) *Miljöopinionen i Sverige*. Lund: Dialogos.
Bennulf, M. & Gilljam, M. (1991) 'Snacka går ju – men vem handlar miljövänligt?', in L. Weibull, & S. Holmberg, (eds.) *Politiska opinioner*. Göteborg: Department of Political Science, SOM-rapport 7.
Bennulf, M. & Holmberg, S. (1991) 'The Green Breakthrough in Sweden', *Scandinavian Political Studies* 13:165–184.
Bennulf, M. & Holmberg, S. (1992) 'Förbättras eller försämras miljön i Sverige?', in S. Holmberg, & L. Weibull, (eds.), *Trendbrott?* Göteborg: Department of Political Science, SOM-rapport 8.
Bennulf, M. & Lundin, U. (1997) 'Gult ljus för bilismen', in L. Nilsson (ed) *Nya landskap*. Göteborg: Department of Political Science, SOM-rapport 19.
Boije af Gennäs, U. (1994) *Miljöprojekt i kommunerna – en sammanställning*. Stockholm: Kommunförbundet.

Cabinet Bills; all to be found in Swedish Parliamentary Record (*Riksdagstrycket*). Stockholm: Swedish Parliament.

Cabinet Office (1997) *Rapport från det tredje partsmötet med klimatkonventionen i Kyoto*. Stockholm: Regeringskansliet, PM December 1997.

Ds 1994:126 (Ministerial Committee Report) *Det svenska miljöarbetet i EU*. Stockholm: Ministry of Environment and Natural Resources.

Ds 1997:26 *Sveriges andra nationalrapport om klimatförändringar, i enlighet med Förenta Nationernas ramkonvention om klimatförändringar*. Stockholm: Ministry of Environment and Natural Resources.

Engen, T. (1995) *På kommunernas villkor. Lokal organisering och interkommunalt samarbete om Västerhavets miljö*. Göteborg: Department of Political Science. Research Report, December.

Från Riksdag & Departement (1995) 'Så fördelades rösterna i EU-valet' # 29:13.

Från Riksdag & Departement (1997) 'Dyrare el och olja' # 14:7.

Gilljam, M. & Holmberg, S. (1990) *Rött blått grönt. En bok om 1988 års Riksdagsval*. Stockholm: Bonniers.

Gilljam, M. & Holmberg, S. (1993) *Väljarna inför 90-talet*. Stockholm: Norstedts.

Gustafsson, J.-E. (1995) *Avriningsområdesbaserade organisationer som aktiva planeringsaktörer*. Stockholm: Royal Institute of Technology.

Hajer, M.A. (1995) *The Politics of Environmental Discourse. Ecological Modernisation and the Policy Process*. Oxford: Clarendon

Holmberg, S. & Gilljam, M. (1987) *Väljare och val i Sverige*. Stockholm: Liber.

Jamison, A. et al (1990) T*he Making of the New Environmental Consciousness. A Comparative Study of the Environmental Movements in Sweden, Denmark and the Netherlands*. Edinburgh: Edinburgh University Press.

Loftsson, E. et al. (1993) *Svensk miljöpolitik*. Lund: Studentlitteratur.

Lundqvist, L.J. (1971) *Miljövårdsförvaltning och politisk struktur*. Lund: Prisma/ Verdandidebatt.

Lundqvist, L.J. (1994) 'Environmental Cooperation among Swedish Local Governments. Professional Networks and the Evolution of Institutions for Collective Action', *International Journal of Public Administration* 17:1733–1766.

Lundqvist, L.J. (1995a) 'Municipal Sewage Treatment in Sweden 1960–1990. From Bans on Bathing to Schools of Salmon', in M. Jänicke & H. Weidner (eds) *Successful Environmental Policy – A Critical Evaluation of 24 Cases*. Berlin: Edition Sigma, pp. 43–59.

Lundqvist, L.J. (1995b) 'Kommunal samverkan: Ett strategival i tiden', in L.J. Lundqvist & J. Pierre (eds) *Kommunal förvaltningspolitik*. Lund: Studentlitteratur, pp. 110–134.

Lundqvist, L.J. (1996) 'Sweden', in P. Munk Christiansen (ed) *Governing the Environment: Politics, Policy and Organisation in the Nordic Countries*. Copenhagen: Nordic Council of Ministers, Report NORD:5, pp. 259–338.

Lundqvist, L.J. (1997) 'Sweden', in M. Jänicke & H. Weidner (eds) *National Environmental Policies: A Comparative Study of Capacity-Building*. Berlin: Springer Verlag, pp. 45–72.

MHTF (1997) *Resurser och organisation för ett effektivt miljö- och hälsoskyddsarbete. Resultat av en enkätundersökning 1997.* Eslöv: Miljö- och Hälsoskyddstjänstemannaförbundet (MHTF).

Ministry of Environment (1997) '5,4 milljarder till hållbara Sverige'. Press Release 44/97, June 18.

Norén, Y. (1996) *Agenda 21 i Sveriges kommuner.* Stockholm: Kommunförbundet.

Persson, G. (1997) 'Så gör vi Sverige till ett föregångsland för ekologisk hållbarhet!' Speech delivered at the SSU and Byggnads miljöseminarium, April 11 (available at *http://www.sb.gov.* under *'Statsministern', 'Tal m.m.'*).

RRV (1997) *Länsstyrelsernas miljövårdsarbete i förändring.* Stockholm: Riksrevisjonsverket, RRV 1997:47.

SCB (1994) Communication to the author, September through November, 1994.

SEPA (1994a) *Hur har det gått? Redovisning av myndigheters arbete mot 9 miljömål.* Stockholm: SEPA.

SEPA (1996) *Forskning och utveckling för bättre miljö 1996.* Stockholm: SEPA.

SOU 1987:32 (Statens Offentliga Utredningar, Investigative Committee Report) *För en bättre miljö.* Stockholm: Ministry of Environment and Natural Resources.

SOU 1993:19 *Kommunerna och miljöarbetet.* Stockholm: Ministry of Environment and Natural Resources.

SOU 1994:128 *Lokal Agenda 21 – en vägledning.* Stockholm: Environmental Advisory Council.

SOU 1994:138 *Rapport från Klimatdelegationen 1994.* Stockholm: Ministry of Environment and Natural Resources.

SOU 1995:96 *Jordens klimat förändras. En analys av hotbild och globala åtgärdsstrategier.* Stockholm: Ministry of Environment and Natural Resources.

SOU 1997:11 *Skatter, miljö och sysselsättning.* Stockholm: Ministry of Finance.

SOU 1997:99 *En ny vattenadministration. Vatten är livet.* Stockholm: Ministry of Environment and Natural Resources.

Uhrwing, M. (1995) 'Intresserepresentation i brytningstid. En studie av intresserepresentation i några miljöpolitiska organ'. Göteborg: Department of Political Science, mimeo.

Vedung, E. (1989) 'Sweden: The "Miljöpartiet de Gröna"', in F. Müller-Rommel (ed.) *New Politics in Western Europe. The Rise and Success of Green Parties and Alternative Lists.* Boulder: Westview Press.

Vem gör vad i Miljösverige 1994 (1994). Stockholm: SEPA and Natlikan.

Chapter 12

The European Union: Environmental policy and the prospects for sustainable development

Alan Butt Philip

The European Union's 15 member states extend across most of the western part of Europe but contain very contrasting natural and man-made environments. The population density of the Benelux countries, with their heavily urban environments, could not be more different from the sparsely populated desolation of the areas north of the Arctic Circle to be found in Sweden and Finland. These very different environmental challenges demand separate as well as common responses to match differences in climatic, topographical and geological circumstances, as well as differences of degree in urbanisation and economic development. Despite over two decades of Community action, the EU's own recent official assessment of Europe's environment, provided by the new European Environment Agency, was on balance pessimistic (Stanners and Bourdeau 1995), echoing its reference in 1993 to the 'slow but relentless deterioration ... of the environment' (CEC, *Towards Sustainability* 1992).[1]

Total carbon dioxide emissions from fossil fuels are still rising in the EU, and while sulphur dioxide emissions per head are falling, in most EU member states nitrous oxide emissions per head are also still rising. The quantity of municipal waste generated continued to rise in the 1980s, but by 1990 substantial quantities of glass and paper board were being recovered (67 per cent of glass and 50 per cent of paper/board in the Netherlands; 21 per cent of glass and 31 per cent of paper/board in the United Kingdom (Eurostat 1995)). The fifth EU Environmental Action Programme, published in 1992, noted that a 20 per cent increase in carbon dioxide emissions was likely to occur between 1987 and 2010 if there were no change in current energy demand growth rates. In the 1990s alone car ownership was set to rise by a quarter within the EC-12, and car mileage by one-sixth (CEC 1992b:23).

The member states also have very different environments to manage and distinct political cultures which determine their separate policy responses, even though they share many of the same problems such as air pollution, waste management, and water quality inadequacies. The extent of their dependence on agriculture varies from 13.5 per cent of GDP (24 per cent of persons employed) in the case of Greece to 1.3 per cent of GDP (2.2 per cent of persons employed) in the case of the United Kingdom in 1991. Population density varies from as high as 366 per square kilometre in the case of the Netherlands to as low as 15 per square kilometre in the case of Finland (1994 estimates). The level of economic development as measured by GDP per head also varies greatly between EU member states, with Luxembourg and Belgium achieving the highest measures, 25,422 and 17,946 Ecu respectively per head while Portugal and Greece had the lowest measured GDP per head, 10,935 and 10,000 Ecu respectively (all figures quoted are at current (1997) prices and purchasing power parities for the year 1993): regional variations in GDP per head are even greater within the EU, at about seven to one.

Accordingly, the European Union's environment policies have to accommodate the problems of highly industrialised and urbanised states, suffering substantial environmental degradation, with individual member states at different stages in accepting and implementing a supposedly common environmental agenda. At the same time other member states are trying to catch up with the leading EU members in economic terms, which feel unable to give the protection of their natural environment the same priority.

EU institutions

No discussion of policy developments in the European Union would be complete without some review of the Union's institutions and its political structure.[2] Policy leadership is most commonly provided by the Commission, headed by a President, nominated by all the member states jointly, and by 19 other Commissioners, nominated by individual member states. This college of Commissioners oversees the work of all the directorates-general and is accountable *en bloc* to the European Parliament. The portfolio for the environment commissioner usually also includes responsibility for nuclear safety and civil protection, since these functions have traditionally been combined in the same directorate-general (DG XI).

The roles of the Commission combine political and administrative functions. It has the sole right of policy initiative, which gives it a crucial agenda-setting role in EU decision-making. As the executive arm of the Union, it carries out EU decisions as required by EU law. The Commission also acts as the watchdog of the EU both to keep the spirit of the treaty base alive within the EU's institutional structure and to ensure compliance with EU law in the member states by governments and other actors.

The Commission may propose legislation for the EU but all such legislation requires the approval or acquiescence of one or more of the other EU institutions. The most important of these is the Council of Ministers, which brings together ministers of all the member states who share the same portfolio responsibility. The work of the Council is organised by a separate Council secretariat based in Brussels and overseen by the member state occupying the Presidency of the Council, which rotates between the members every six months. The Presidency acts as spokesman for the Council, chairs all meetings of the Council and its committees, determines the agenda of such meetings, and mediates between states when there are disputes. Below the ministerial level is the committee of permanent representatives (COREPER) which brokers most of the deals between national governments and which provides day-to-day continuity, whatever subjects are discussed and whatever governmental changes at national level may bring. Below COREPER sit up to 200 working groups of national officials giving line-by-line consideration to legislative proposals. It is in the various committees of the council that the scrutiny of legislation is most closely exercised, much more so than in the European Parliament (EP).

The EP comprises members directly elected by voters in all 15 member states. The EP is elected for a five-year fixed term and meets in plenary session for one week every month in the calendar year except August, its committees also meeting monthly. One of these committees is devoted to environmental issues. Members of the EP (MEPs) sit by loosely arranged party groups rather than by nationality, and it is the party-group structure that dominates the organisation of the Parliament's work. As no one party group has a majority among MEPs, the EP is mostly run by a cabal of representatives of the two largest party groups, the Socialists and the Christian Democrats.

The EP's powers now include the right to dismiss the college of Commissioners en bloc, the right to reject the EU's annual budget and to determine some of its expenditure in detail, the right to block some types of EU decisions and to make others jointly with the Council (such as many environmental policy decisions).

The European Court of Justice (ECJ) is the fourth major EU institution. Its functions are to interpret the founding treaties of the EU and all the decisions flowing from them. This occurs either through cases which the Court hear directly, these being initiated often by the Commission, or by reference from national courts hearing cases where a preliminary ruling from the ECJ is needed to help determine points of European law relevant to each case. In effect, the ECJ acts as the supreme court of the EU and, where there is a void in the EU's legislation armoury, the Court may choose to develop policy case by case on the basis of principles clearly articulated in the treaties.

Other institutions also play minor but not necessarily insignificant roles in the decision-making system of the EU. The Court of Auditors, for example, scrutinises in detail the way money is spent from the Union's budget and whether value for money is achieved in terms of meeting the EU's declared policy objectives.

The EU treaty base

The treaty base of the European Union exercises vital and enduring control over its operations. The EU is based upon a binding legal order to which all member states must agree before accession to the Union and which they alone can change significantly by means of treaty amendments, which have to be agreed upon by the EU institutions themselves and ratified by every member state. The treaty on European Union, signed at Maastricht in February 1992, and finally implemented in November 1993, sets up what is commonly known as a three-pillar structure of governance for the Union. The first pillar is occupied by the European Community (EC): decisions made within this pillar are subject to the complex decision-making rules and the scrutiny of all the institutions listed above. A very high proportion of all policy making by the Union occurs within this framework, including the environment, transport, energy and single-market policies, and economic and monetary union. The intergovernmental second and third pillars cover common foreign and security policy and co-operation on judicial and home affairs respectively.

The treaty base of the EU's predecessors did not recognise the specific category of environmental policy until the Single European Act was implemented in 1987. Up to that point, the broad objectives set for the EC placed primary emphasis upon economic integration among the member states and the economic growth this was expected to bring about. Even so, a 'continuous and balanced' expansion was set as the EC's goal and some recognition of specific regional problems, improved living and working conditions, and transport policy needs was included. The principal legal basis used for the development of EC environmental legislation was, however, the residual Article 235 of the EEC Treaty which authorises the adoption of specific measures necessary to further the common market, and also Article 100 of the same treaty which provides for harmonisation measures to be adopted. Decisions using either article as the legal base required unanimity among the member states up to 1987, and Article 235 decisions still do.

The Single European Act (SEA) provided, for the first time, an explicit competence for the EC to deal with environmental issues in Articles 130R, S and T. These spell out the principles that should underlie EC environmental policy, accept that member states should not be required to lower their existing environmental standards in the interests of harmonisation, and apply the test of *subsidiarity* to whether the Community should be legislating at all in a given field. The Community is only permitted to take action in situations where policy objectives are likely to be better attained by actions taken at the Brussels level than at national levels.

The Treaty on European Union (the Maastricht Treaty) generalises this subsidiarity provision so that it covers EC decision-making in all areas where the Community does not have exclusive competence. It restates the principles which must be respected by EC environmental policy, first set out in the SEA, sets objectives for this policy, enables the EU to make binding international agreements on behalf of the member states and confers additional powers to decide environmental policy upon the European Parliament. The Maastricht Treaty commits the EC to promoting throughout its territory a harmonious and balanced expansion of economic activities and 'sustainable non-inflationary growth respecting the environment ...' (Article 2). The Community is also now charged with ensuring that its environmental policy objectives are applied consistently in other policy areas.

The treaty of Amsterdam, signed in June 1997, finally committed the European Union to the principle of sustainable development and also made it easier for individual member states to maintain higher environmental standards than those set for the EU as a whole. More powers of co-decision on environmental policy issues were handed to the European Parliament.

The scope and development of EU environmental policy

Environmental policy in the EC, and now in the EU, does not have a strict definition and frequently cuts across other policies which the Union pursues. Although the emerging treaty basis for environmental policy makes the scope clearer, in the late 1960s, the EC was already justifying environmental legislation on the basis of furthering the internal market or of seeking a general improvement in living conditions. Since that time, there has been a continuing tension between the economic and environmental policy objectives of the emerging European Union which has yet to be fully reconciled.

It is evident that the European Union's capacity to shape and to influence environmental policy in the member states is of relatively recent origin. Member states, upon accession to the Union, have brought into this community national environmental attitudes and objectives at widely different stages of development. In a few cases, member states had almost no corpus of public policy initiatives on environmental questions to contribute and they have had to accept EU policies *en bloc*.

The EC made its first decisive step towards building its own environmental policy at the Paris Summit in October 1972 when it declared that, as referred by Wilkinson (1990), economic expansion is not an end in itself. The Community's firm aim should be to enable disparities in living conditions to be reduced. As befits the genius of Europe, particular attention will be given to intangible values and to protecting the environment, so that progress may really be put at the service of mankind.

This then led to the adoption of the first environment action programme (EAP) in 1973 (CEC 1973) which proposed three sets of actions; to minimise and prevent pollution, to improve the existing European environment, and to pursue EC policy objectives at other international levels.

The shift in direction which these decisions marked reflected the growing world-wide consciousness and concerns for general environmental degradation, climate change and the consumption of irreplaceable natural resources such as oil and other fossil fuels in the cause of economic gain. These had come to a head at the level of policy with the convening of the UN Conference on the Human Environment in Stockholm in 1972. There was also a recognition that national environmental policies could distort the operation of the common market. The Commission presented its proposals in April 1973 and these were adopted as a legally non-binding declaration later that year.

The first EAP, which has been followed up by a series of five-year plans, was based largely on Article 2 of the EEC Treaty, using the member states' vague commitment to 'harmonious development' as a justification for environmental action. This was an insubstantial legal foundation, and had to be made politically acceptable by linking environmental policy to the concept of helping 'to harness expansion to the service of man' (CEC 1974:235); in other words, economic growth remained the priority. However, despite the declamatory nature of the EAP, the expression of political will which it represented allowed the Commission to bring forward substantive legislative proposals in the knowledge that, as long as they stayed within the bounds of the politically acceptable, there was a good chance that they would be turned into EC law. The first EAP was, therefore, 'most important' (Johnson and Corcelle 1989), or, as the Commission put it, 'an important political act' which 'defined the terms of a Community policy on the environment' (CEC 1974:235)[3].

The first EAP set out the three major principles of EC environmental policy: the necessity of preventative action; the responsibility of the polluter for environmental damage and its rectification; and the need for action to be taken at the most appropriate level (an early prototype of subsidiarity). Eleven principles for action were also laid down; these were based on agreements reached by national environment ministers at an intergovernmental conference in Bonn, held in October 1972 (Haigh 1989).

Despite the talk of prevention, however, the EAP was largely curative in nature, and in its emphasis on improving the quality of life, rather than on the environment, it did nothing to distract attention from the overriding goal, that of economic growth (Liefferink, Lowe and Mol 1993:3).

The second EAP was launched in 1977 (CEC 1977). It was little more than a continuation and extension of the first EAP, but was adopted by a Council Resolution, rather than a Declaration, representing a slightly higher legislative status.

The third EAP duly followed in 1983, also adopted by means of a Resolution (CEC 1983). By now, despite the weakness of environmental policy's legal base, over 100 legislative measures had been adopted. The uncertain legal status of environmental Directives had also not prevented the European Court of Justice (ECJ) from hearing Article 169 actions by the Commission against member states who were guilty of non-implementation. Furthermore, controversy over the EC's competence in dealing with external-relations aspects of environmental policy seemed 'to have disappeared silently' (Koppen 1988). In fact, 'in many critical areas of environmental policy-making Brussels had come to play as important a role as the nation state' (Vogel 1993).

However, basing legislation almost entirely on Articles 100 and 235 forced the Commission to adopt an economic, customs-union-related approach to the environment, and the unanimity rule in the Council made it difficult to get effective measures agreed. Nevertheless, a very substantial body of legislation came through. While the third EAP could do nothing to remedy this situation, it was notable for placing more emphasis on the prevention of environmental degradation, and on a more integrative approach to the environment and the economy (Johnson and Corcell 1989:12, Koppen 1988:21). The EAP stressed that the preventative side of environment policy must be strengthened, and the environmental dimension integrated into other EC policies. It also foresaw that environmental policy initiatives could contribute to economic growth and job creation in non-polluting industries and services.

Importance was also attached to the complementing of regulation, by the use of other types of measure: for example, the idea of environmental impact assessment (Liberatore 1991). And it was suggested that environmental policy could have a positive effect on other policy areas, especially via job creation and industrial innovation. The third EAP can, therefore, be seen as indicating a gradual shift towards a policy which sought to affect fundamental economic behaviour, in favour of a more environmentally aware and sustainable future, and this led inexorably in the direction of policy instruments that were less rooted in 'command-and-control' systems.

However, it was clear by 1983 that the EC's approach to the environment was fundamentally flawed. Environmental policy as an adjunct to the common market meant patchy, *ad hoc* development. Where economics and environment overlapped in a mutually supportive manner, progress could be made: elsewhere, advances were hard to come by (although not unknown, as directives on wild bird protection demonstrated). There was immense difficulty in drawing unequivocal conclusions on the basis of existing scientific knowledge, and yet industry would frequently demand categorical evidence before it would concede ground on environmental legislation. As a result, a constant and arbitrary trade-off had to be made between short-term costs and long-term benefits, with the latter usually losing out. In addition, the diversity of national approaches to the environment, and of national institutional and administrative structures, made creating a policy which could, or would be implemented uniformly across the EC virtually impossible (Rehbinder and Stewart 1995; CEC 1984). Above all, unanimity for environmental decision making in the Council led either to no decisions at all, or decisions based on the lowest common denominator. This generally made it possible to legislate for broad framework objectives only, leaving enormous scope for member states to interpret and to implement directives as they wished.

In signing up to the SEA, the EC's member states had set the course of future environmental policy for the Community by defining its three founding principles: that prevention, not cure, should be the preferred course of policy; that rectification of problems should occur at source; and that the 'polluter pays' principle should be applied.[4] The fourth EAP was adopted shortly after the entry into force of the SEA (CEC 1987). The sequence of SEA and EAP was important: environmental policy had in some ways 'come of age' (Koppen 1988:24).

The fourth EAP signalled an intention to make environmental policy more coherent, and to use Article 100a wherever possible, to force legislation through (and to involve the EP via the co-operation procedure), and subsequently to enforce implementation. Whether this would be politically feasible was another question, but by 1987 environmental policy had been transformed from the ragtag policy of the early 1980s into what appeared to be the brink of a dynamic new phase in its development, adopting an increasingly cross-media approach to make the environment both a policy in its own right and an integrative part of all other EC policies (Liberatore 1991). Priority areas for action were identified, such as waste management, biotechnology, atmospheric and marine pollution, and the enforcement of EC environmental laws.

At the beginning of the 1990s, EC environmental policy was very largely regulatory in nature, and tended to shrink from controversy: it was flexible, but at the price of coherence, and avoided imposing costs on Community industry (Liberatore 1991) There were powerful arguments for reforming the Treaty once again to give environmental policy a proper, single legal basis, with adequate administrative, political and financial support (Wilkinson 1990). Also there was a rising groundswell of opinion in favour of a shift away from command and control remedies to policy instruments which would enable markets to work in a more environmentally sensitive manner.

The weaknesses in environmental policy were manifest. The EC could be guaranteed to act effectively only where environmental and economic issues, and member states, converged. But the political leaders of the EC who came together in 1991 for the intergovernmental conference (IGC) which led to the Maastricht Treaty had other priorities (Verhoeve, Bennett and Wilkinson 1992:11). The resulting Treaty on European Union (TEU) failed to resolve many of the ambiguities: when the Community was faced with a tension between environmental protection and economic integration, 'it chose the latter over the former' (Vogel 1993:127). The TEU did, however, enshrine the principle of 'sustainable ... growth respecting the environment' (revised Article 2), and gave the Court of Justice the power to impose fines on member states for non-implementation. On the other hand, the TEU's attachment to subsidiarity had the potential to form the basis for member state opposition to further EC action. Sustainability was ill-defined, and there was no indication as to how the environment was to be integrated into other EC policies (Verhoeve, Bennett and Wilkinson 1992:16). The range of decision-making procedures which could now be applied to different types or aspects of 'environmental' legislation presented policy makers with a legal minefield.

Nonetheless, environmental policy was now much stronger. The launch of the fifth EAP, to run from 1993 to 1999 and significantly entitled *Towards Sustainability*, opened with a frank admission that current measures were inadequate. Sustainable development 'will require significant changes from current patterns of development, consumption and behaviour' (CEC, *Towards Sustainability* Vol I, 1992). The Commission, in its report on the state of the environment, (*ibid.*: Vol II.) indicated that environmental problems were growing, and that the time had come to attempt a far more rational and coherent approach to the issues: a 'qualitative step forward' (Ripa di Meana 1993:3). The fifth EAP would, therefore, aim primarily to be preventative and proactive, and would attempt to induce:

> changes in current trends and practices which are detrimental to the environment, so as to provide optimal conditions for socio-economic well-being and growth for the present and future generations. (CEC, *Towards Sustainability* Vol I., 1992:5)

In other words, the EC was aiming for sustainable practices which would remove the causes of future pollution. Five priority areas for action were identified (industry, energy, agriculture, transport and tourism) and a broader range of instruments was to be used to facilitate change. As well as regulation, the EC would now make much greater use of market-based instruments, with the intention of internalising costs without introducing comparative disadvantage. This would enable a more 'bottom-up' approach to be taken to the protection of the environment, involving 'all levels of society' (*ibid.:* 9). One of the Commission's primary aims was to campaign for information and awareness-building among the general public (*ibid.:* 5). This, combined with the desire to involve the public more by means of market-based instruments, led logically to the adoption of policy instruments such as the EU eco-labelling scheme (*ibid.:* Vol II:69).

Current environmental politics

It would be easy to see the entry into force of the Maastricht Treaty in November 1993 as the main watershed in the power relationships between national governments and the EU institutions, as this was the formal moment when the requirement for unanimity among member states before adopting environmental legislation was significantly weakened. Yet this would wholly misread the balance of power between actors in the environmental policy game. In reality, the Commission has been the driving force behind the development of EU environmental policy and until the mid-1990s, the Commission for the most part was able to get through the measures it promoted, despite the need for unanimity in the Council of Ministers. The Commission, its room for manoeuvre in many policy areas blocked by disagreements among the member states, notably in realising the common market itself, found that it was pushing at an open door in developing an environmental agenda for the EC in the 1970s. Many member states had no environmental policies to speak of in place, but a few in northern Europe were fast developing their own. These few states wanted to generalise their national policies across all member states to minimise the possible loss of competitiveness arising for home industries from meeting the additional costs of compliance with new environmental regulations. There were international obligations to be fulfilled, especially to the United Nations, where the legal order of the EC offered the best possible means of realising such commitments across much of the continent of Europe in a transparent way. Public opinion, in survey after survey, has recognised since the early 1970s that environmental protection in Europe ought to be organised through the EC/EU because of the inherently cross-border nature of the problems to be addressed (e.g. pollution of air and water) given the close proximity of member states to each other.[5] The legitimacy of EC activities in environmental policy was rarely questioned in the early days of its development, even when measures to protect wild birds, habitats and the quality of drinking and bathing water were proposed, where cross-border issues were minimal. Indeed, the Commission can usually expect support for stronger EU-wide environmental policies from the European Parliament, the majority of whose members often press the Commission to be bolder in its initiatives, and from the governments of member states anxious to establish their green credentials.

The direct elections to the European Parliament have demonstrated that green issues have a substantial resonance with voters. The 1989 elections saw the vote for green parties rise to as high as 15 per cent in the United Kingdom and 11 per cent in Belgium, with major blocks of Green MEPs elected to represent France, Germany and Italy. By 1996 the Green group of MEPs could count on 25 supporters drawn from nine member states in the 626-member parliament, although support for green causes is strong among MEPs from most of the political groups.

Among the member states in the Council, there is a solid group of usually six member states (Germany, Netherlands, Denmark, Sweden, Finland and Austria) that favour strong environmental policies. The other nine states show varying degrees of ambivalence about the feasibility, desirability and cost of adopting and complying with such legislation.[6] Typically the least economically developed states (such as Spain, Portugal and Greece) have the greatest reservations about strengthening EU-wide environmental regulations, partly on grounds of cost (and the cost to competitiveness) and partly because they do not have to face serious domestic political pressures from environmentalist movements The six green states, in contrast, are anxious to respond to the greener demands articulated within their political systems, without hurting their international competitiveness, so it is in their economic interests to spread the higher costs of compliance by means of stricter environmental laws across the whole European Union. The support of pan-European multinational firms is also often forthcoming for such an approach, since it is usually in their operational interests to have only one set of EU rules to comply with rather than 15 sets of national environmental regulations (Porter 1995:44–76, 194–234). In the middle of this 'leader-laggard' spectrum of states stand the United Kingdom, France, Luxembourg and Belgium, which find it harder to strike a clear balance between environmental and economic development objectives. Some of the divisions between member states occur not so much because of the economics/ environmentalism divide but because national environmental policies built up prior to EU involvement in environmental policy-making are based upon different and sometimes incompatible philosophies, e.g. German use of uniform emission standards in regard to pollution control compared with the British preference for environmental quality standards. Equally German policy makers are happy to apply the precautionary principle, and if necessary to make policy judgements in the absence of scientific data, while British policy makers prefer to be governed by what science can justify. Member states can attempt to use their temporary tenure of the presidency of the Council to push forward their preferred proposals (and to delay others): German efforts to change the integrated pollution prevention and control directive in 1992, while in this position, however, were largely unsuccessful (Schnutenhaus 1994). Major pieces of environmental legislation are subject to substantial intergovernmental bargaining at the Council stage, as was the case over the 1994 packaging and packaging waste directive and, in contrast to the 1970s and 1980s, the negotiations are taken very seriously by member states.[7] No longer are these deals left to junior ministers to conclude; the mid-1990s witnessed a shift of emphasis towards cost-benefit analysis (using specialist expertise from outside Brussels to train administrators) and better implementation of existing policies. Concerns about over-regulation and administrative simplification have also been injected into policy development at the behest of member states such as Germany and Britain.

The European Court of Justice has also played a key role in the recent development of European environmental policy. It has made rulings to allow member states and local authorities to introduce or retain their own environmental states in the absence of EU legislation (notably in the Danish bottles case of 1988) despite the adverse effects on the single market.[8]

The Court has been asked to judge how far environmental regulations can be introduced using single market treaty articles which avoid the requirement for unanimity, on occasion agreeing to this stratagem, and it has increasingly asserted that EU environmental legislation should be backed up by proper implementation in the member states and commensurate penalties for non-compliance that are appropriate, proportionate and dissuasive. In 1996, the United Kingdom was, for example, fined for allowing more pesticides into the water supply than permitted by the EU's drinking water directives.

Interest groups play a considerable part in the evolution and administration of the EU's environmental policy. Although there are about 20 Europe-wide groups lobbying exclusively for environmental causes in Brussels out of approximately 1000 interest groups (Butt Phillip and Gray 1996), they are still able to be effective in their interventions, especially in the European Parliament and when co-ordinating their efforts with other EU pressure groups and with national-level member organisations (Mazey and Richardson 1993; Porter 1995). The most significant such groups are Friends of the Earth, the World Wide Fund for Nature and the European Environmental Bureau. But lobbying on environmental issues is dominated by industrial and commercial interests organised in sectoral groups and in specific advocacy coalitions on particular issues. They have far greater resources than the green lobby, even though the latter sometimes receives some financial support from the Commission. Occasionally, corporate and environmentalist interests will make common cause, for example, in support of a single regulatory standard to apply across the whole EU. But on many issues, green groups find their views marginalised as corporate interests are able to mount broad-based lobbying campaigns backed by considerable practical expertise while playing on fears of increasing regulatory costs and decreasing European competitiveness.

Public opinion, meanwhile, solidly supports EU-wide measures to protect the environment, ranging well beyond strictly single-market issues to include the protection of flora and fauna and climate change. But there are real worries in some states that EU environmental legislation will lead to reduced national standards (a fear that contributed to the negative Norwegian vote on full EU membership), while in others Brussels is seen as presenting a hostile approach to traditional country pursuits (such as hunting and shooting) and to strong local fishing interests.

It remains to be seen whether the new provisions of the Treaty of Amsterdam, reinforcing the ability of national governments to set higher environmental standards than the Union as a whole, will overcome these suspicions.

New approaches to environmental policy

The EC ventured into environmental policy using a 'command-and-control' approach to policy development. This had the advantage of ensuring that the costs of compliance were largely borne by national and local governments and by individual economic actors rather than by the relatively small Community budget. The regulatory approach from Brussels also played to the strength of the supranational legal order constructed by the new community and, by requiring the member states to implement directives, obscured the real sources of policy development. Member states, for their part, at least in the 1970s and 1980s, seemed content to approve many ambitious environmental directives, safe in the knowledge that they would control the implementation process. By the mid-1990s some 200 directives had been adopted by the EU along with more than 300 other measures. In general terms, these have dealt with issues such as air pollution, chemical substances and industrial hazards, water, noise pollution and nature conservation. The

EU has also developed a waste management strategy and a requirement that major public-sector building projects should be subject to an environmental impact assessment. The latter policy, agreed in 1985, represents the first major change of approach away from vertical command-and-control measures towards horizontal measures, of general cross-sectoral application. This has been followed by the adoption of a voluntary eco-management and audit scheme, a voluntary eco-labelling scheme and a mandatory directive on packaging and packaging waste containing ambitious targets for member states in the recovery and recycling of packaging materials. Further initiatives are anticipated in regard to civil liability for environmental damage, the recycling of end-of-life vehicles, and landfill of waste.[9]

The change in the nature of legislative proposals emanating from Brussels, and the reduction in their number, has occurred in parallel with a rapid expansion of lobbying efforts on environmental questions by industrial and commercial interests. Environmental issues appear in the 1990s to be viewed increasingly in terms of the logic of the single market and of the impact of EU proposals upon long-term competitiveness of the European industrial and commercial base. The plan for the imposition of a tax upon carbon dioxide emissions, first prepared in 1991, has made almost no progress because of conflicts within EU institutions and fears about loss of global competitiveness raised by the most affected economic interests. Other far-reaching proposals, such as the 1991 urban waste-water treatment directive, have been surrounded by controversy over their final cost to governments and industry, a concern heightened by the approach of economic and monetary union and the necessity for many member states to curb public spending in order to meet the eligibility criteria for participation in the EU prescribed in the Maastricht Treaty. The Commission's programme of work for 1996 contained only two legislative proposals in the environmental field, a far cry from the 1980s when up to 14 environmental directives a year were being adopted by the EC institutions.

It may be suggested that DG XI has been captured by certain powerful economic interests which have frustrated environmental policy development, but this does not take account of the subtle change to a more cautious mood of national governments who previously might have been expected to support a strong proactive Brussels-led expansion of policy measures. Nor does such a view give sufficient weight to the enormous problems of lack of implementation of EU legislation thrown up by the 'command-and-control' approach. In addition, the possibility of reconciling economic and environmental policy objectives through 'ecological modernisation' has won increasing acceptance in the Commission as well as among many corporate interests lobbying the EU institutions. Policy making in the environmental sphere has continued strongly in Brussels long beyond the deadline of the single market programme. The Commission has argued that the development of environmental technologies in response to regulatory pressures and of environmental services and expertise is a positive development for the Union, both in terms of job creation and competitiveness.[10] Significant amounts of the EU research and development budget have been allocated for research in these sectors.

Administration and implementation of environmental policies

The main responsibilities for carrying out the EU's environmental policies are effectively shared between the Commission and the member states, each of which has its own individual administrative and legal structures for dealing with environmental issues and its own distinct political culture in which such issues are perceived. The Commission is itself fragmented by strong vertical lines of division separating the different parts of the administration. The environment directorate-general (see Table 12.1) is one among 25 located in Brussels/Luxembourg and it has significant overlaps of policy responsibility with other directorates-general. Its remit is very wide, but it has few day-to-day responsibilities for directly applying existing EU environmental policies: this is usually the task of national or regional governments. With some 400 staff in five directorates, it is among the larger directorates-general, although it is located slightly out of the way within Brussels in comparison with the other DGs, and few are concerned directly with the implementation of EC law.

It would be a mistake to concentrate attention on DG XI alone since several other directorates-general have a considerable say on aspects of environmental policy, e.g. DG VII (transport), DG XVII (energy), DG VI (agriculture) DG III (industry) and DG XII (research and development). Frequently there are differences of view between the different directorates-general on environmental issues, for example over the proposed tax on CO_2 emissions, and it is open to outside interests to exploit such differences of approach in order to achieve the best possible hearing for their own point of view. Member states have also been known to press for changes within the Commission on approaches to environmental policy. A Franco-British paper of 1993 demanded a review of 22 items of adopted or proposed EC environmental legislation on the grounds of complexity, cost and adherence to the principle of subsidarity. The Commission agreed to 'simplify' some of the earlier directives, notably the water quality directives. When DG XI was reorganised at the end of 1994, it was widely believed that the senior official in charge of implementation of EU environmental law was moved after protests from national governments about the enthusiasm he had brought to his role.

COMMISSIONER
DIRECTOR GENERAL
DEPUTY DIRECTOR GENERAL

CABINET

Directorate A General and international affairs	Directorate B Environmental instruments	Directorate C Nuclear safety and civil engineering	Directorate D Quality of the environment and Natural resources	Directorate E Industry and the environment
A1 Inter-institutional relations	B1 Economic analysis and environment forward studies	C1 Radiation protection	D1 Water protection, soil conservation and agriculture	E1 Control of industrial installations and their emissions
A2 Budget finances and contracts	B2 Management and coordination of financial instruments; environmental impact assessments	C2 Security of nuclear installations	D2 Nature protection, coastal zones and tourism	E2 Chemical substances and biotechnology
A3 Information and communication	B3 Legal affairs, legislative activities and application of EC law	C3 Management of radio-active waste	D3 Air quality, urban environment, noise, transport, energy	E3 Management of waste
A4 International affairs, trade and environment	B4 R and D, relations with EEA, statistics, education, training and health	C4 Civil protection	D4 Global aspects of the environment: climate change, biosphere and geosphere	E4 Industry, internal market, products and voluntary approaches
A5 Technical co-operation with third countries				

Table 12.1: Organisational structure of Commission DG XI (1996) Environment, Nuclear Safety and Civil Protection

The decision to establish the European Environment Agency was taken in 1990. The purpose of the Agency initially was to co-ordinate and to harmonise information about the environment emanating from each member state. Only when similar sets of data become available can cross-border comparisons be sensibly made. The Agency (with a staff of over 60) monitors the state of the environment, promotes information exchange on best available technologies, harmonises methods of measurement and disseminates information about the environment. Disputes about the siting of this Agency, finally chosen as Copenhagen, delayed its setting up by more than three years and its real impact on policy has yet to be made. Many of the Agency's original supporters, in the European Parliament and elsewhere, hoped that once it had established its role in monitoring and evaluation data, and providing analysis of existing policies, the Agency could be called upon to play a much more sensitive role in investigating compliance records in the member states and, possibly, taking responsibility for enforcement as well. As it stands, the Agency is a pale shadow of its federal US counterpart, the Environmental Protection Agency, which does combine these functions and which disposes of far more resources.

The record of member states in implementing EU environmental legislation is a matter of great concern to several governments and many economic interests. Not only is the transposition of directives into national law regularly behind schedule: at the close of 1993, Italy had transposed only 81 per cent and Greece 84 per cent of directives due for full implementation at that time (Butt Philip 1994). But there are also numerous instances of incorrect implementation as well as non-implementation. Indeed environmental directives have the worst record of all policy areas for this: at the start of 1994, nearly one-third of all environmental directives due for implementation at the end of 1988 were the subject of infringement proceedings initiated by the Commission, and seven of the 15 least well implemented of these directives fell into the environmental policy category (Butt Philip 1994). Although many reasons can be found for this enduring situation, common to other policy areas, two particular problems deserve to be highlighted.[11] First, the implementation deficit is frequently encountered in member states (such as Germany, Belgium and Spain) where sub-national governments are responsible for environmental policy but national governments conduct the negotiations on their behalf in the Council of Ministers. Secondly, there is a tendency for member states to adopt legislation at EU level which they know they cannot implement properly in practice for a variety of technical, administrative and financial reasons. In effect, this leaves the governments of the member states with considerable leeway about the timing and the keenness with which they seek to conform with EU norms. It is the member states which have to find the administrative resources to achieve full implementation and enforcement, and they retain the right to decide penalties for non-compliance by individuals. In some cases it has become clear that member states have adopted environmental legislation mainly for symbolic reasons rather than really to change practice on the ground (Butt Philip 1994). Given the limited resources available to the Commission for direct oversight of the implementation and application of European rules, it is very dependent upon affected interests; individual firms, local amenity groups, pan-European pressure groups or private persons, making complaints direct to Brussels about cases of suspected non-compliance. Friends of the Earth and local groups brought eight major road schemes in the United Kingdom to the attention of the Commission in 1990 as suspected cases of non-compliance with the Environmental Impact Assessment directive: the Commission upheld most of their complaints and forced the UK government to negotiate a settlement which allowed most but not all of the schemes to go ahead. In 1987, designated by the EC institutions as the European Year of the Environment, the Commission organised a series of seminars in all the member states to inform local environmental groups about the EC's environmental policies and the mechanisms for bringing allegations of non-compliance in individual member states to the attention of the Brussels authorities.[12] In the early 1980s fewer than 50 such allegations concerning environment policy each year were subjects of the formal complaints and inquiries procedure; in the first half of the 1990s around 500 such allegations each year were being similarly dealt with (Butt Philip 1994).

The EU and the international dimensions of environment policy

One of the problems the EU encounters in managing and developing environmental policy is that the EU is not synonymous with Europe, or even the whole of the western part of that continent. Inevitably, EU environmental policy has implications internationally as well as for domestic politics in the member states. Concerns over climate change have, for example, led to several EU trade policy and aid policy measures to help protect the Amazonian rain forests. The EU's own high environmental standards, if imposed on its trading partners or taken into account by Brussels in deciding how much access to EU markets to allow to a third country, can all too easily be perceived as barriers to trade, for example the strict controls on the imports of Amazonian hardwoods or baby seal skins from Canada.[13] Sometimes for reasons of the EU's inappropriate geography, member states will prefer to settle environmental issues on an intergovernmental basis, e.g. the periodic conference on the North Sea (which includes non-member states such as Norway) but whose conventions may then be carried forward into EU legislation (such as the 1991 urban waste water treatment directive). More generally, the EU is represented, although sometimes only after a struggle for recognition, at intergovernmental meetings on environmental matters such as the Rio Earth Summit of 1992. The Commission regards itself as bound to seek to implement through EU measures, if appropriate, the Conventions on biodiversity and climate change, and the Montreal Protocol on ozone-depleting substances, by virtue of the fact that all EU member states are signatories. The EU's willingness to enlarge its own membership also leads it into trying to improve the standard of care for the environment in applicant countries in central and eastern Europe. Moreover, the EU's very vulnerability to major environmental accidents, such as the discharge of radioactive material from the Chernobyl nuclear reactor in the former USSR in 1986, has caused it to take a continuing interest in nuclear safety developments across the continent, particularly since the collapse of the Soviet Union in the early 1990s. The Maastricht Treaty amended Article 130R of the EEC Treaty to enable the EU to contribute to the pursuit of its avowed environmental objectives by 'promoting measures at international level to deal with regional or world-wide environmental problems'. Thus the EU has an explicit role on the international stage, one that was first confirmed by member states in 1987, but which is exercised concurrently with the member states acting independently. However, member states may also find that any unilateral action on environmental standards taken by them not only has to be reported to the Commission (under Article 130T) but may be refused authorisation by the Commission on grounds of incompatibility with the EU's international obligations (Hession and Macrory 1994).

The EU played the leading role in brokering a deal between developing and developed economies at the Kyoto summit talks on climate change held in December 1997. It is now left with the task of delivering its side of the bargain, in particular a reduction in greenhouse gas emissions of eight per cent from 1990 levels by the year 2010. The UK presidency in the first half of 1998, however, made progress on a range of measures to limit air pollution including emissions of greenhouse gases, one of its central objectives, with some effect.

The interplay between policy and policy organisation

The European Commission moved quickly in the early 1970s to establish the leadership of the European Community in environmental policy. Despite the requirement that environmental policy measures be adopted unanimously by the member states, hundreds of EC policy measures were agreed under this major constraint. The initial policy approaches were characteristically legalistic and regulatory in nature (Majone 1994). That reflected the unusually strong legal order for an international organisation with which the European Community is endowed and the fact that it also commands a very small budget (around 1.3 per cent of EC-GDP) in comparison to those of the governments of the member states. The EC therefore continued to use the powers that it did dispose of, essentially to regulate, to research and to inform, until the member states, in the Council of Ministers, began to complain. This highly active regulatory approach continued unabated until 1991–92, at which point some member states and industrial interests began to press strongly for alternative policies (using more market-based mechanisms where possible), more subsidiarity and less prescriptive and onerous legislation.

The primary instrument for EU legislation has been the directive, which is intended to give flexibility to member states to implement EU decisions. This has been exploited by several member states to enable incorrect application of EU measures, and occasionally no application of directives at all. Awareness of this growing 'implementation deficit' led to increasingly detailed directives being agreed by member states in the Council, but ultimately produced legislative and financial indigestion. This contributed to the change of climate away from 'command-and-control policies', and led Jacques Delors early in 1992 to offer to return responsibility for the implementation of environmental standards to the member states. Interestingly, this idea did not find favour either with environmentalist interest groups, which saw it as leading inevitably to a dilution of green standards, or with large multinational firms, which foresaw a growing inconsistency of rules governing the environment emerging from such a decentralisation, thereby serving to frustrate the single market. Subsequent deregulation initiatives have been promoted by, notably, the British and German governments and have largely concentrated on compliance costs, implications for global competitiveness, and the lack of full scientific justification as the basis of their case for questioning existing EU environmental laws. The Commission for its part would like to simplify and consolidate some legislation, while also seeking to raise environmental standards as economic prosperity and new technologies allow. In this they have been encouraged both by the terms of the SEA and the Maastricht Treaty, and by a broad swathe of opinion in the European Parliament. The result of such contradictory pressures has been a major slowdown in policy development, best exemplified by the stalemate on the proposed tax on CO_2 emissions, and a determination to take stock of current EU environmental regimes and to concentrate on achieving better implementation of existing policies. By 1995 it would seem that the activities of DG XI of the Commission had been reined in, and that the narrower single market agenda, rather than the search for sustainability, was its primary focus.

The rules of the political game in regard to environmental policy have been the subject of important changes, but these changes have reflected rather than shaped events so far. With the implementation of the SEA from July 1987, it became possible occasionally to seek to vote through environmental legislation using a single market legal base (and hence qualified majority voting in the Council, instead of unanimity): a good example is provided by the 1994 packaging and packaging waste directive which was ultimately adopted under Article 100(a) (single market) and not under Article 130 (environment rules). The Maastricht Treaty specified four different decision-making procedures for different aspects of environment policy, with unanimity of the member states reserved for measures primarily of a fiscal nature; town planning, land use (other than waste management and general measures) and management of water resources; and measures significantly affecting a member state's choice between different energy services and the general structure of its energy supply (Article 130s). Member states were to continue to be responsible for the finance and implementation of EU environmental policy although if a particular EU measure was deemed to involve costs 'disproportionate for the public authorities of a Member State' the Council shall make appropriate provisions either in the form of temporary derogations and/or financial provision from the newly established Cohesion Fund.

But the major problem facing policy makers in the European Union remains how to integrate the Union's environmental policy objectives with its economic and other policy objectives. The objective of high self-sufficiency in agricultural production is not reconcilable with the EU's environmental protection and anti-pollution objectives. The single market programme during 1985-92 sought to boost cross-frontier trade, implying much more use of transport services and infrastructure which would have adverse effects on the environment in the member states. The commitment to the economic development of the poorer regions of the EU, mainly through use of the structural funds, can hardly avoid adverse impacts on some of Europe's most unspoiled environments as a result of new industrial and infrastructure projects which the EU itself has financed. The incoherence of the pursuit of the EU's environment policy objectives is underlined and perpetuated by the fragmentation of policy making in the Commission. Although this problem is well recognised in Brussels, it is unlikely to be overcome because environmental policy cuts across so many different existing policy interests at all levels of government. The subordination of all these other policy interests to EU environmental policy objectives is not a realistic prospect and would only occur if the EU abandoned the primacy it has long accorded to the objective of economic integration.

Conclusion

The high EU involvement in the development and application of environment policies is a product of serious political and economic pressures, and institutional opportunism. Policy making for the environment by EU institutions has been intensive and often tortuous. EU measures have shied away from provision of financial aid until the advent of the Cohesion Fund in 1993, and have been primarily regulatory and legalistic in nature. They have frequently been over-ambitious, given the substantial implementation deficits that have been documented, and at times clearly 'symbolic' in nature: member states may will the ends of much EU environmental policy but they find difficulties in making available the means to their achievement. The diversity of environmental problems, the divergence of economic conditions in the member states, and the frequent conflicts between economic and environmental policy objectives all lead to

Hession, M. and Macrory, R. (1994) 'Maastricht and the Environmental Policy of the Community: Legal Issues of a New Environment Policy', in D. O'Keefe and P. Twomey (eds) *Legal Issues of the Maastricht Treaty*. Chichester: J. Wiley, pp. 151–67.

Johnson, S. and Corcelle, G. (1989) *The Environmental Policy of the European Communities*. London: Graham and Trotman.

Koppen, I. (1988) *The European Community's Environmental Policy: From the Summit in Paris, 1972, to the Single European Act, 1987*. Florence: EUI.

Kramer, L. (1992) *Focus on Environmental Law*. London: Sweet and Maxwell

Liberatore, A. (1991) 'Problems of Transnational Policy-Making: Environmental Policy in the European Community', *European Journal of Political Research*, Vol. 19, Nos. 2 & 3, pp. 281–305.

Liefferink, J. Lowe, P. and Mol, A. (1993) *European Integration and Environmental Policy*. London: Belhaven Press.

Macrory, R. (1992) 'The Enforcement of Community Environmental Laws and Some Critical Issues', *Common Market Law Review*, Vol.29, pp. 347–69.

Majone, G. (1994) 'The Rise of the Regulatory State in Europe', *West European Politics*, Vol.17 No.3, pp. 71–101.

Mazey, S. and Richardson, J. (eds) (1993) *Lobbying in the European Community*. Oxford: Oxford University Press.

Nugent, N. (1994) *The Government and Politics of the European Union*. Basingstoke: Macmillan.

Porter, M. (1995) *Interest Groups, Advocacy Coalitions and the EC Environmental Policy Process*. Bath University: Unpublished Ph.D. Thesis.

Rehbinder, E. and Stewart, R. (1985) *Integration through Law: Europe and the American Federal Experience*, Vol 2: Environmental Protection Policy. Berlin: De Gruyter.

Ripa di Meana, C. (1993) 'Introduction: Towards Sustainability: A Program of Action for the Environment', in *EC Environment Guide 1992/93*. Brussels: EC Committee of the American Chamber of Commerce in Belgium.

Schnutenhaus, J. (1994) 'Integrated Pollution Prevention and Control: New German Initiatives in the European Environment Council', *European Environmental Law Review*, vol. 3 No.11, pp. 323–8.

Sidendorf, H. and Ziller, J. (eds) (1988) *Making European Politics Work*. London: Sage.

Single European Act, Article 25 introducing a new Article 130R into the EEC Treaty.

Stanners, D. and Bourdeau, P. (eds) (1995) *Europe's Environment. The Dobris Assessment*. London: Earthscan.

Verhoeve, B., Bennet, G. and Wilkinson D. (1992) *Maastricht and the Environment*. Arnhem/London: Institute for European Environment Policy.

Vogel, D. (1993) 'The Making of EC Environmental Policy', in S. Andersen and K. Eliassen (eds) *Making Policy in Europe*. London: Sage, pp. 115–31.

Vogel, D. (1995) *Trading Up: Consumer and Environmental Regulation in a Global Economy*. London: Harvard University Press.

Wallace, H. and Wallace, W. (1996) *Policy-Making in the European Union*. Oxford: Oxford University Press.

Weale, A. and Williams, A. (1992) 'Between Economy and Ecology? The Single Market and the Integration of Environmental Policy', *Environmental Politics*, Vol.1 No.4, pp. 45–64.

Wilkinson, D. (1990) *Greening the Treaty: Strengthening Environmental Policy in the Treaty of Rome*. London: Institute for European Environmental Policy.

Chapter 13

Environmental challenges and institutional changes

An interpretation of the development of environmental policy in Western Europe

Alf-Inge Jansen, Oddgeir Osland and Kenneth Hanf

> And the end of all our exploring
> Will be to arrive where we started
> And know the place for the first time.
> T.S. Eliot

Introduction

Returning to our point of departure – the assumption that the development of a country's environmental policy and practice is the result of the interaction between the strategic activities of the various actors and the institutional structure within which they act – we here give an overview and interpretation of significant developments in environmental policies in Western Europe. Firstly, we describe the emergence and the development of environmental policy. Secondly, we address the principal changes that took place around 1970, and the governmental responses in the period when environmental policy was established as a policy field and the environmentalist conceptualisation of the relation between man and nature was put forward as a challenge to the economic and political institutions of Western Europe. Thirdly, we focus on the expansion of the environmental policy field from the mid-1980s, a period characterised by a widening of the scope of environmental policy and the second wave of popular environmental concern as well as by a profound internationalisation of environmental policy. Furthermore, we examine how the environmentalist conceptualisation of the environmental *problematique* was redefined through the notion of sustainable development and further reconstructed and transformed into the policy strategy of ecological modernisation. Lastly, we address the question of the extent to which this policy strategy has been put into practice. The emergence and authority of this policy strategy is interpreted in terms of the overriding political project of policy elites for restructuring the relation between the different institutional orders of Western European societies.

On the emergence of environmental policy

Undoubtedly it was scientists, naturalists and professionals like medical doctors and architects who, during the nineteenth century, first constructed and promoted a certain conception of man and nature as a relationship according to which effects related to industrialisation were defined as undesirable. Certain activities that intervened in and changed nature were seen as having particularly damaging effects. These pioneers mobilised public attention and support, and enlisted allies in governmental circles for taking action, e.g. developing policies and taking measures to prevent or reduce these interventions or to mitigate or repair their damaging effects.

What were defined as the problems and the policies adopted and measures taken to solve these problems can be grouped according to three traditions. Notwithstanding that the names may differ in the various countries, these traditions refer to dealing with threats to (1) hygiene and public health, (2) nature areas, animal and plant species and natural monuments, and (3) the cultural heritage of the nation.

Already by the first half of the nineteenth century the health conditions due to the pollution and poor sanitary conditions in the rapidly growing urban areas in countries such as Britain and France were causing concern. Because of the active involvement of members of the medical profession and of scientists, public health and hygiene became issues in public debate. Some health regulations were adopted by central authorities and regulatory areas like sanitary engineering and town planning were developed. The first nation to industrialise and urbanise was Britain, and as pointed out by Carter and Lowe (Chapter 2), Britain's nineteenth century governmental as well as civic leaders were the first to grapple with appalling health and amenity problems of urban-industrial society. In the second half of the century, in response to unsanitary conditions and polluted drinking water, and concomitant incidents of cholera epidemics, most of the countries examined in this book had adopted health laws and regulations, and established local administrative bodies to implement them.

Similarly, most countries during the nineteenth century adopted *factory acts* designed to regulate damage, danger and nuisance associated with industrial and manufacturing activities and processes. In France the government had already by 1810 issued a decree on establishing a service for registering and inspecting industrial plants. These measures were later incorporated in the Industry and Trades Law of 1917, which also defined standard distances to be maintained between industrial and residential areas. The Prussian *Gewerbeordnung* (adopted by the German *Reich* in 1869) was another well-known example of this kind of factory law. In 1895 the German *Reich* also adopted the *Technische Anleitung Luft* (TA) which can be regarded as the start of a German clean-air policy. The definition of required environmental standards was, in accordance with the rules of German federalism, left to the discretion of local authorities. Consequently, institutional characteristics of

Germany opened the way for great variation between the *Länder* as well as within the same *Land*. The pioneering effort in legislating against air pollution was, however, the British Alkali Act of 1863. The authority set up to implement it, the Alkali Inspectorate, was the world's first governmental 'environmental agency'. This agency practised regulation in a co-operative style based on 'advice and friendly admonition' (Ashby and Anderson 1981:27). An informal, consultative and negotiatory practice that relied on persuasion and gentlemen's agreements was institutionalised as the British style of clean-air policy implementation. This implementation style was consistent with the general relationship of trust and confidence that in the UK had been developed between administrative authorities and their officials on the one hand and societal interests, like those of rapidly growing industry and its representatives as well as those of professions of increasing authority and status (medical doctors, lawyers and engineers), on the other. There was no ideological barrier separating the representatives of the state and the representatives of these societal interests, much less an ideological conflict between state and society.

As to the second tradition, it has been noted that in Britain the Victorian middle class sought to protect wildlife from the onslaught of urbanisation. In France the study of habitats of species and natural space has a long history. Larrue and Chabason note (Chapter 4) that the establishment of the National Museum for Natural History in 1791, together with local museums for natural history and botanical gardens, are events in the institutionalisation of this tradition. At the turn of the century, naturalists argued for the creation of natural parks and reserves as well as for the protection of endangered species. German university botanists and naturalists were actively promoting the cause by setting up associations and enlisting the support of the public authorities, and nature reserves were set up and forested areas were preserved as parks. The first efforts of organising nature conservation in Denmark, Norway and Sweden were to a great extent initiated by scientists who were inspired by and co-operated with German colleagues. In 1909 the Swedish Parliament decided to set aside a string of 'national parks' in accordance with principles developed by scientists (Lundqvist 1971:23).

The tradition of cultural heritage and the preservation of historic monuments, sites and landscapes was to an extent established in countries like Britain, France, Germany and the Netherlands before the turn of the century. These tasks had been taken on board as tasks of the state in most of the countries before World War I.

The time of emergence, the extent to and the manner in which these three traditions were developed in each of these countries, i.e. the distinctive profile of the historical development of the antecedents of environmental policy, was influenced both by the country's geographical and socio-economic characteristics as well as by the character of its institutions. Moreover, it was a characteristic feature of this development that policy content, in terms of acts and programmes, as well as the organisation and technology for the implementation of these acts and programmes, took place in relation to three different and to a great extent mutually independent traditions. That is to say, acts and policies were directed to some specific subclass of what we today call environmental problems. The administrative bodies that implemented these policies were set up and anchored in different policy sectors (such as health, industry, agriculture, culture and education). Different professions, with their specific knowledge and technical competence, were recruited into the different agencies and bodies. Only the first of these three traditions, the one dealing with the threats to hygiene and public health, had a significant effect on public policy. None of these traditions significantly challenged, dislodged or supplanted policies for economic modernisation and growth in sectors like agriculture, industry, mining and transport.

The environmentalist construction of the environmental challenge — the environmental movement and governmental response

The late 1960s and early 1970s ushered in a new era, in which a transformation of the established conception of the man-nature relation in public debate can be observed. This relation between man and his external natural surroundings was reconstructed into a concept of the environment as being both the complex interrelated reality surrounding us and including us, as an interacting whole. Concomitant with this new concept was the belief that the existence of this interrelated reality of which mankind and its social organisation are parts, was threatened by human activities and their organisation. A new society-nature relationship was constructed, and it was popularised by means of metaphors and models that illustrated the interdependence and vulnerability of mankind and mother Earth (e.g. 'Spaceship Earth', 'Limits to Growth') and was promoted vigorously in the public debate in all Western European countries.[1]

A number of biologists, in particular Americans, were principal claim makers (Rachel Carson, Barry Commoner, Paul Ehrlich, Garret Hardin, René Dubos), but other types of scientists followed suit (Kenneth Boulding, J.W. Forrester, Dennis L. Meadows, E.F. Schumacher) in constructing this new society-nature relationship. The influence from the American debate was important for public debate in Western Europe, but the European environmentalist mobilisation was anchored in the European situation. Dramatic events, like the wreck of the tanker *Torrey Canyon,* that in 1967 spilled 120,000 tons of crude oil into the English Channel, facilitated mobilisation of popular opinion and, to an extent, the reorganisation of many of the old conservationist organisations as well as the founding of new environmental organisations with roots and links to the student rebellion of 1968 and the anti-Vietnam war movement.

The environmentalist construction implied assumptions and represented ideological positions that could be seen not only as strong criticism of existing policies and administration but also as challenges to core institutions of the existing political order. Notable was that ecology and economic growth were seen as polarities embodying strongly contrasting sets of values and assumptions regarding man and nature (Caldwell 1970:11). A dominant view in the new environmentalist movement was that the relation between environmental protection and economic growth was one of a zero-sum game. It was in this context that Commoner (1971:44–45) introduced his famous fourth law of ecology: 'there is no such thing as a free lunch'. He stated that the energy and environmental crises reveal the truth about the 'deep and dangerous fault in the economic system' (Commoner 1976:235–36). He insisted, therefore, on a 'rational ideal' that makes the 'production system' conform to the 'ecological system' and the 'economic system' conform to the 'production system' (Commoner 1976:2 and Rubin 1994:69ff.).

Various authors have pointed to similarities in the way governments in the OECD area responded to the development of an environmentalist movement and its demands, and that these similarities reflected policy imitation across national boundaries. However, comparing the year of establishment and formally defined characteristics of administrative units and environmental legislation (see Weale 1992:14) may be rather superficial if we want to understand how the governments of Western Europe responded to the environmentalist definition of the environmental challenge. For instance, the setting up of an environmental ministry in one country may be less significant in terms of environmental policy than the establishment of a central agency in another country. The British Department of the Environment (the DoE), established in 1970, gave a false impression of being an environmentally oriented 'super ministry'. In reality fragmentation prevailed. Only a minor part of the DoE dealt with environmental issues, and this part experienced serious difficulties in gaining high priority for environmental issues on the agenda of the DoE. In contrast, a central agency like the Swedish Environmental Protection Agency, set up in 1967, was given wide responsibilities with both great legal and professional authority and became a highly influential actor in Swedish environmental policy (Lundqvist 1996:276).

Similarly, in the 1970s environmental phenomena were responded to by legislation that was relatively similar in formal terms, but was quite different in terms of operational goals and instruments. An important example is the difference in operational definitions of pollution. The British tradition has been to see pollution as an effect, i.e. as the undesirable consequence, rather than the presence of pollutants. The implication here is that undesirable materials present in low concentration, widely dispersed or transformed by natural processes, may be harmless. The objective is, therefore, not to avoid emissions irrespective of cost, but to define the 'least-cost' use of the environment, which can differ depending on local factors, the cost of emission-avoiding technology, and the economic situation of the firm concerned (Héritier, Knill and Mingers 1996:78). Consequently, this definition of pollution emphasises the need for convincing evidence of environmental damage before taking action (Boehmer-Christiansen and Skea 1991:14–15). The other view, defining pollution as the presence of undesirable pollutants, which is consonant with the precautionary principle, has been adopted by a number of countries, among them Germany. The British Health and Safety Act and the German Federal Pollution Control Act were in many respects quite different policy responses to the environmentalist wave of the late 1960s and early 1970s. While the Federal Pollution Control Act confirmed the legal requirement that plants in FRG had to comply with emissions and pollution concentration standards and meet the 'Best Available Technology' requirement (BAT), the British legislation focused on damaging environmental effects and made state intervention contingent on demonstrable detrimental impacts from air pollutants being more harmful and costly than the potential costs of avoidance measures.

These differences in legislation are partly due to the fact that the British Isles have far more favourable prevailing wind conditions for dispersing air pollution than Germany. But they also reflect different traditions of state intervention and administrative practice. Similarly, the significant differences as to changes in environmental policy due to the setting up of the Swedish central agency for nature protection compared with the effects of the establishment of the British DoE reflect the different tradition of state intervention and other institutional characteristics of the British and Swedish politico-administrative systems. In other words, to understand what Western European countries had in common in terms of governmental response to the new definition of the environmental challenge, we must go beyond comparisons of formal characteristics of organisation and policy.

The new environmentalist movement undoubtedly gave a significant impetus to the upsurge of environmental awareness in most Western European countries, but the extent of mobilisation of people for the environmentalist cause varied greatly across countries. Particularly in the three Scandinavian countries, but also in the Netherlands, one can discern a mobilisation and a reorganisation of a movement that increasingly demanded more environmentally oriented policies and a restructuring of public administration. In the other countries there was no significant political pressure from specific strata or large groups or from the population at large for a general environmentalist policy. In countries like France and West Germany such demands developed at the local level in certain parts of the country. However, it was not until the mid-1970s that the environmentalist movement had a national basis in these countries. At the other extreme, the military dictatorship in Greece (1967–74) prevented voluntary environmental mobilisation.

Without doubt, the environmentalist movement of this period had effects in Denmark, Norway and Sweden. In Sweden, new administrative bodies to deal with environmental issues were established and more environmentally oriented pollution control policies were adopted and implemented. Both Denmark and Norway set up Ministry of Environment (ME) in the early 1970s. In Norway, the new environmentalism injected fresh energy and character to the conservationist movement which was transformed into a much more environmentally oriented political movement. The Norwegian environmental movement became a political movement that overwhelmingly co-aligned with the social movement against Norwegian membership in the EC. The EC was seen as a political organisation that would be instrumental in increasing the exploitation of Norwegian natural resources (e.g. oil, watercourses, natural areas) in the pursuit of economic growth. In the mid-1970s the environmental movement in Norway was a political force with considerable clout. The resistance against exploitation of watercourses was still the constitutive issue of the movement but its platform had been widened. When preparations were made to lay down the basis for production of nuclear power, a new round of environmentalist mobilisation was forthcoming. However, the reservoirs of petroleum on the Norwegian continental shelf gave the dominating coalition (the Labour government, the Labour Party and the Confederation of Trade Unions together with the Conservative Party and the Confederation of Industries) room for political manoeuvrings. Consequently it was able to anchor its growth policies in the oil economy and to avoid confrontation on the issues of nuclear power and of exploitation of watercourses. In this way, the environmentalists' potential for mobilisation was reduced.

The development in West Germany was quite different. Here policy makers saw nuclear energy as a key industry in the modernisation of the economy. To counterbalance the country's dependence on imported energy (underlined by the oil crisis of 1973–74) the SPD-FDP coalition under Chancellor Schmidt launched a vast long-term programme for nuclear energy production. Procedures for the approval and selection of sites remained the preserve of a rather small and closed policy community. No local organisations, let alone citizen action groups, were expected to play an active role. The building of nuclear power plants was seen, and presented, as a *Sachszwang*, an imperative for a competitive and expanding German economy (Dyson 1982:27–28). However, as a growing number of sites were named, other aspects of nuclear energy achieved a new visibility. Resistance against *Der Atomstaat* became the most constitutive issue in the mobilisation of the loosely coupled but fast-growing German environmental movement and a necessary condition for the electoral success of the Greens. Moreover, as a result of the German system of state subsidies for parties winning over 0.5 per cent of the vote, the income that the alliance of the Greens received from state subsidies for campaigning in the 1979 European elections made it practical for them to become a political party in 1980. Another crucial condition for the Greens' success was the characteristic of the German electoral system. When in 1983 the Greens won 5.6 per cent of the vote, they passed the five per cent threshold, and won 27 seats in the *Bundestag*. The launching of a policy for nuclear energy production had significantly mobilised the German environmental movement while the country-specific characteristics of the German electoral system had contributed to transforming this support into representation in the *Bundestag*.

The Greens' electoral success changed both the party system and the dynamics of environmental politics. As Pehle and Jansen point out (Chapter 5), since even marginal electoral swings can have significant impacts in a party system that favours government by coalition and since all parties and governing coalitions are engaged in continuous election campaigns at *Bund* and *Land* level, the success of the Greens made the other parliamentary parties very sensitive to the changing public mood in favour of environmentalism.

Public concern over *Waldsterben* and SO_2 pollutants made the governing coalition act. The Ordinance on Large Combustion Plants, the 1985 amendment to the Federal Pollution Control Act and the 1983 and 1986 amendments to the technical Guidelines on Air Quality (*TA Luft*) led to the further development and consolidation of the characteristic BAT requirement. These policy measures have been characterised as representative of the (ideal-typical) 'synchronous' interplay between the development of environment law and environment technology '…to protect the environment not only by regulating what comes out of the pipe but also intervening legally in the production process to regulate material inputs and production procedures.' (Héritier, Knill and Mingers 1996:53) A significant consequence of this policy of stopping emissions at the source has been the technological progress on environmental protection in the FRG and the development of a large new industry and market for environmental technology. The consistent application of legal instruments combined with the approach of combating emissions at the source in pollution-control policy reaffirm what is seen as the strength of particularly legal values mediated and practised by officials trained in the Roman-law tradition of coherence and consistency and practised in accordance with the norms of the *Rechtstaat*. Concurrently, in line with the long German tradition of emphasising *der Stand der Technik*, the application of the BAT requirement reaffirms the authoritative role of the professions of engineers and lawyers in the area of pollution control.

Despite simplification of the formal procedure of implementation, all the rules and the thorough *Sachlichkeit* by which officials decide on the various industrial projects make authorisation a cumbersome and time-consuming process. The detailed legislation limits the scope for co-operative negotiation and informal agreements, but a 'sounding out' and preliminary negotiation between licensing authority and industry nevertheless takes place in many cases. Environmental organisations have only marginal access to the routinised decision-making process between licensing authority and emitter. At the local level, for instance, in development and urban planning, these organisations and groups try to put pressure on local government and politicians. Hearing procedures that have been legally institutionalised are, however, limited to those directly affected by the proposed measures (Derlien 1995:85–86).

As it was for environmental policy in Germany, the decision to develop an unprecedented nuclear energy programme was crucial for the development of French environmentalism and environmental policy. But the development and implementation of this programme had quite different consequences in

France than in Germany, primarily due to institutional and other country-specific features. After the first oil-price shock, the French government moved quickly to adopt a large-scale nuclear energy programme. This led to the mobilisation of a loosely organised anti-nuclear energy movement and isolated protests against nuclear power plants. There was also a growing voter sympathy for Green politics. In the 1977 municipal elections Green lists won more than 10 per cent of the votes in a number of towns, including Paris, and 4.4 per cent at the European elections in 1979. Because of characteristics of the electoral system, this relatively strong showing did not result in any significant representation in the legislative assemblies. Still the Greens' share of the vote made the Socialists grant 'green' concessions.

In the 1980s the support for the Greens appeared to have stagnated, and attention was gradually drawn to how nuclear power contributed to a significant decrease in SO_2 emissions. France had no problem in signing the Helsinki protocol on the reduction of sulphur emissions or the 1987 Montreal protocol on CFC emission abatement, and only minor problems in meeting the objectives of a EC directive on large combustion plants.

French complacency with its nuclear power programme was demonstrated in the case of the Chernobyl disaster. Compared with the reaction in Germany and Italy, the most striking feature of the French reaction was the almost total lack of response, in terms of policy measures and social reactions (Liberatore 1995:82). As certain areas of France were as highly contaminated as certain areas of Germany and Italy, this difference in response was not due to objective differences, but was shaped by institutional factors. During the 1980s, the nuclear energy industry and its role in French economic life had become 'taken for granted' – it had become one of the institutionalised characteristics of the French political order. The case of the French nuclear energy programme highlights one of the principal points made by Stephen Krasner (1988:71): Once the choices of developing and implementing the programme were made, they constrained future policies. In tandem with the relative closure of French legislative assemblies that hampered minority interests like those of the environmentalists in gaining representation, they strongly limited the range of options perceived by the policy makers. Consequently, decisions and policies made later were in line with the interests of the French nuclear industry.

A further important feature of French environmental policy during the 1970s and 1980s was the relative weakness, in terms of jurisdiction and resources, of the core environmental organisation. The ME of 1971 was a lightweight trans-sectoral structure that was given an inter-ministerial role '…without unduly disturbing the already-existing overall administrative organisation.' (Larrue and Chabason, Chapter 4, p. 60) It had no executive powers and was dependent on the sector ministries and their field administration. Three of these sector ministries, the Ministry of Public Works, the Ministry of Agriculture and the Ministry of Industry, were each staffed, and also dominated, by one of the three main *corps*. As three of the four directorates of the ME each was staffed with the same *corps*, it follows that each of the three sector ministries had its corporate counterpart within the ME. In other words, one of the institutional characteristics of French public administration, the predominance of the *corps*, was also reproduced in the field of environmental administration. The regionalisation laws of 1982 laid the ground for an increased role of elected assemblies at the region and department level, and step by step these bodies (*conseils régionaux* and *conseil généraux*) have developed into political actors of consequence. Combined with the great variety of instruments and measures that are available in French environmental policy as well as with the mechanisms employed to realise implementation based on consensus, partnership and negotiation, a diversified environmental policy has been developing from the bottom up, partly to counterbalance the policy standards issued in Paris.

British environmental organisations and groups differ greatly from the environmental movement in Germany and France. British environmental organisations focus primarily on conserving nature and the countryside, traditional Victorian values. Confrontation and demonstrations have not been part of these organisations' repertoire of action; they rather play by the institutionalised rules of confidentiality and discretion.

There was no significant change of British environmental policy in response to the environmentalist threat definition. Carter and Lowe (Chapter 2) sum up the environmental policy process as characterised by accretion of common law, statutes, agencies, procedures and policies. The highly devolved and decentralised regulatory structure of British environmental administration reflected and, in turn, perpetuated, the fragmented and piecemeal character of the country's environmental policy. In contrast to German legalistic regulation, environmental control in Britain continued to be pervaded by administrative rather than judicial procedures, while legislatively prescribed standards and quality objectives were avoided. Legislation continued to be broad and discretionary, leaving regulatory agencies considerable latitude in their enforcement. The approach of the authorities was deliberately to foster co-operation and to achieve the objectives through negotiation and persuasion. The application of a notion like 'Best Practicable Means' (BPM) ensured that regulatory authorities would be sensitive to the economic constraints of societal interests. When environmental policy impinged on major economic interests, policy making, in accordance with the British tradition, took place in relatively closed policy communities. British environmental policy in the period up to the late 1980s was characterised by continuity and complacency, and subordinated to the Thatcherite policies of a general restructuring of British society in order to promote entrepreneurial spirit and economic competitiveness.

Greece, Italy and Spain are often grouped together in the environmental policy literature, for a number of reasons, which are other than their sharing a similar climate and geography. As described in the chapters on these three countries, public awareness of the environmental challenge developed later here than in other countries dealt with in this volume. The first piece of Spanish national environmental legislation, the Air Pollution Control Act of 1972, was in many regards a response to the Stockholm Conference. Characteristically, environmental policy during the 1970s and the 1980s was fragmented among eight ministries, and in almost all of them, and certainly in the government as a whole, environmental policy was subordinated to considerations of economic development and modernisation. At the regional and local levels of government, policy content was ambiguous and its administration was ineffective. In practice, environmental considerations and interests were systematically subordinated to the interests favouring industry and employment.

In Italy and Greece, neither internal pressure nor stimuli from abroad led to any governmental offensive in environmental policy. In Italy, a major piece of environmental legislation concerning water pollution was finally passed in 1976, and it took another ten years of 'filling the legislative tool box', as Lewanski puts it (Chapter 7). Until the establishment of the ME in 1986, environmental administration was very fragmented. At the central level it was characterised by inter-ministerial rivalries in which the interests of the economy and industry dominated. Of great importance to the Italian environmental policy system are the three other governmental levels: the regions, provinces and municipalities. Lewanski characterises the environmental policy field as crowded with an increasing number of actors who are more concerned with trying to stake out their own areas of influence than with coping with substantive problems. As he puts it, the distribution of powers is perceived by actors as a zero-sum game. From this perspective, centre-periphery relations are seen in terms of a 'layer cake' rather than a 'marble cake'. With the exception of some parts of northern and central Italy, local authorities have had a rather poor record in terms of administering environmental policy. Implementation turned out to be extremely difficult, largely due to the high degree of organisational fragmentation referred to earlier, which is a major cause of the well-known 'implementation deficit' in Italian environmental policy.

In Greece, the process of setting up an environmental administration was long and difficult. By 1980 over 15 ministries had developed a special unit with responsibilities for environmental protection. After prolonged inter-ministerial rivalries, the present Ministry of Environment, Physical Planning and Public Works was set up in 1985. The interaction between the two parts of the ministry is described by Spanou (Chapter 6) as limited, and environmental priorities have not appeared to have a central position in public works policy. Field services are fragmented and have overlapping responsibilities. The municipalities were rather hesitant to extend their responsibility to environmental matters. They lack technical as well as financial resources and predominantly pursue policies of economic development and employment. The framework law on environmental protection was approved in 1986 after many years of debate, and only after it had been adjusted to accommodate the vigorous resistance from industrialists. This law is characterised by

Spanou as 'a list of wishful thoughts'. In general the development of environmental policy up to the late 1980s was symbolic, fragmented and reactive in character. The lack of effective implementation and control in Greek environmental policy reflects the low commitment of successive governments in this policy field.

The Netherlands represents a contrast to the situation in Greece, Italy and Spain. From 1970 to 1975 the Dutch environmental movement underwent, as pointed out by Hanf and van de Gronden in Chapter 8, a period of unprecedented growth and was transformed into a more politicised social movement that mobilised public opinion against the government's environmental policy and vigorously participated in more general socio-political criticism. The government responded to the upsurge of environmentalism by introducing a number of laws which typically covered one environmental policy area and took the form of framework laws. In the case of pollution control, each act had its own set of regulatory procedures, and companies and individuals therefore had to apply for several different permits and in most cases to deal with different regulatory authorities. The oil crisis and the oil boycott of the Netherlands that accompanied it effectively led to the subordination of environmental policy to core concerns of economic policy. From the mid-1970s the environmental movement stagnated. Economic recession was the number one issue, and increasing criticism against the environmental laws and environmental policy had more political weight. Industry, for its part, claimed that the licensing procedures were far too time consuming and that there was insufficient co-ordination between the various authorities involved in the licensing process. In addition criticism grew against the lack of coherence and integration of environmental policy.

The Directorate General for Environmental Hygiene was set up in 1971 as a part of the Ministry of Public Health and Environmental Hygiene. It had a modest start in terms of competencies and both financial and human resources. The new Ministry of Housing, Physical Planning and Environment (here called the ME) of 1982 became an organisation of some consequence. At the same time, the context of environmental politics changed as a centre-right coalition came to power, elected on a programme for deregulation and rolling back the state.

The new Minister of the Environment, Dr Pieter Winsemius, changed the ME's internal organisation and its relations with its surroundings. By organising policy making according to themes, geographical areas, flows of materials and target groups, a more integrative approach was adopted. By extensive co-operation with other ministries as well as with provincial and municipal authorities, the ME aimed at both more cross sectoral and vertical co-ordination. Notable was its comprehensive new regulatory regime. Characteristic was the effort to change the detrimental social practices through seeking to bring about a mental change (*verinnerlijking*) within the target groups, e.g. bio-industry, chemical industry, utility industry and households. Environmental renewal (*Milieuvernieuwing*) was to be based on a new partnership between industry, households and government. The traditional 'hierarchical system of command and control' was to be supplanted by a partnership, where the ME had to take account of the interests of industry. Most Dutch environmental NGOs became negotiating partners for environmental departments at the national, provincial and municipal levels, and thereby became integrated into routine environmental politics and policy-making. The new environmental policy strategy was in line with the government's commitment to economic growth and to rolling back the state, and it did not challenge any basic institutional interests of the Dutch political order. It was explicitly stated that environment and economy could be mutually reinforcing, provided they were managed correctly. This policy strategy was anchored in the characteristic accommodative tradition in Dutch policy making, and reaffirmed the country's institutionalised bias towards co-operation between the leaders of government, politics and industry.

The discussion so far can be summed up as follows: the new conception of man-nature relationship, or society-nature relationship, constructed primarily by biologists in the late 1960s, increased the awareness of environmental deterioration in these ten societies and mobilised people of different ages to join the new environmental movement. However, only in a minority of these countries was the pattern of politics significantly influenced by this movement, and from the last half of the 1970s and into the 1980s, environmental politics in the great majority of these countries was overshadowed by employment and economic policies. However, from the mid-1980s onwards, as the economies of all these ten countries started gaining momentum, the environmentalist cause experienced a comeback.

During the 1970s a characteristic governmental response was increasingly to use categories, notions and references that were part of or compatible with the symbols of the biologists' construction in environmental policy documents. The development of environmental policy generally took the form of accretion in terms of organisation and legislation, i.e. setting up additional organisational units and adopting additional laws for regulating human activities within a specific environmental area (such as water, air or soil pollution control). Characteristically, the dominating coalition of actors in all ten countries saw to it that this process of accretion was given energy and direction that was compatible with institutionalised interests, in line with the tradition of state intervention and the institutionalised policy style of the country.

It should be pointed out, however, that this development was an outcome of political processes and, therefore, not always a smooth passage in calm waters. In some countries it occurred that the dominating coalition had a rough crossing. The Federal Republic of Germany was, because of the far-reaching consequences of the implementation of the dominating elite's policy strategy of developing nuclear power plants, an outstanding case. The implementation of this policy strategy contributed to different but interrelated events as the downfall of the SPD-FDP coalition government and to the 1983 breakthrough of the Greens that changed both the German party system and the dynamics of German politics. These events, in combination with ecological accidents, propelled environmental issues onto the public agenda. The emergence of political pressure as a precondition for moving towards a more substantial environmental policy, can, therefore, to a significant extent be interpreted as an unintended consequence of the dominating coalition's pursuit of the nuclear power strategy. The implementation of the nuclear power programme in France had a different unintended consequence, such that pollution did not become a salient issue in national politics. In Norway the exploitation of petroleum resources made it possible for the dominating coalition to stop the planning for developing nuclear power plants and still vigorously pursue its policies for economic growth while making acceptable compromises on the issue of exploitation of watercourses. In this way it was possible to pacify significant parts of the environmental movement and even, as in most of the other ten countries, integrate them into routine politics.

One can conclude that during the 1970s and the first half of the 1980s there were significant differences in the governmental response to the 1970 biologists definition of the environmental challenge. These differences were primarily a reflection of characteristic features of country-specific political and administrative institutions. One prominent example is the differences between German and British responses in the area of pollution control. The importance of political institutions is also highlighted both by the significance of the lack of democratic institutions in Greece and Spain in the late 1960s and early 1970s and by the differences in the electoral systems of Germany and France. Similarly, the significance of institutional characteristics of national public administration is illustrated by the development in Greece and Italy where characteristic features of public administration have been reproduced in the area of environmental policy and represent a major obstacle for the implementation of such policy. Although strong formal standards have been introduced they are not enforced, which is part of the dynamics of a vicious circle characterising a political culture in which lack of trust between state and society is prominent. This could be contrasted with the 'rule-deferential' political culture in the Nordic countries in which, and partly due to which, formal rules are implemented despite 'soft' implementation practices (Christiansen and Lundqvist 1996:361).

However, the responses of the various governments in these ten countries had the following characteristics in common: although many of these governments became vigorous actors in the 'politics of environmental symbols', they never accepted the basic assumption of a zero-sum game between economic growth and environmental protection. Few of these governments' responses in terms of policy and measures had significant effects on limiting the overall deterioration of the environment. None of them met the requirements for comprehensive and co-ordinated action implied by what we have above referred to as the biologists' definition of the environmental challenge. However, by the mid-1980s, there were indications, such as in the Netherlands and Germany, of a revitalised recognition of the gravity of the environmental challenge and of the emergence of a new general and comprehensive policy strategy.

Bridging the gap: ecological modernisation as a policy strategy

From around the mid-1980s, ecological disasters (in particular Chernobyl), the scientific consensus over the issues presented as the 'ozone hole' and 'global warming', as well as the almost apocalyptic mood characterising the Western mass media's coverage of these issues, led to renewed increased attention on environmental issues among the general public as well as among the politico-administrative elites. Environmental issues gained higher saliency in national politics all over Western Europe and were also given increased attention by organisations such as the OECD, the EC and the UN. During this process, leading members of the policy-making elite in Western European countries reconstructed the environmental *problematique* and put forward a new general policy strategy.

In this new phase of environmental politics, the Brundtland Commission (WCED) was an important actor, and its report, *Our Common Future*, was of great significance. The publication of the report in 1987 and the promotion of its recommendations coincided with the increased public attention to environmental issues. The report had the role of midwife for a new approach towards environmental problems, and for the lifting of this new approach into the limelight of public attention and political discourse. The Brundtland Commission's understanding of the character of environmental challenges was in some ways similar to that of the biologists around 1970, but the Commission effectively restructured the approach to these challenges, particularly by its application of the concept of sustainable development.

The concept of sustainability originally refers to ecological sustainability, i.e. harvesting and managing renewable resources in such a way as not to damage future supplies (Lélé 1991:609, Baker *et al* 1997:7). However, whereas the Commission's definition of sustainable development; development that '... meets the needs of the present without compromising the ability of future generations to meet their own needs.' (WCED 1987:8), may be seen to be in accordance with the original meaning of the concept, the application of the concept in various parts of the report implied an extension and redefinition of this meaning. The concept of sustainability was not only applied to the use of non-renewable resources, but also linked to economic development and growth. This redefinition may be interpreted as an attempt to bridge the gap between those actors advocating and promoting economic growth and those arguing for environmental interests. The report provided '... a slogan behind which first world politicians with green electorates to appease, and third world politicians with economic deprivation to tackle, could unite.' (Brenton 1994:129)

In this way the Brundtland Commission Report offered a reconstructed conception of the society-natural environment relationship that combined key elements of different and partly contradictory conceptions of this relationship which had been promoted by competing actors within the environmental policy sector. On the one hand, its description of environmental threats implied an acceptance of the definition of the gravity of environmental problems and of the necessity to solve them that had been offered by the biologists and the environmentalist movement around 1970.[2] On the other hand, the Commission did not support the assumption that policies of economic growth and of environmental protection were necessarily contradictory. On the contrary, the Commission not only emphasised that environmental quality and economic development were interdependent and mutually reinforcing in third-world countries, it was also optimistic about the prospects of economic growth in the industrialised countries. The Commission saw environmental protection and economic growth as inexorably linked, and operationalised its general position by explicitly referring to prospects of annual economic growth rates of three to four per cent for industrial countries and states.[3] In other words, the Brundtland Commission redefined the relation between economic growth and environmental protection from that of a zero-sum game to a positive sum game. As is alluded to in its report, the Commission, by this redefinition, supported efforts and recommendations made by multilateral institutions, such as the OECD and the EC.[4]

If we turn to the policy recommendations of the Brundtland Commission, these were rather few, and vague, and did not represent a consistent programme, let alone a theory, for achieving sustainable development. However, these recommendations can be seen as consistent with recommendations that since the first half of the 1980s had increasingly been offered by the OECD and a little later also by EU officials. Although the recommendations are not always fully consistent with each other, they are, nevertheless, based on some common assumptions and key elements that are inter-related to an extent that makes it reasonable to consider them as constituting a new general strategy to solve the environmental problems. This is the policy strategy which a number of social science scholars have characterised as ecological modernisation,[5] a strategy that aims at a 'greening' of key European institutions, the market economy and those of the state. Based on certain assumptions this policy strategy represented a clear choice as to the kind of policies to be pursued and types of instruments to be used to influence societal actors as well as to the organisational alternative for making environmental protection '... an integral part of the mandates of all agencies of governments,...' (WCED 1987:312)

Firstly, the assumption that the relation between economic growth and environmental protection can be considered to be a positive-sum game is based on the premise that technological innovations should not be regarded only, and not even primarily, as the cause of environmental problems, but rather as the 'solution' for these problems.[6] Technological innovations, 'clean technologies', are seen to be a necessary condition for delinking economic growth from both increased pollution and from increased energy-consumption, and a number of 'paradigmatic examples' have been provided that support the feasibility of this option (Hajer 1996:249).

Secondly, the ecological restructuring of the economy is not, as it was in the environmentalist conception from around 1970, seen to be in conflict with the institutional logic of the market economy – production for a market with the purpose of making a profit. On the contrary, rather than aiming at a radical restructuring of the institution of the market, this policy strategy emphasises an extensive use of the market mechanism. Pollution is seen as a symptom of inefficiency in industrial production (WCED 1987:220), and reduced pollution ('increased efficiency') can be achieved by ensuring that environmental considerations are incorporated at an early stage into all decisions in all sectors of the economy as well as by the internalisation of environmental costs in the price of the goods in the market. In fact, as part of this strategy, a principal policy at the macro-level is to make use of the market mechanism in combination with the establishment of conditions that make it favourable for firms to implement environmental measures. Such a policy should pursue the development of incentives and measures that encourage actors in the market (firms and individual consumers) to be both aware of and responsible for the environmental impact of their decisions. This is believed to be an effective way to promote the competitiveness of environmentally sound production processes and products. Consequently, at the micro level, it will be both possible and desirable for the firm to integrate environmental considerations into its standard operating procedures.

Based on this construction, 'that environment and technology, environment and competition, have become brothers and sisters'[7], proponents of the ecological modernisation strategy argue that the market in many cases is not only a more efficient but also a more effective institution for attaining sustainable development than is regulation by the state. In terms of environmental policy, the deficiencies of the state are primarily understood as stemming from its bureaucratic character, i.e. being '...inflexible, economically inefficient and unjust, a brake on rather than a motor of technological innovations, unable to monitor and control the billions of material and energy transformations taking place each day, and incapable of stimulating companies to adopt more progressive environmental behaviour.' (Mol 1995:46)

Consequently, the advocates of this new policy-strategy argue that the deficiencies of the state as well as the character of new environmental problems call for developing alternatives to the regulatory approach. Instead of relying on top-down regulations, one should make use of instruments that both promote environmental considerations and are compatible with the predominant type of rationality characterising the market economy. Actors in a market economy are assumed to act in a calculating manner in order to maximise monetary gains. The choice of instruments in environmental policy should, therefore, be based on the recognition that firms have to make money at the same time as they make improvements that reduce the environmental damage of their production. Similarly, in the case of utility-maximising consumers, the alternatives that are less harmful to the environment should also be the cheapest. Therefore, environmental regulations should be more flexible, cost-effective and sensitive to the logic of the market. Consequently, more extensive use should be made of types of instruments that do not suffer from the limitations of the regulatory approach. Analytically we can here distinguish between two types of instruments that have been put forward by proponents of this policy strategy:

1. Governments should make more extensive use of economic instruments, in particular environmental taxes/charges,[8] in order to promote cost-effectiveness and to guide producer and consumer behaviour.[9]
2. Governments should seek to introduce instruments that encourage internalisation of environmental values among economic actors, that is, make them feel responsible for the environmental impact of their action.

The first of these two types of instruments aims at internalising environmental costs in the prices of the goods. Its recommendation is based on the assumption that it is feasible to promote actions with less damaging consequences for the environment by changing the costs of the different alternatives open to economic actors. In other words, this type of instrument does not imply that economic actors have to take environmental considerations into account when they make their decisions. They do not follow 'the road to sustainable development' because they have changed their motivation and values, but rather because they find it, according to their calculation of costs, profitable to do so.

While the first type of instrument aims at internalising environmental costs in prices of the goods, the second type of instrument, in principle, aims at internalising environmental values and goals, i.e. getting economic actors to take environmental considerations more systematically into account as part of their standard operating procedure. This type of instrument aims at developing a kind of environmental inner-directedness among economic actors. Covenants between representatives of industry and of the state are assumed to be an instrument of this type. Environmental criteria and considerations are not intended to supplant the type of rationality that is predominant among economic actors. On the contrary, based on the assumption that there is a win-win relationship between economic growth and environmental protection, the intention is to ensure that standard operational procedures are designed and evaluated in accordance with *both* economic *and* environmental criteria.

Choosing these types of instruments points to a restructured role of the state in environmental policy. The preference for market-sensitive instruments is based on the belief in the importance of market dynamics and the role of economic actors as entrepreneurs in ecologising the economy. In the case of environmental taxes, the assumed primary role of governmental agencies is to calculate the costs of environmentally damaging activities and to decide on the right level of environmental quality to be aimed at. In the case of self-regulation, the primary role is to provide the conditions for developing ecological rationality through discussions and deliberation. The application of these two types of instrument is, therefore, believed to lead to a reduced need for an expensive bureaucracy responsible for implementation and control of environmental policy.

The Brundtland Commission Report, as well as OECD policy documents, also addressed the problem of how to ensure that environmental protection becomes 'an integral part of the mandates of all agencies of governments'. The Commission specifically recommended that these agencies '...must be made responsible and accountable for ensuring that their policies (...) encourage and support activities that are economically and ecologically sustainable.' (WCED 1987:312) Translated into organisational terms, these recommendations represent a choice of organisation for dealing with the trans-sectoral dimension of environmental policy making and administration.[10] By this choice, referred to as the sector responsibility approach, each ministry (and its agencies) is explicitly given the responsibility for integrating environmental considerations into the policy process of its sector. As phrased by Weale (1992:124–25), it aims at improving the operating software of the government machine, rather than reconfiguring the existing hardware. To the extent that the role of the ME and the environmental agencies is laid out in this approach, it is noted that they '...will be called upon to advise and assist central economic and sectoral agencies as they take up their new responsibilities for sustainable development.' (WCED 1987:319) As we see, there is a correspondence between this approach to the 'greening of government' and the approach to the 'greening of the market economy': both approaches have 'self-regulation' as an overriding tenet, and they are based on the assumption that it is possible to integrate environmental considerations into standard operating procedures of political and economic actors without changing the logics of their respective institutional spheres.

Logically, another organisational solution for integrating overriding considerations into public policy trans-sectorally would be to establish the ME as a super-ministry, i.e. as a ministry that is given the mandate and the resources needed to be able to enforce environmental policy goals in the various policy sectors, against the will, if necessary, of sector ministries. In the literature, such a ministry has been referred to as a co-ordinative structure that has no operative functions.[11] The establishing of such a super-ministry for the environment would logically be compatible with the biologists definition of the nature and the overriding significance of the environmental *problematique*. The sector responsibility approach, on the other hand, is compatible with the understanding of the relation between ecology and economy as a positive-sum game. The latter pictures 'integrating policies' as an analytical process, as a question of problem-solving on the basis of relevant information and not as a process of subordinating other interests to environmental interests. Therefore, a key element in the organisational part of the ecological modernisation strategy is that of providing the relevant information for making 'informed choices',[12] and this way of 'turning government green' logically makes two types of changes necessary:

1. The necessity of improving information about 'the state of the environment': economic and environmental values can and should be integrated on the basis of analyses and calculation, and therefore reliable data on the state of the environment as well as on aspects of the production process are decisive.
2. The necessity of implementing procedural changes to ensure that such information is included as premises in the policy process: for instance, by including in the policy process organisations that possess such knowledge (e.g. through interministerial committees, hearing procedures) and/or by making such information part of the standard operating procedures of the sectoral agencies and ministries.

In the following we shall address the question of to what extent key elements in the policy strategy described above have been put into practice, as well as the significance of these changes.

Institutional continuity and institutional change: the ecological modernisation strategy in practice

When addressing characteristic features of empirical developments in these countries it is necessary to take note of the changes in the context of environmental politics, policy making and administration that took place during the last half of the 1980s. Firstly, there was a significant widening of the scope of environmental policy (for example, through the inclusion of issue areas like those of depletion of the ozone layer and climate change). This expansion of the environmental policy field implies that new actors are (potentially) affected by environmental policies and, therefore, encouraged to participate in policy making. Secondly, to some extent as a part of this process, there has been a profound internationalisation of environmental policy (e.g. the UN and its sub-organisations have become more important as arenas for environmental politics and policy-making of consequence for most countries). Of great importance during the late 1980s was the role the EU came to play in environmental policy. As Butt Philip points out in Chapter 12, the EU's fourth environmental action programme (EAP) signalled an intention to make EU environmental policy more coherent, and the fifth EAP, to run from 1993 to 1999, was significantly entitled *Towards Sustainability*.

On the interplay between member states and the EU

Until the beginning of the 1990s, the EC's environmental policy was dominated by a regulatory approach, and by the mid-1990s about 200 directives and over 300 measures had been adopted by the EU. There is a group of states that in general promotes stronger environmental regulations than the ones adopted by the EU (Netherlands, Germany, Denmark, Austria, Sweden and Finland). Facing stronger domestic pressure in favour of environmental regulations, government in these countries sees the introduction of such regulations by the EU as a way of reducing the possibility that such national measures will weaken their international competitiveness, and, in some cases, they are even seen as a way to gain future competitive advantages. These initiatives are often opposed by the countries that are regarded as least economically developed (Portugal, Spain and Greece). Their reservations are related to the costs to the national budgets as well as to the competitiveness of domestic industry. The conflict over EU environmental policy also reflects the tendency of national governments to promote policy alternatives that are in accordance with the institutional characteristics of their countries. This is clearly demonstrated by conflicts over a number of important EU directives.

The conflict over the EU directive on Integrated Pollution Prevention and Control (the IPPC Directive) reflects the earlier rivalry between the policy alternatives of substantive emission standards and ambient air quality standards. Germany, the most important proponent of the first alternative, has found that her persistent pursuit of a 'synchronous' interplay between the development of environmental law and environmental technology has paid off. Between 1980 and 1990 Germany realised a '…remarkable achievement in reconciling economic growth and environmental objectives.' (OECD 1993: 205) During the same period, Germany consolidated her predominance in the environmental technology markets. This development contributed to reactions from other countries. In particular, in the environmental policy making process of the EU, repercussions of the German technologically oriented approach, and its concomitant commercial success, have been manifest. Not surprisingly, Britain has established herself as Germany's chief opponent. In the case of the IPPC Directive, the German position, supported by the Netherlands and Denmark, was that EU-wide emission limits should be defined and related to the BAT-requirement. The British were strongly opposed to EU-wide emissions standards. Characteristically, they favoured the 'site-specific best practicable environmental option'. The positions were 'best available techniques versus local environmental situation' (Héritier, Knill and Mingers 1996:244). Proponents of the former position feared that air-quality standards would permit two different levels of industrial pollution rights; i.e. in underdeveloped areas in countries like Spain and Greece, industry would have greater pollution rights and, therefore, lower environmental costs than in highly industrialised countries like Germany. Correspondingly, from the perspective of Spain and Greece, it is not reasonable to expect industry in 'underdeveloped' areas in these countries to implement emission standards that are necessary in highly industrialised Germany and to do so by buying the technical equipment necessary – from Germany. The southern European members supported the British position. In 1995, after offering some compensation to the southern members, the BAT clause was accepted and a compromise struck.

Politics and decision making on a number of other EU directives, such as the EU directive on the Freedom of Access to Information on the Environment (the FAIE Directive), the EIA-Directive[13] and the EU Eco-Audit regulations, reveal similar conflicts over emphasis on procedural measures *versus* substantive policy measures, over different types of regulatory systems and over state-society relations. In these conflicts, Germany and the UK were the principal contesting parties. In another case the EU Commission saw the adoption of an EU CO_2/energy tax as an instrument for demonstrating 'environmental leadership' globally and for enhancing its competence in energy policy. Germany, Denmark, the Netherlands and Luxembourg were in favour of an EU environmental tax, but the southern member states feared that such a tax would slow economic development and, therefore, their approval was contingent on appropriate compensation. Again we find Britain in the forefront of the opposition, because of the assumed economic effects and the loss of national autonomy such a tax would represent. The UK was in line with European industrialists who were intensely lobbying in Brussels against such an EU-wide tax. The EU Council in December 1994 decided that there would be no tax set at EU level.

As these cases illustrate, there is no unidirectional power exerted by the EU on member states. All member states are interested in influencing the EU policy. The three big members – Germany, France and Britain – have been the major contestants in trying to put their stamp on EU environmental policies. On the above-mentioned issues where Britain and Germany have had contrary positions, France's role has been characterised as that of a 'friendly onlooker and coalitionist' (Héritier, Knill and Mingers 1996:262ff.).

All member states have been influenced by EU policy. Although EU policies have been adjusted and redefined in the process of national implementation, even in the biggest states they have transformed existing practices and authority patterns. Not only has EU legislation displaced national legislation in many areas and changed the locus of decision from individual countries to Brussels, but also state-society relations have been affected in most of these countries.[14]

The EU environmental directives have been particularly influential in Italy, Greece, Spain and France. Morata and Font (Chapter 10) point to Spain's entrance into the EC in 1986 as a determining factor for the development of the country's environmental policy. Larrue and Chabason (Chapter 4) note that the increased impact of EU environmental policy may have reinforced the authority of the French ME in relation to industry and to other ministries, although the highly sectoral structure and the power of constituencies of the sector ministries are still the most prominent characteristics of the French environmental policy system. Spanou (Chapter 6) emphasises that in Greece the pressure for complying with EU directives has been important for the adoption of national regulation as well as for improved implementation. From an environmentalist point of view, EU directives fill the void of a rather weak environmental movement, and environmental NGOs' use EU procedures and directives as a 'resource' in national environmental politics. In many cases, EU initiatives and legislation are constrained and redefined in national implementation processes, shaped by the institutional characteristics of the different countries. For example, in the case of Italy, the 'implementation deficit' continues to be a major problem, particularly when it comes to putting EU directives into practice. These problems in establishing an effective environmental policy are closely linked to the characteristics of the Italian polity (e.g. particularism, clientelism and corruption).

The case of Britain represents a profound example of the EU's competence to make binding international agreements on behalf of the member states and its capacity to shape and influence environmental policy throughout its territory. Until the late 1980s, the British Government was able to implement most EC environmental directives under existing laws with only minor administrative adjustments required. As observed by Carter and Lowe (Chapter 2), with the gathering pace of European environmental policy making, there were increased pressures for the integration of British policy into a European framework and the administrative insularity, in terms of a system of environmental regulation that had evolved in an *ad hoc*, pragmatic and piecemeal manner was challenged. EC directives on bathing areas and drinking water, and in particular the large combustion plants directive of 1988, implied the establishment of uniform air and water quality standards and the standardisation of pollution-control procedures that upset the 'gentlemanly' British style.

Legislative measures for integrated pollution control (based on emission orientation, standards and the requirement of 'Best Available Techniques Not Entailing Excessive Cost' (BATNEC) to supplant the BPM[15]) were put forward by the British government in 1988 and integrated in the new Environmental Protection Act (EPA) of 1990. This law created the framework for an integrated approach that from 1996 was implemented and supervised by the Environment Agency (which brought together Her Majesty's Inspectorate of Pollution and the National Rivers Authority). Statutory substantive emission standards and the establishment of public registers, which contain data on permitting and operation and on monitoring of emissions, have given the controlling authorities less room for manoeuvre in their interaction with industry. This development has fundamentally challenged the informal, consultative and negotiative standard operating procedure in the tradition of the Victorian-era Alkali Inspectorate and has contributed to a politisation and a growing openness of the process of British pollution control.

This change of rights of access to environmental information was in line with a fundamental tenet of Thatcherism and the Conservatives' neo-liberal philosophy – the active consumer who according to her utility calculations makes her choices. Citizens are seen as individual utility maximisers, who according to quality criteria choose among alternatives on the market as well as among the services offered by the authorities. Logically, to make such rational choices, citizens must have access to correct information. Access to information and openness in public policy making were also seen as necessary conditions for holding government authorities accountable. In this way, both the introduction of market-type relationships into the public sector and privatisation required new regulatory frameworks and generated institutional changes. The resulting emphasis on and the growing openness within the policy process enabled environmental groups to gain access to previously closed policy communities. This development also partly changed the role of the British government in EU politics. For instance, it has supported the EU Commission's effort to depart from its original German-influenced position of adopting substantive environmental policies and implementing them from the top down and to develop a strategy that seeks to generate 'pressure from below' (Héritier, Knill and Mingers 1996:233). This is illustrated in the case of the FAIE Directive referred to above. The Tory Government was keenly aware that the new EPA would provide foreign industry access to the British public registers on emission-related matters and was afraid that without EU-wide legislation, British industry might suffer a competitive disadvantage.

Together with the Thatcher regime and the subsequent Major government, the EU has contributed to greater centralisation of UK policy making. In this way developments in environmental policy are part of a general development which has moderated aspects of the characteristic role that many relatively autonomous organisations and groups have played in British policy making and implementation that have given UK policies and the political order their traditional pluralistic character. During the 1990s, Britain has become a more centralised and unitarian state.

The development of national environmental policy – towards a 'greening' of political and economic institutions?

As pointed out above, improved information on the environmental impact of political and economic decisions is a key element in the strategy of ecological modernisation. Indeed, we find that in several respects, the institutionalisation of the application of environmental information has represented one of the most significant changes in the environmental policy field. Carter and Lowe (Chapter 2) point to the introduction of new and more extensive sustainability indicators and the publication of annual reports as one of the most important developments in British environmental policy since the late 1980s. Another prominent example is the Dutch case, where the review of environmental consequences of 'politics as usual' was an important factor that contributed to the National Environmental Policy Plan (NEPP) of 1989.

The number of environmental monitoring and reporting systems that have been established, both nationally and internationally, since the mid-1980s is closely linked to the internationalisation of environmental issues. International agreements, recommendations from international organisations (such as the

OECD) and standards and directives adopted by the EU lead to pressure for establishing and harmonising data-collection and analysis across countries. Environmental monitoring systems not only cover a growing number of environmental areas and variables, but environmental information is also increasingly presented in statistical terms. This reconceptualisation, 'the quantification of the environment', is both a result of and a prerequisite for attempts at formulating environmental policy goals in terms that can be easily operationalised and evaluated.

The definition of goals and the choice of programmes and organisations that are to monitor and evaluate the means and measures taken to achieve these objectives have, therefore, become important elements in both international and inter-sectoral negotiations. The significance of these processes is related to the fact that as the application of environmental information is institutionalised, it tends to establish a continuous pressure for governmental action. Words (declared goals and commitments) are never free, and now they are no longer even cheap. As environmental information reveals a discrepancy between a government's declared goals and commitments on the one hand, and the existing state of the environment on the other, this information becomes a resource that MEs as well as environmental NGOs may draw upon in public discourse.

The institutionalisation of information on the state of the environment may, therefore, facilitate public debate and deliberation. Another consequence is, however, that environmental issues are increasingly being transposed into technical and analytical problems and dominated by actors within scientific communities and public administration. Consequently, the relations between environmental NGOs and governmental agencies are being altered: as the modes of solutions become that of calculation and administration, and the technologies applied become those mastered by the professionals (i.e. technical experts in economics and engineering), the dynamics of issues and politics fizzle out and the logic of routine and standardisation takes over. In response, the environmental NGOs change their strategies from mobilisation of the public to a stronger role as advisers, consultants and lobbyists, and restructure their organisations in line with this role through processes of professionalisation and bureaucratisation.

Procedural changes intended to deal with the situation in which the MEs and the core environmental administration have been excluded from the policy processes in the traditional policy sectors, have to different degrees been put into practice in the various countries. For example, in the case of Greece, the exclusion of the ME continues to be a problem, whereas in the Nordic countries there is a development towards increased intergovernmental co-operation as well as mandatory environmental impact assessments (EIAs). However, also in these countries the demand for EIAs has not been sufficient to enforce the precautionary principle (for example, in the transport sector).

In other words, even though these changes (the establishing of publicly stated goals and commitments, the institutionalisation of the application of environmental information and the procedural changes that give the ME the formal right to participate in the policy processes of other sectors) are significant changes, they are not sufficient to ensure that environmental considerations have any significant effect in these policy processes. 'Intersectoral co-ordination' in many cases implies that environmental interests may be weighed in the balance and found wanting. As sector agencies increasingly participate in the environmental policy process and are assigned responsibility for the integration of environmental values in the policy processes within their own jurisdictions, this also implies that environmental issues are (re)defined and (re)conceptualised in accordance with the symbolic systems and social practices that are predominant within the respective sectors. The traditional, well-established policy sectors (such as transport, agriculture and industry) have been established in order to promote and represent specific interests and values (such as developing infrastructure, promoting employment and production). These are the values and interests that have been institutionalised as part of the conceptual systems and practices of professionals and bureaucrats within these sectors (engineers, agronomists, economists, for example). In the cases when the conceptual systems and values predominant in sector agencies are compatible with environmental interests and values, the transfer of responsibility of environmental matters to sector agencies may be seen as unproblematic or even positive from an environmentalist point of view (such as in the role of the Dutch water boards, see Andersen 1994:146–168). However, in many cases the traditional values and interests predominant in sector agencies are in conflict with environmental goals and values.

Against this background we can see the significance of the political role, in terms of jurisdiction and authority, of the ME in the politico-administrative system. After the latest reorganisation in Spain in 1996, all EU countries have set up a ME, but these MEs vary in several respects. In some countries (such as Britain and Greece) the organisational solution has been to assign the responsibility for environmental issues to a ministry with a much wider jurisdiction, i.e. only a minor part of the ministry's portfolio is environmental matters. As a consequence, actors in the ministry that promote environmental values have experienced difficulties in getting these to the top of the ministry's agenda. In contrast, the setting up of a ministry that exclusively deals with environmental issues not only tends to give these issues a formally higher rank within government, but, as the case of Germany shows, leads to a transformation of conflicts within the central administration from intra-ministry to inter-ministry conflicts, which implies that the potential for socialisation of conflicts and mobilisation of the public as well as environmental interest organisations is extended.

In countries where the ME deals exclusively with environmental issues, we can also observe variation in terms of which areas of environmental policy are included in the portfolio of the ME. In many countries, the establishment of a ME was a long and cumbersome process, during which the well-established ministries acted strategically in order to restrict the jurisdiction of the new ME. The expansion of the jurisdiction of the MEs typically has taken place in periods of increased public and political attention to environmental issues (in particular in the late 1960s/early 1970s and the mid to late 1980s). As jurisdictions are defined and institutionalised through a process of negotiations during the first years succeeding the setting up of the ministry, the established inter-ministerial structure gives priority to certain actions and policies and exclude others.

This structure tends to allow for stronger initiatives (such as legislative reforms) and more substantial impacts in areas within the jurisdiction of the MEs than in areas outside this jurisdiction. For instance, in Italy the capacity and the authority of the ME is still rather low, not the least because the new ministry (which was largely composed of personnel that were transferred from other ministries) was given limited jurisdiction. Moreover, in areas of overlapping jurisdiction, the authority has been hampered by the necessity to secure agreement from other ministries which tend to protect certain societal interests. In Norway, the two central environmental agencies have recently and jointly stated that the development of environmental policy has been most successful in areas within the jurisdiction of the ME. This statement may be interpreted not only in terms of inter-bureaucratic politics, but also as a lucid observation of the consequences of the institutionalised distribution of authority in the Norwegian environmental policy system. The results of the various changes in formal organisation and interaction patterns that have been introduced vary between sectors as well as in different areas of environmental policy. But, as noted by Jansen and Mydske in Chapter 9, the ME has lost most battles in relation to the Ministry of Transport, the Ministry of Industry and Energy and the Ministry of Finance. An exception to this general pattern may be the recently established co-operative relation between the ME, on the one side, and the Ministry of Agriculture and the farmers' organisations on the other, in the area of agricultural pollution.[16]

This indicates that the extent to which the jurisdiction of the ME matters is dependent on 1) the extent to which the ME has its own field organisation[17] and 2) the characteristics of the ME and this field organisation (e.g. in terms of types of professions predominant in the ME and its field organisation, the type of knowledge or technology these organisations make use of, the degree to which the professionals identify with environmental interests and values and to the extent to which they have professional autonomy).

In Norway, the ME has its own agencies both at the central and the regional level, and these are all parts of a hierarchical structure subordinated to the ME. The Norwegian case is also illustrative of the second dimension. Admittedly the core environmental administration is characterised by a significant number of professionals trained in the key technologies of other policy sectors (engineers, agronomists, economists and lawyers), but in addition, biologists and other professionals trained in natural sciences have been recruited. The expansion of the core environmental administration has led to the inclusion and institutionalisation of a new type of technical competence in public administration and also to an environmental administration that to a great extent is staffed by civil servants who look upon themselves as being advocates for environmental interests and values.

The development in France is illustrative of an extreme position at the other end of these dimensions. The French ME has no field organisation but is dependent on the sector ministries and their field organisations. It has no profession of its own, but is dominated by the same *corps* that dominate the sector ministries. Moreover, while French environmental policy organisation at the national level seems to reflect typical features of a centralistic unitary state, the political structure at the regional and local levels increasingly have permitted participation by the public and some degree of influence by local interests. Local environmental action groups are included in the implementation process and typically in this process the interaction between the environmental authorities and industry is, as pointed out by Larrue and Chabason in Chapter 4, one of dialogue, negotiation and partnership. The decentralisation laws of 1982 also made it possible to set in motion a process that transformed regional and department councils into significant actors in environmentally important policy (e.g. regional planning), and that consequently further eroded the Bonapartist character of the French state.

The significance of the lack of an effective field organisation is also illustrated in the case of Greece. Despite the organisational reforms around 1990 aiming at developing an environmental administration in the regions and within the prefectures, Greek environmental administration continues to be weak. This weakness creates problems both for the implementation of environmental policies and for the monitoring and control of environmental quality. These problems are not only linked to lack of staffing or financial resources, but also to the fact that institutional characteristics of the Greek state (such as particularism and personalism) still persist and also shape the environmental policy sector.

The distribution of jurisdictions and responsibility between ministries is also of significance in inter-ministry negotiations and conflicts. If a matter is defined as part of the jurisdiction of a certain ministry, it gives this ministry a high degree of control in the policy process, in some cases even 'a right of veto'. The choice and application of the sector responsibility approach is therefore of crucial importance for the development of environmental policy. Assigning responsibility for the integration of environmental considerations into the policy processes to the relevant sector ministries/agencies implies that such sector ministries/agencies (with their specific symbolic systems and social practices) become the 'centres of gravity' for environmental policy making in the areas in question. Ministries are embedded in wider socio-economic contexts, and they can in many cases be seen as functioning like bridgeheads or even as agents for certain societal interests and actors. Consequently, these societal interests are included in the policy process whereas others are excluded.

Seen in this perspective, in the cases when environmental interests and values are in conflict with the predominant values of a traditional policy sector, the prospects of the former are therefore rather poor when the sector responsibility approach has been chosen as the organisational alternative. This may be outweighed if the ME maintains control over environmental policy goals in the process during which national goals and targets are 'broken down' into specific goals for sectors and/or target groups as in the case of the Netherlands. In the Netherlands, the NEPP identified eight themes and nine target groups, and established overall objectives for environmental policy. As noted by Hanf and van de Gronden in Chapter 8, the target groups, which included both sectoral administration and societal actors, were strongly involved in the development of the plan and were also given responsibility for achieving the emission reduction targets. The process of establishing these targets required analytical skill and capacity and this comprehensive process of planning has characteristics similar to, and was probably facilitated by, the tradition of economic macro-planning that was established in the Netherlands in the post-World War II era.

Although there may be cases where environmental goals are used as authoritative premises for the development of sector policy, there are numerous examples showing that alliances between sector ministries and organised societal actors predominant within these sectors have restricted and defeated initiatives from the ME or other environmental actors to give priority to environmental goals and values both in national policy and in sector policy as well as to reshuffle authority relations in favour of environmental interests. Such examples are typical in the area of climate policy which during the last decade has been put at the centre of environmental policy.

In the case of Britain, climate policy illustrates both new and traditional features of British environmental policy. The British programme of 1994 for reducing greenhouse gas emissions passed through a long consultative process with a wide range of invited interests (O'Riordan and Rowbotham 1996:235) which may be seen as reflecting the new openness within the policy process. A carbon/energy tax was ardently resisted by the policy communities around ministries like the Treasury, Trade and Industry and Transport, which fought against it because it would influence industrial competitiveness. Consequently it was ruled out. Emissions from cars are the fastest growing contributor to CO_2 in the UK (O'Riordan and Rowbotham 1996:255–256), but there is still no integrated transport policy. Also, in Germany, conflicts erupted over policy initiatives taken by the ME on the national target for reduction in CO_2 emissions and on the application of instruments. These inter-ministry conflicts demonstrate that authority relations between ministries are embedded in an economic and political context, in which the ministries' relations to different societal interests and the 'weight' of these interests in the political order are of great significance. In general, the initiatives of the ME have had little impact on the Ministry of Economic Affairs and the Ministry of Transport. In the case of the introduction of a German CO_2-tax, the government solved its inter-ministerial conflict by tying such a tax to an EU-wide solution (Pehle and Jansen in Chapter 6).

Both these cases reveal that the introduction of such taxes has been postponed after opposition from sector ministries and representatives of big industry. Norway is one of the countries where such taxes have been introduced, but only 60 per cent of the CO_2 emissions are covered by a tax, since exceptions are made due to reasons of competitiveness. Also in Norway attempts to increase taxes or remove exemptions have been blocked. The position of the dominating coalition in climate policy (Labour and the Conservatives) has been that the increase of such taxes should be made conditional on developments in other OECD countries. This position is explicitly legitimised by the principle that policies should be cost-efficient across climate gases, sectors and countries. The adoption of this approach logically reduces the credibility of the front-runner role in environmental policy. If Country A does not introduce environmental taxes, the introduction of such taxes in Country B would not be cost efficient. The consequences of this approach may be interpreted as the precautionary principle in reverse, i.e. that national goals and the introduction of policy-instruments are to be considered on the grounds of competitiveness. The practical consequence, then, is that a government which in general considers itself to be a front-runner in environmental policy postpones the introduction of concrete measures in climate policy and makes them dependent on international agreements. Thus, while awaiting the establishing of an effective international climate regime, 'the rear party sets the pace'. This means that when the principle of cost-effectiveness is used as a premise for environmental policy, advocates of stronger national environmental policy measures are manoeuvred into a 'Catch 22'-situation: 'If you are to introduce stronger policy measures, you should make use of economic instruments rather than legal ones. However, you should not introduce economic instruments, either in a policy sector or in a nation unless these instruments are introduced in other sectors and nations.'

This highlights the fact that the common criteria for comparing environmental and economic values and goals that have been developed with the declared purpose of serving environmental interests are to a great extent rooted in the concepts and categories of economics. The inclusion of and redefinition of the environment into economic analyses and calculation implies that whatever the intentions, the environment is perceived in monetary terms and ultimately is deprived of any moral value. This process, the calculation of pros and cons within economic analyses, is therefore well suited for laying down lines for state intervention in accordance with the governing principles of the market economy, including de/prescribing the circumstances in which such intervention should not take place, e.g. in the cases of trade distortions and loss of competitiveness.[18]

The tension between economic and environmental goals can also be observed in the EU. EU environmental policy has, like the Roman god Janus, two faces. As we have seen, EU environmental policies have influenced legislation and the choice of instruments and measures in all these ten countries, and have been an impetus for stronger governmental responses in several member states. In many cases, the introduction of new directives, which promote institutional changes and convergence in national policies, will be seen to be in accordance with the logic of the single market. That is to say, they reduce the inconsistency of rules and regulations between member states and thereby promote free trade. The other face is that there is a tension between the economic and environmental policy objectives of the EU, a tension that is illustrated both in cases when new measures are postponed due to reasons of global competitiveness and in cases when the primacy of economic integration and growth within the single market leads to environmental deterioration (e.g. increased cross-frontier trade).

The same tension can be observed in the case of proposals for making more use of environmental taxes in general, even in the Nordic countries, which seem to be in the forefront in this area. On the one hand, the recommendation for increased use of economic instruments has taken root in the national environmental policy strategy in these countries. From the late 1980s, several environmentally motivated taxes have been introduced (e.g. taxes on CO_2 emissions, artificial fertilisers and pesticides) together with first initiatives towards a 'greening' of the whole tax system. The 'green tax exchange' (*'skatteväxling'*) of the Swedish tax reform in 1990, where radical cuts in income taxation were covered by increased taxation of environmentally damaging products, may be considered a paradigmatic example of such a comprehensive strategy. The breakthrough in Denmark came with the 1993 tax reform, which implied a gradual increase of environmental taxes (e.g. on water and non-renewable resources). On the other hand, the introduction of 'green taxes', has, in some areas of environmental policy, been met by strong opposition from societal interests and governmental agencies. Andersen, Christiansen and Winter's statement in Chapter 3 (p. 43) on the development in Denmark is illustrative of this: 'Despite widespread recommendations of economic instruments, they seem to be very difficult to introduce in the business sector.'

The tax reforms that have been planned and implemented in these countries are largely a result of a more general policy strategy for reducing the tax on labour and capital in order to limit distortions in the economy and increase employment. Interest in 'greening' the tax system coincided with the dominating national coalition's interest in a general overhauling of the tax system to make it more efficient and to meet the requirements set by reduced public support for high levels of taxation. The reputation of these countries as forerunners in making use of environmental taxes can, therefore, partly be interpreted as a result of the abilities of the policy elite to play the international game of symbolic politics.

The development in the use of environmental taxes reveals a striking discrepancy. The national governments' response has been to pay lip service to and to second recommendations for the use of economic instruments, which implies that one makes use of norms and principles that are supposed to regulate economic action (self-interest, calculation of costs, profitability) rather than aiming to change or limit the extension of these norms and principles (such as through legislation). On the other hand, this type of instrument has only been put into practice to a limited extent.

In many cases, actors oppose such taxes due to economic self-interest. But there are other sources of opposition. Proposals for using such instruments often evoke resistance in scientific and professional communities where the assumptions that an environmental tax will influence the actors to reduce pollution in the way assumed in economic theory are questioned. For instance, agronomists have questioned the assumption that taxes on fertilisers would lead to reduced nitrogen emissions. Such proposals also often evoke 'ideological' resistance: environmental taxes are seen as being a way for polluters to ransom themselves. In many cases we find that all three sources of oppositions are activated, and the introduction of economic instruments is opposed by policy communities that resist such instruments due to economic, scientific (professional) and ideological reasons.[19]

The opposition from representatives of industry is of particular interest since the principal position – that economic instruments should be preferred to regulatory instruments – is consistent with the view of predominant scientific communities as well as with ideological positions of the dominant policy community in industry. In practice, however, the introduction of environmental taxes has been postponed and/or important parts of industry have been exempted from such taxes for reasons of international competitiveness. In other words, in these cases environmental taxes, which in general are not only consistent with but also make actively use of the rationality of *Homo Oeconomicus,* are ruled out because they are seen to reduce the chances of making national industry profitable, i.e. the overriding imperative of any enterprise in a market economy.

In many countries, representatives of organised industrial interests have argued that covenants or agreements should be introduced as an *alternative* to environmental taxes. As noted by Larrue and Chabason in Chapter 4, in France, branch contracts aimed at reduction of pollution date back as far as the early 1970s and can be interpreted as mainly an offshoot from the French tradition of central planning. After the mid-1980s, a new wave of covenants or agreements as a specific policy instrument occurred in countries that have been characterised by having both an ambitious environmental policy and a tradition of consensual and co-operative relations between state and industry, in particular the Netherlands and Denmark.

Dutch environmental policy has been far ahead of the other nine countries when it comes to the introduction of voluntary agreements or covenants. The widespread Dutch application of covenants confirms and reproduces institutional characteristics of the relation between government and societal actors in the Netherlands, in particular the institutionalised values of co-operation and consultation. Government, industrial branches and individual companies have a mutual understanding of the necessity to sustain the competitiveness of Dutch firms, and there is a 'gentlemen's agreement' that direct regulations and environmental taxes will not be introduced by the government when agreements are concluded, which provides an incentive for firms and branches to participate in this new partnership (ECON 1996a:32).

In Denmark, agreements were first introduced in the late 1980s, aiming at reduction of specific polluting substances (e.g. PVC, VOC) (Mol *et al* 1996:11). As to the effects of these agreements, the interpretation is that they have been rather successful in some areas (e.g. used tyres) and rather poor in other areas (e.g. the battery covenant) (Holm, Klemmensen and Schrama 1996:6). The introduction of agreements in new areas such as climate policy has been closely linked to the green tax reform. Firms that conclude agreements will obtain reductions in the energy/CO_2/SO_2 charges. Firms that do not fulfil an agreement can be forced to refund the reduction in charges that they have obtained (ECON 1996a:25–26).

It should be noted that both in the Netherlands and Denmark agreements were introduced by liberal-conservative governments as part of programmes for deregulation, and that '...these innovations have abandoned to some extent their initial conservative deregulation-privatisation connotation.' (Mol *et al* 1996:20) Agreements have not supplanted regulations or taxes, but have become institutionalised as an alternative to further regulations and as part of the ecological modernisation strategy in these countries. Recently, agreements have also been introduced as an alternative to regulation in countries like Germany and Spain.

In order to avoid new extensive regulations and in particular increased charges (e.g. the proposed CO_2/energy tax) representatives of German industry, proposed in the early 1990s that this industry should conclude voluntary agreements with the authorities on reductions of CO_2 emissions. This was, however, seen as a way of obtaining exemption from charges, and therefore interpreted as being in contradiction with principles of the German constitution, particularly the principle of equal justice. Nevertheless, in 1995 the authorities reacted positively to the statement from 15 branch organisations that before the year 2005 German industry and trade was prepared to reduce specific CO_2 emissions by 25 per cent, with 1987 as the base year. Negotiations subsequently followed. Notably no formal agreement has been signed by the parties; the negotiations are rather a continuing dialogue between the parties aimed at increasing the level of ambition during the years to come (ECON 1996b:6–10).

In general, also, environmental policy in Spain has continued to rely on a regulatory approach, a strategy that is consistent with its closed and bureaucratic policy style. The policy-making process has been characterised by a general lack of institutionalised co-operation and consultation between the state and societal actors. This 'statist institutional design' has been reproduced in the area of pollution-control policy, where the relationship between government and industry has been limited to personal contact on specific issues (Aguilar-Fernandez 1994:104). However, as both large industries and government experience the need to comply with EU directives, the government has seen agreements as an alternative to regulations, thus taking initiatives that may lead to a transformation of this traditional relationship between government and industry. Since the regions continue to be the most important actors, agreements have been developed at the regional level, particularly in highly polluted areas.

Although these descriptions show that it is a characteristic of covenants or agreements that they have no well-established definition as a policy instrument, we still assume it is useful to see them as a distinct class of policy instrument. Logically, using this class of policy instrument implies taking a position on a number of significant dimensions of politics, policy and polity, and we can observe that these countries have chosen different positions on such dimensions. Consequently, they have applied different types of agreements as instruments in environmental policy.

One crucial dimension of agreements is which subjects are open for negotiation. Agreements may represent occasions only for choosing measures for reducing the negative environmental impact of the activity of the firms while the government sets the targets with which the industrial branches/firms are expected to comply. At the other extreme, agreements may represent arenas or occasions for negotiation over the environmental quality objectives and/or targets for the reduction of pollutants emitted and/or the application of alternative policy instruments (e.g. regulations or economic instruments).[20] In the Netherlands, representatives of branches of industry and big firms without doubt influenced the setting of goals and targets as well as measures to be taken, but long-term reduction targets were not negotiable. In several cases (e.g. the Dutch and the Danish VOC-agreements) negotiations have been used by business organisations to postpone regulation and weaken environmental goals (Holm, Klemmensen and Schrama 1996:18).

Another dimension is the distribution of tasks of monitoring and control. One extreme may be that government agencies are able to counterbalance the asymmetric relation between themselves and firms when it comes to technical competence and information through extensive monitoring both of the general state of the environment as well as the measures taken by industrial branches and individual firms and the effects of these measures. The other extreme may be that only the industry or the firms monitor the effects of measures taken. This implies that firms will increase their technical competence and professional authority in relation to government agencies. In the Netherlands, a professional and independent agency, *Novem*, is assigned a principal role in monitoring the results of the agreements, while in Germany it is the branch organisations of industry that are to play a crucial role in the follow up of the agreements.

Agreements also vary as to how mandatory they are and as to type of sanctions available to a party if the other party does not comply with the agreement. At one extreme, branches/firms may be obliged to comply with the agreement to the extent that they are liable to be taken to court by government agencies for failing to do so. At the other extreme, compliance can be purely voluntary, with no recourse to sanctions by the authorities. For instance, the agreements in Denmark are more firmly anchored in legislation than are those in Germany. The German covenant framework has been characterised by the authorities as an 'open-ended process'. The agreements are not contracts in legal terms. In addition to the complexities and difficulties in connection with verification, there are few sanctions that can be applied against branches or firms that do not fulfil an agreement.

Another dimension is the number and types of actors included in the processes of consultation and negotiation. It may be an open and deliberative process in which many actors representing a variety of interests are involved or, on the other extreme, it may be a closed process within a small policy community of representatives of central sectoral agencies and of representatives of industrial branches and big firms (e.g. Italian central authorities and Fiat).

In general the agreements are being criticised for the lack of democratic control (Mol *et al* 1996:10–11). Moreover, as such agreements are made (whether anchored in law or in gentlemen's agreements and understandings) they curtail the possibility of future policy initiatives by parliament and government. It should be recognised that the use of agreements represents a (re)structuring of the relation between the state and industrial interests. Increased use of this type of instrument involves a (re)structuring of the environmental policy sector: some actors and interests are included while others are excluded and, subsequently, new patterns of authority and influence over environmental policy are established.

Without doubt, agreements may give organised interests the chance of influencing and negotiating environmental policy goals as well as their implementation. They may even give such interests the chance to postpone or prevent the application of alternative instruments. Negotiations and conflicts over what type of agreements should be adopted are significant because as practices such agreements constitute institutional features that shape politics and strategies. As such they will make a difference for future interaction between the state and these interests. When established as rights and statuses for industrial interests, to be allowed to participate and to be heard as well as to influence and have the right to negotiate, they are institutional characteristics that give agents of industrial interests an advantage in the future game of power. As they are established, these practices become part of the arrangement of power and authority in environmental policy, and therefore not only between state and industry, but also relative to other societal interests. It is because of their function of 'relating' state and society and in doing so also ranking societal interests that such agreements become part of the political institutions of a society (Wolin 1960:6–7).

The general finding is that the strategy of ecological modernisation has not supplanted the application of traditional regulatory instruments. The impact of this policy strategy is primarily in the new areas of this policy field (e.g. climate policy and the expanded area of international environmental policy). This policy strategy has not been pursued and implemented as a coherent strategy. Rather, there is great variation as to which and to what extent the different countries have adopted its various elements. Characteristically, the various elements have been (re)defined, redirected and in some cases have been defined as politically feasible policy as they have been confronted with and adapted to the institutional characteristics of the different countries. In many cases, policy communities within sectors have opposed environmental taxes, in particular when such taxes have not been earmarked and thereby reallocated within the same sector, and given their support to various types of self-regulative instruments. The increased use of various elements of the policy strategy of ecological modernisation is therefore highly dependent on whether it is consistent with the institutional characteristics of the various countries and the interests and values that they promote: the increased use of environmental taxes in the Nordic countries, in particular on consumers, has been justified in the context of the traditional institutional logic of the welfare state, i.e. that such a burden is necessary to promote the common welfare. The long and well-established tradition of public planning in countries like the Netherlands, Norway and Sweden has also contributed to making it reasonable to use environmental information as an important element in current public planning in these countries.

The significance of this policy strategy in the new areas of environmental policy lies not primarily in the sense that the measures and instruments that have been seen to be in accordance with this strategy have been put into practice, but rather in the sense that alternatives that have been seen to be inconsistent with this strategy have been ruled out. Both environmental taxes and agreements, which often have been proposed as competing policy alternatives, may be interpreted as parts of the reservoir of policy instruments of the same market-oriented policy strategy. This market-oriented strategy has focused on the deficits of 'command-and-control' instruments which are to be ruled out because they are both 'inefficient' and incapable of stimulating companies and other actors to adopt more progressive environmental behaviour.

The major accomplishment of the policy strategy of ecological modernisation is, then, its significance for the efforts of policy elites to redefine and redirect initiatives in the public environmental debate, in particular during the period of the expansion of environmental policies from the mid-1980s. From around 1970, these policy elites were confronted with the environmentalist definition of the environmental *problematique* which pointed to the necessity of a radical restructuring of the core institutional orders of modern western society, in particular the market economy and the established state apparatus. The environmentalist movement represented a radical critique and an attack from below on what was seen as the 'growth-promoting state industrial complex'.

The promotion of the policy strategy of ecological modernisation may, therefore, be interpreted as an attempt both to neutralise the critique of the shortcomings of the market economy, and to capitalise on the critique of the flaw of environmental regulations. As such, this policy strategy has now been successful. The ecological modernisation discourse has become hegemonic in the environmental policy discourse (Hajer 1995). In other words: the policy strategy of ecological modernisation has served as an instrument for harnessing the social forces that were set in motion by the mid-1980s through public reactions to experienced and anticipated environmental disasters. By framing the expansion and internationalisation of environmental policies in terms that stress particular types of instruments based on the institutional logic of the market economy, environmental policy instruments anchored in other types of institutional logics are organised out as applicable alternatives.

An interpretation of strategy and practice

To understand why the ruling elites of Western Europe currently are increasingly pursuing the policy strategy of ecological modernisation this strategy has to be interpreted in relation to the imperatives to which these elites have to respond. Pressure from citizens who want to protect their environmental quality is only one among numerous other 'heavyweight' premises for action by these elites. The core of these imperatives are those that today follow from the structural position of the liberal-democratic state in a global market economy. The status of the ruling elites is dependent on their ability to deal with these conflicting imperatives. The defining features are that *access to political power* in liberal-democratic states is dependent on the extent of support from the voters in free elections. However, *the ability to use the power gained*, in terms of being able to implement policies and programmes, is dependent on finances derived from various modes of taxation upon private wealth and income, i.e. on capital generated through private accumulation. Therefore, there is an institutional self-interest of the state, that is to say an interest of all those who wield state power (in particular, the elite of politicians and officials), to safeguard the competitiveness and vitality of the country's economy. Phrased in terms of the logic of government action: if the government fails to secure economic stability and growth, the result is falling tax revenues for policy programmes (e.g. for welfare services which in many states now have been taken for granted) that the government wants and may have promised the voters to pursue, and, consequently, the government will be less popular and possibly suffer electoral defeat.[21]

All ten country chapters refer more or less directly to the necessity of achieving economic stability or growth as a tenet of governmental action, irrespective of what political party is in power. Seen in terms of this imperative of accumulation, to which governments must accommodate themselves, we see how radical, how revolutionary, were the means and policies to solve the environmental threat proposed by the biologists and environmentalists around 1970. We also see that for the political elites of Western Europe, and of the OECD countries in general, the strategy of ecological modernisation offers an alternative by which these elites logically can be politically on the offensive in relation to environmental problems while they act to meet what is required of them in terms of economic rationality. By adhering to the strategy of ecological modernisation, governments can logically pursue policies of environmental protection without producing a negative impact on the confidence of capitalist investors. Industry and business do not have to be regulated in a way that intrudes on the terms of equal competition or interferes fundamentally with the logic of action characterising the market economy. An additional bonus is achieved when the proponents of this strategy succeed in gaining acceptance among firms for the belief that the environmental quality of products and processes of production gives them a competitive edge in the market. The crucial assumption is that if this policy strategy is put into practice internationally, governments can pursue environmental protection while they pursue policies that safeguard their countries' competitiveness and place in the world's capitalistic order.

In this perspective, the overriding 'project' of the policy elites of Western Europe over the last quarter of this century has been to respond rationally to the above imperatives, which during the last decade have been emphasised by the increasingly felt dictates of international regimes for free trade and finance in a globalised economy. In all these ten countries we can, in particular from the middle of the 1980s, follow how this project has been operationalised into policies and programmes that were aimed at reducing the role of the state and expanding the domain of the market. These efforts took place in several policy areas:

a) *Competition policy*: All ten countries have made efforts, partly driven by the EU, to strengthen competition legislation and policies (e.g. price liberalisation, deregulation of sectors previously sheltered from foreign competition, liberalising public monopolies, liberalising regulation on mergers and acquisitions).

b) *Employment policy*: To deal with imperfections of the labour market, employment legislation has been 'liberalised' (e.g. reducing unemployment benefits and the period of eligibility, reducing restrictions on part-time work, relaxing limits on working hours, relaxing the monopoly on placement of the public employment services).

c) *Deregulation of the financial system*: In most of these countries the financial markets had been regulated to fund public deficits and specific sectors of the economy that had been targeted by the political authorities. Deregulation meant significant changes, such as the freeing of interest rates, an end to state credit controls and liberalisation of capital movements. It also meant a profound restriction on the role of the state. In the financial system, prudential supervision of market participants was to be strengthened and the superior status of the state was to come to an end. The liberalisation of capital movement, and the associated banking and financial services, both nationally and in the European single market, was soon followed by a similar liberalisation internationally. Concomitantly stability-oriented monetary policies were pursued.

d) *Fiscal consolidation policy, limiting the size of the public sector*: Most of these countries experienced that persistent pressures for spending led to large governmental debts and fiscal imbalances, which proved to be a significant weakness in the increasingly competitive international marketplace. In all these ten countries policies and programmes for cutting the cost and increasing the efficiency of the welfare state were launched in combination with renewed efforts to reduce budget deficits and limit the growth of the public sector.

e) *Policies for administrative reform and public-sector management*: These policies aimed at changing the size and structure of the public sector as well as the ways in which government activities are carried out (Ormond 1993:4). A number of new types of instruments were introduced or increasingly used, such as (1) privatisation, (2) assigning a new role to central management bodies, (3) market-type mechanisms (e.g. internal markets, user-charges, vouchers, franchising, contracting out) (4) mechanisms for promoting 'managerial mentality' (OECD 1987:125) (e.g. corporatisation, performance-based pay).

These policies represent a coherent policy strategy to increase rationalisation in terms of the institutional logic of the market economy, and consequently, to expand the institution of the market into spheres of activity, in particular state activity, that previously had been more or less characterised by the logic of other types of institutions. In line with our assumptions, these policies have, therefore, been contested and opposed, e.g. by trade unions, traditional professions and professions of welfare state services as well as by representatives of agricultural interests. The extent to which they have been adopted, pursued and implemented has varied among the ten countries. Without doubt, the UK has most consistently and vigorously pursued and implemented all these types of policy. Since the Thatcher Conservative government first launched its programme aimed at a thorough restoration of the logic of the market and market discipline in British society, Conservative governments from 1979 onwards have implemented policies in the areas of a), b) and c) above to the extent that there today is a lower degree of regulation in this country than in the other nine countries (Koedijk and Kremers 1996:446–447). Similarly, the UK stands out with regard to the above policy areas d) and e). From the late 1970s, the Thatcher government implemented the transfer of one million jobs from the public to the private sector in the subsequent decade (Hood 1996:37). What followed was a general 'corporatization' of the civil service. There was a strong reliance on performance-related pay, compulsory competitive tendering procedures, an explicit move away from a standard service-wide set of process rules towards a structure in which process rules were negotiated *ad hoc* and greater attention was placed on outputs expressed in terms of new accounting procedures and management information systems (Hood 1996:39–40; Pollitt 1990:181). This restructuring and managerialism was particularly clearly demonstrated in the health service sector (Stewart and Walsh 1992:502). Authority relations were significantly changed: the authority and leadership of professions like medical doctors, lawyers and teachers were reduced and supplanted by that of leaders trained in management techniques and often hired on fixed-term contracts. In institutional terms, these policies were revolutionary. They represented a revolution from above, legally implemented.

In general the changes in deregulation and competition policies took place more gradually in continental Europe. Germany, however, took the lead in pursuing a restrictive monetary policy accompanied by a restrictive fiscal policy. Although the ruling Christian Democrats of the Kohl coalition that came to power in 1982 were inspired by their counterparts in the UK and the USA, there were significant differences between the CDU and the neo-conservatism of Thatcher and Reagan. Their 'turn around programme' (*die Wende*) was primarily a programme to eliminate the excesses of state and interest-group intervention in the market in order to re-establish the basis for sustainable economic growth (Krüger and Pfaller 1991:205). The federal government sold the shares in some of the corporations it controlled and there were other signs of a withdrawal of the state from society, in particular in terms of changing substantive policy programmes that were implemented by public enterprises or were heavily regulated. However, the structure of the administrative system as well as the general pattern of decision making and management remained unchanged (Derlien 1996:157–158, 165).

In most of these policy areas, the UK and Germany represent two contrasting cases of policies for rolling back the state and restructuring its role in line with the logic of institutional order of the market economy. This contrast follows from the policy elite's strategies and power as well as depending on specific institutional characteristics of the two countries. The radical policies that were implemented to improve entrepreneurship and competitiveness in British society and to change the public sector to be compatible and aligned with the institutional logic of the market economy was of course dependent on the will and power of a central actor, like Mrs. Thatcher. But it was also dependent on country-specific institutions of the UK, in particular the two-party system and the Westminster model of government that gives the governing majority such a commanding power to adopt and implement its policies. In Germany, the constitutional principle of the social welfare state sets effective limits to federal government activism. Moreover, the constituencies of the CDU-FDP coalition ensured that the traditional policies of co-operation and consensus between the important interest groups of German society were continued, and that institutional changes took place more gradually.

In France, during the last 15 years the state was gradually and moderately rolled back by privatisation, deregulation, supply-side economics. Its somewhat restructured role was further accentuated by a more vigorous administrative modernisation policy after 1988 (e.g. by a more consistent introduction of management tools like pay-for-performance (Rouban 1994:98, 1995:54–58, 61) and an increased application of concepts like 'service', 'the market', 'customers' or 'corporatism' in the public sector (Montricher 1996:261–265). The changes that have taken place have made French public administration, in terms of structure and role, less distinctive and specific (Rouban 1995:39). The movement away from a state-directed to a more market-oriented economy is obvious. After 1990 French governments have not had the capability to be *dirigiste*. This has, however, not meant an end to state influence: governments have not stopped seeking to guide business, albeit in more indirect ways, more in keeping with the new international economic environment and the dictates of the European Union. They as always play a primary role in deciding the direction of economic growth and the shape and organisation of economic activity, even as they engineer the retreat of the state. (Schmidt 1996:400)

Countries like Greece, Italy and Spain have more reluctantly, but gradually, followed in implementing deregulation, competition and fiscal consolidation policies. Programmes for restructuring the public sector and changing its management have also been adopted. The Netherlands and the Nordic countries have even more vigorously pursued deregulation and fiscal consolidation. As to programmes for administrative reform and public-sector management, the Netherlands has during the last 10 to 15 years implemented three 'efficiency operations' in which privatisation and contracting out as well as administrative modernisation policies have been pursued. In the Nordic countries similar programmes have been adopted. The implementation style has been oriented towards compromise. However, significant changes in terms of privatisation and rationalisation of public monopolies (such as in telecommunications and postal services) as well as in terms of the structure and role of public administration, have gradually taken place. New management techniques have been introduced and applied in manners similar to how they are practised in business and industry. Furthermore, it is noteworthy that the highly centralised negotiation system in the state sector has been transformed into a system of more decentralised bargaining. Together with the introduction of pay-for-performance, this has contributed to a departure from the traditional egalitarian wage policy based upon 'solidarity' within the state sector (Christensen 1994, Lægreid 1994, Sjølund 1994).

As in the case with the implementation of the strategy of ecological modernisation, there is significant variations among the ten countries with regard to the implementation of policies in these five areas. However, there is no doubt about the general expansion of the logic of the institutional order of the market in all ten societies. Different kinds of studies report such findings and numerous articles in the OECD and the EU literature (e.g. *The OECD Observer*, *OECD Economic Studies*, *OECD Economic Surveys* and various reports and studies in the series of *European Economy* from the European Commission) specifically point to this development. The OECD and, particularly, the EU are exerting pressure to ensure that this institutional logic is consolidated and is further expanded into spheres of activity that in these countries have been predominantly state activity and characterised by institutional logics that differ and may partly be contradictory to that of the market economy. A most relevant case is the convergence criteria that must be satisfied prior to the establishment of the European Currency Union. All these criteria are part of the EU policy elite's strategy to rationalise the EU economies in terms of the institutional logic of the market. At present this is particularly well illustrated by the criterion that member states' fiscal deficit should be no larger than three per cent of GDP. In a majority of these ten countries this requirement is seen as quite contradictory to the kind of instruments and resources the state needs in order to pursue counter-cyclical economic policies and to fight the battle for jobs. The establishment of the European central bank, with unprecedented legal authority that is beyond the control of politically responsible government, is another significant case that illustrates the withdrawal of the state and concomitantly the undermining of the institutions both of popular control and representative government.

The interpretation that the driving force behind this institutional development are relatively homogeneous European policy elites, who concentrate on responding rationally to what they see as imperatives of any liberal-democratic state in a globalised economy, assigns relatively little significance to traditional political ideologies and party cleavages. A test of this interpretation would, therefore, be what the Labour government in the UK is doing on these issues. One would expect that such a government would reverse the Thatcher revolution. But, based on the interpretation offered here, the Blair government will, rethoric to the contrary notwithstanding, mainly follow in the tracks of the Conservatives. Similarly in France, we expect that the government of the political left, headed by Lionel Jospin, will, after some initial ideological window-dressing, arrive at the moment of truth and accordingly pursue policies much in line with the policies of the centre-right administration.

It is against this background that we can understand the significance for these policy elites in succeeding in redefining the environmental *problematique* and in redirecting initiatives in the public debate on environmental issues by framing them in the policy strategy of ecological modernisation. This policy strategy of ecological modernisation is not only compatible with, but may even be seen as part of the overriding project of the ruling policy elites to expand the logic of the institutional order of the market.

We have shown that the key assumption of the policy strategy of ecological modernisation – the assumption that the relationship of economic growth and environmental protection is a positive-sum game – is compatible with what the ruling elites see as the rational response to the current imperatives of government in Western Europe. It is this compatibility that primarily has made this assumption the *Leitmotif* of these elites. In other words, it is (political) credibility rather than (scientific) validity that explains its authoritative status as the environmental policy strategy in these ten countries.

The question now is whether or not it is reasonable to assume that governments' pursuit of the strategy of ecological modernisation will result in a system of governance that can realistically meet the environmental challenge. In answering this question, we should note, on the one hand, that significant results have already been achieved. Some types of damaging pollution have been greatly reduced, e.g. industrial pollution, sulphur emissions. Some countries, among them the most pronounced proponents of ecological modernisation, like Germany and the Netherlands, have achieved substantial reduction of industrial pollution while achieving significant economic growth. On the other hand, it is notable that there is a general agreement that none of the most active promoters of ecological modernisation will achieve its goals in terms of the central categories and criteria of environmental policy (e.g. national targets for reduction of CO_2 emissions and energy consumption). More importantly, however, fundamental developments have demonstrated the weaknesses of the policy strategy of ecological modernisation and the validity of the construction of the biologists of around 1970. The defining characteristics of the latter construction, such as the finite character of the Earth and the interdependence and vulnerability of mankind and mother Earth, are being demonstrated by long-term consequences of ecological accidents and the encroachment on open space (not only in the Amazon and the Antarctica but in all types of Western European countries as well). These characteristics are also at the core of vital issues like global warming and genetic engineering. There are many scientists and thoughtful citizens who corroborate concepts like Risk Society (Beck 1992) and The End of Nature (McKibben 1989), and even those reluctant to accept the validity of these concepts acknowledge that the phenomena which these concepts claim to denote are unintended and probably irreversible side-effects of man's attempt to control and manipulate nature. Hence, there emerges in political and scientific discourse, with increasing frequency but at irregular intervals, the suspicion that mankind is carrying out a large-scale experiment on our planet and, in fact, on the human species itself.

As of now, no definite answer can be given to the question of whether the pursuit of the ecological modernisation strategy can lead to a solution of environmental problems. This, of course, is due to the fact that future developments which will give the answer go to the core of political life in our societies, and because of their fundamentally political character, they may transcend any calculation or logical prediction. As a concluding note, however, we offer an observation and a somewhat normative point. When reading the environmental policy statements regularly made by members of the political elite of Western Europe at environmental conferences in Brussels and other capitals (e.g. Berlin, London, Paris, Oslo) we cannot help but being reminded of the Hopi Indians and their rain dance. The purpose of the Hopi rain dance was to bring rain, and this it did not do. Robert K. Merton and other sociologists have, however, taught us that the important function of the Hopi rain dance ceremony was to provide an occasion on which members of a group assemble to engage in common activities, to emphasise common symbols and generate support for solutions to perceived common problems.

320 *Environmental challenges*

Notes

1. The first UN Conference on the Human Environment held in Stockholm in 1972 had as its theme 'Only One Earth' and may be seen as symbolising the official emergence of the environment as an object of world-wide international concern.

2. The WCED's position is illustrated by the following statement: 'In the middle of the 20th century, we saw our planet from space for the first time. (...) From space, we see a small and fragile ball dominated not by human activity and edifice but by a pattern of clouds, oceans, greenery, and soils. Humanity's inability to fit its activities into that pattern is changing planetary systems fundamentally. Many such changes are accompanied by life-threatening hazards. These new realities, from which there is no escape, must be recognized – and managed.' (WCED 1987: 308)

3. 'Such growth rates could be environmentally sustainable if industrialized nations can continue the recent shifts in the content of their growth towards less material- and energy-intensive activities and the improvement of their efficiency in using materials and energy.' (WCED 1987:51) On this basis the Commission saw 'the possibility for a new era of economic growth,...' (WCED 1987:1)

4. Notably the OECD Environmental Committee had promoted such a development for a decade, and in the conclusion from the big OECD International Conference on Environment and Economics in 1984 it is stated that '... the environment and the economy, if properly managed, are mutually reinforcing; and are supportive of and supported by technological innovation.' (OECD 1985:10)

5. The concept of ecological modernisation was introduced by the German social scientists Joseph Huber and Martin Jänicke in the early 1980s, and it has been used to categorise both a social theory and a new political programme/ideology/discourse (Spaargaren and Mol 1992, Weale 1992:66–93, Mol 1995:28ff., Hajer 1995:25ff.).

6. As stated by the Brundtland Commission, 'With careful management, new and emerging technologies offer enormous opportunities for raising productivity and living standards, for improving health, and for conserving the natural resource base.' (WCED 1987:217)

7. Mr Laurens Brinkhorst, the Director-General of the environment directorate of the European Commission, speaking about the ECs Fourth Environmental Action Programme before the British House of Lords Select Committee on the European Communities in 1987 (Weale 1992:77).

8. Terms like taxes/charges have different connotations in some of these countries. In some countries the distinction between taxes and charges refers to that of general versus earmarked purposes of taxation, whereas in other countries this distinction refers to that of direct versus indirect taxation (TemaNord 1994:22). If not otherwise stated we shall here use these two terms interchangeably.

9. It should be noted that this new policy strategy implies a much stronger emphasis on a certain type of economic instruments: in many countries subsidies have been widely used, e.g. in order to promote technological improvements in the firms, but environmental taxes/ charges are seen to be preferred because they, unlike subsidies, do neither lead to an increase in governmental expenditure nor to trade distortions, and, if properly designed, are assumed to be in accordance with both the principle of cost-effectiveness and the polluter-pays principle. Note

that the application of other economic instruments (such as tradable permits and subsidies) and market-sensitive instruments like eco-labelling will not be reviewed here.

10. It should be pointed out that this organisational approach follows more as an implication of this strategy's mooring base in economic categories and logic than it represents an organisational alternative anchored in organisation theory. The lack of an explicit argument/justification of this approach in organisational terms is striking.

11. See the special issue of *Public Administration Review*, Vol XXVIII, July/August 1968, edited by Lynton K. Caldwell. When discussing a 'super-department for the environment and natural resources', Professor Caldwell several times around 1970 pointed to the manner in which '...the Department of Defense relates to the Department of the Army, the Navy and the Air Force'. It has also often been noted that such an alternative has been applied for integrating overriding concerns to promote economic growth (Ministry of Economic Affairs) and to manage the state financial resources (Ministry of Finance). An illustrative case is the discussion on the role and responsibility of the Norwegian ME before it was set up in 1972 (Jansen 1989).

12. This perception of the problem is explicitly expressed in a publication prepared for the Meeting of Ministers of the Environment of OECD Member countries, Paris 1979. In the section that addresses difficulties to be overcome in order to include environmental considerations in the decision-making process, it is explicitly referred to 'ignorance' as one such difficulty and the summarising statement is: 'The main causes underlying the above-described attitudes and difficulties are the lack of training, the inadequate provision of information, operation of the administrative agency, the inappropriateness of classical economic concepts, and the inadequacy of economic and technical data.' (OECD 1980:69)

13. In 1985, the Council of Ministers adopted the directive on Environmental Impact Assessment (the EIA Directive), but as to its implementation the member states were given discretionary scope of action which resulted in EIA practices varying extremely in level of aspiration (Héritier, Knill and Mingers 1996:294). Also, in this case, implementation was problematic in Germany where the cross-media perspective of the EIA Directive contradicted the established medium-specific approach. Procedural rules like those in the EIA Directive play a minor role in Germany where emphasis is on technical standards. Moreover, the provisions of more extensive public participation were in conflict with German tradition of public administration. In formal terms this directive has now been transposed into German law, but in terms of requirements for integrated administrative authority procedure, for substantive standards and for public participation, it can be reasonably argued that such requirements have been watered down during the process of implementation (Héritier, Knill and Mingers 1996:297–8).

14. The case of the CO_2/energy tax also illustrates how the policy-making process of the EU has affected the pattern of interaction between authorities and societal interests, *in casu* between the EU authorities and representatives of 'big industry', and, therefore, also the dilemmas and problems of how to develop the EU into a democratic polity.

15. On BATNEC, BPM etc. see Franklin, Hawke and Lowe (1995:37–49).

16. The results in this area are linked to the fact that the ME entered the agricultural policy sector in a period of deinstitutionalisation of and reduced support for traditional agricultural policies in government and parliament (Jansen and Osland 1996:241). The interest of the dominating alliance of the agricultural policy sector in enhancing its 'green' legitimacy and its need for allies are factors that explain why the ME has become an actor of consequence within a policy sector that traditionally has been characterised by a strong, and rather closed, policy community.

17. With public policy being made and implemented at a number of governmental levels and throughout the territory, the bottom line is that '...the more a bureaucracy is territorially grounded, the more powerful it is... .' (Montricher 1996:251)

18. These remarks do not imply that we consider economics as a unified theory that represents or is an ideology for certain societal interests. On the contrary; the relation between science, public administration and societal interests in this sector and its influence on environmental policies represents a research area of profound interest and significance. Although recognising the relevance of intra-disciplinary controversies for the policy process, we are here primarily addressing the relevance of inter-disciplinary relations, the point that economics can be analysed as a certain type of perspective or approach (in a broad sense) that one can distinguish from other perspectives (say biological or judicial) and that these perspectives are based on different assumptions, make use of and evoke certain types of concepts and models.

19. In one of the few countries where taxes on the use of fertilisers have been imposed (Norway) this was done against stiff resistance from farmers' organisations and the Ministry of Agriculture as well as from agricultural scientists. It has been argued that the latter ones opposed the introduction of this policy instrument because it was incompatible with 'the agronomist world view' (Vedeld and Krogh 1996). Due to interdisciplinary controversies and interest mobilisation, these taxes have not been significantly increased after they were introduced in the late 1980s.

20. Of great importance is the kind of concepts or symbols used to denote environmental policy goals or targets. For instance, whether or not the goals are transposed into the goal predominant in (the) economy, that of efficiency. If the goals of voluntary agreements are reformulated from concepts or categories denoting the quality of the environment into concepts or categories denoting goals like reduced environmental impact of each unit of consumption/production (e.g. reduced energy use, pollution), the latter goal may be achieved, while the former goal of environmental quality may be offset by increased production.

21. This necessity for the government of such states to accommodate the accumulation imperative has, for more than two decades now, been recognised by political scientists as different as Claus Offe and Charles Lindblom.

References

Aguilar-Fernandez, S. (1994) 'Spanish Pollution Control Policy and the Challenge of the European Union', in S. Baker, K. Milton and S. Yearly (eds) *Protecting the Periphery*. Ilford: Frank Cass, pp. 102–117.

Andersen, M.S. (1994) *Governance by green taxes*. Manchester: Manchester University Press.

Ashby, E. and Anderson, M. (1981) *The Politics of Clean Air*. Oxford: Clarendon Press.

Baker, S., Kousis, M., Richardson, D. and Young, S. (eds) (1997) *The Politics of Sustainable Development. Theory, Policy and Practice within the European Union*. London: Routledge.

Beck, U. (1992) *Risk Society*. London: Sage Publications.

Boehmer-Christiansen, S. and Skea, J. (1991) *Acid Politics*. London: Belhaven Press.

Brenton, T. (1994) *The Greening of Machiavelli*. London: Earthscan.

Caldwell, L.K. (1968) 'Environment: A New focus on Public Policy?' Special issue of *Public Administration Review* XXVIII, July/August.

Caldwell, L.K. (1970) *Environment: A Challenge for Modern Society*. Garden City: The Natural History Press.

Christiansen, P.M. and Lundqvist, L.J. (1996) 'Conclusions: A Nordic Environmental Policy Model?', in P.M. Christiansen (ed.) *Governing the Environment: Politics, Policy and Organization in the Nordic Countries*. Copenhagen: Nord 1996:5 pp. 337–61.

Christensen, J.G. (1994) 'Denmark: Institutional Constraint and the Advancement of Individual Self-Interest in HPO', in C. Hood and B.G. Peters (eds) *Rewards at the Top. A Comparative Study of High Public Office*. London: Sage Publications, pp. 70–89.

Commoner, B. (1971) *The Closing Circle: Nature, Man and Technology*. New York: Bantam Books.

Commoner, B. (1976) *The Poverty of Power: Energy and the Economic Crisis*. New York: Knopf.

Derlien, H.-U. (1995) 'Public administration in Germany: political and societal relations', in J. Pierre (ed.) *Bureaucracy in the Modern State*. Aldershot: Edward Elgar, pp. 64–91.

Derlien, H.-U. (1996) 'Germany: The Intelligence of Bureaucracy in a Decentralized Polity', in J.P. Olsen and B.G. Peters (eds) *Lessons from Experience. Experiential Learning in Administrative Reforms in Eight Democracies*. Oslo: Scandinavian University Press, pp. 146–79.

Dyson, K. (1982) 'West Germany: The Search for a Rationalist Consensus', in J. Richardson (ed) *Policy Styles in Western Europe*. London: George Allen & Unwin, pp. 17–46.

ECON (1996a) 'Erfaringer med avtaler som klimapolitisk virkemiddel'. ECON Senter for økonomisk analyse. Rapport 21/96. Oslo.

ECON (1996b) 'Erfaringer med avtaler som klimapolitisk virkemiddel – fase II'. ECON Senter for økonomisk analyse. Utkast til rapport. Oslo.

Franklin, D., Hawke, N. and Lowe, M. (1995) *Pollution in the UK*. London: Sweet and Maxwell.

Hajer, M. (1995) *The Politics of Environmental Discourse. Ecological Modernization and the Policy Process*. Oxford: Clarendon Press.

Hajer, M. (1996) 'Ecological Modernisation as Cultural Politics', in S. Lash, B. Szerszynski and B. Wynne (eds) *Risk, Environment and Modernity*. London: Sage Publications, pp. 246–69.

Héritier, A., Knill, C. and Mingers, S. (1996) *Ringing the Changes in Europe. Regulatory Competition and Redefinition of the State. Britain, France, Germany*. Berlin: Walter de Gruyter.

Hood, C. (1996) 'United Kingdom: From Second Chance to Near-Miss Learning', in J.P. Olsen and B.G. Peters (eds) *Lessons from Experience. Experiential Learning in Administrative Reforms in Eight Democracies*. Oslo: Scandinavian University Press, pp. 36–70.

Holm, J., Klemmensen, B. and Schrama, G. (1996) 'Covenants – A New Way for International Ecological Policy?' Paper for the 5th Greening of Industry Conference, Heidelberg 24–27 November 1996.

Jansen, A.-I. (1989) *Makt og miljø*. Oslo: Universitetsforlaget AS.

Jansen, A.-I. and Osland, O. (1996) 'Norway', in P. M. Christiansen (ed.) *Governing the Environment: Politics, Policy and Organization in the Nordic Countries.* Copenhagen: Nord 1996:5, pp. 179–256.

Koedijk, K. and Kremers, J. (1996) 'Market opening, regulation and growth in Europe', *Economic Policy* 23, pp. 443–60.

Krasner, S. (1988) 'Sovereignty: An Institutional Perspective', *Comparative Political Studies* 21, pp. 66–94

Krüger, M. and Pfaller, A. (1991) 'The Federal Republic of Germany', in A. Pfaller, I. Gough and G. Therborn (eds) *Can the Welfare State Compete?* London: Macmillan, pp. 187–228.

Lægreid, P. (1994) 'Norway', in C. Hood and B.G. Peters (eds) *Rewards at the Top. A Comparative Study of High Public Office*. London: Sage Publications, pp. 133–45.

Lélé, S.M. (1991) 'Sustainable Development: A Critical Review'. *World Development* 19 (6) pp. 607–21.

Liberatore, A. (1995) 'The Social Construction of Environmental Problems', in P. Glasbergen and A. Blowers (eds) *Perspectives on Environmental Problems*. London: Arnold, pp. 59–83.

Lundqvist, L.J. (1971) *Miljövårdsförvaltning och politisk struktur*. Stockholm: Bokförlaget Prisma.

Lundqvist, L.J. (1996) 'Sweden', in P.M. Christensen (ed.) *Governing the Environment*. Copenhagen: Nord 1996:5, pp. 257–333.

McKibben, B. (1989) *The End of Nature*. New York: Random House.

Mol, A.P.J. (1995) 'The Refinement of Production'. *Ecological Modernization Theory and the Chemical Industry*. Utrecht: International Books.

Mol, A.P.J., Lauber, V., Enevoldsen, M. and Landman, J. (1996) 'Joint environmental policy-making in comparative perspective'. Paper prepared for presentation at the 'Greening of Industry' Conference, Heidelberg, 24–27 November 1996.

Montricher, N. de (1996) 'France: In Search of Relevant Changes', in J.P. Olsen and B.G. Peters (eds) *Lessons from Experience. Experiential Learning in Administrative Reforms in Eight Democracies*. Oslo: Scandinavian University Press, pp. 243–77.

OECD (1980) *Environment Policies for the 1980s*. Paris.

OECD (1985) *Environment and Economics*. Paris.

OECD (1987) *Administration as Service. The Public as Client*. Paris.

OECD (1993) *Environmental Performance Reviews. Germany*. Paris.

O'Riordan, T. and Rowbotham E.J. (1996) 'Struggling for Credibility. The United Kingdom's response', in T. O'Riordan and J. Jäger (eds) *Politics of Climate Change*. London: Routledge, pp. 228–67.

Ormond, D. (1993) 'Improving Government Performance'. *The OECD Observer*, No. 184 pp. 4–8.

Pollitt, C. (1990) *Managerialism and the Public Services. The Anglo-American Experience.* Oxford: Basil Blackwell.

Rouban, L. (1994) 'France: Political Argument and Institutional Change', in C. Hood and B.G. Peters (eds) *Rewards at the Top. A Comparative Study of High Public Office.* London: Sage Publications, pp. 90–105.

Rouban, L. (1995) 'Public administration at the crossroads: the end of the French specificity?', in J. Pierre (ed) *Bureaucracy in the Modern State.* Aldershot: Edward Elgar, pp. 39–63.

Rubin, C.T. (1994) *The Green Crusade.* New York: The Free Press.

Schmidt, V.A. (1996) 'The Decline of Traditional State *Dirigisme* in France: The Transformation of Political Economic Policies and Policy-making Processes', in *Governance* 19 (4) pp. 375–405.

Sjølund, M. (1994) 'Going against the Cultural Grain', in C. Hood and B.G. Peters (eds) *Rewards at the Top. A Comparative Study of High Public Office.* London: Sage Publications, pp. 120–32.

Spaargaren, G. and Mol, A.P.J. (1992) 'Sociology, Environment and Modernity: Ecological Modernization as a Theory of Social Change'. *Society and Natural Resources* 5, pp. 323–44.

Stewart, J. and Walsh, K. (1992) 'Change in the Management of Public Services', *Public Administration* 70, pp. 499–518.

TemaNord (1994) *The Use of Economic Instruments in Nordic Environmental Policy.* Nordic Council of Ministers. Copenhagen: TemaNord 561.

Vedeld, P. and Krogh, E. (1996) 'Rationality is in the eye of the actor. Economists and agronomists and a discourse over environmental taxes in agriculture'. Ås: Norwegian Agricultural University, Department of Economics and Social Sciences.

WCED – World Commission on Environment and Development – (1987) *Our Common Future.* Oxford: Oxford University Press.

Weale, A. (1992) *The New Politics of Pollution.* Manchester: Manchester University Press.

Wolin, S.S. (1960) *Politics and Vision.* Boston: Little, Brown and Co.

List of Contributors

Mikael Skou Andersen is Associate Professor in Political Science at the University of Aarhus.
Alan Butt Philip is the Jean Monnet Reader in European Integration in the School of Management, University of Bath.
Neil Carter is Lecturer in Political Science at the University of York.
Lucien Chabason, former senior official of the French Ministry of Environment. Currently he is Co-ordinator of the Action Plan for the Mediterranean Area.
Peter Munk Christensen is Associate Professor in Political Science at the University of Aarhus.
Nuria Font is Assistant Professor in the Department of Political Science at the Autonomous University of Barcelona.
Egbert van de Gronden is a Consultant in administration and organisation with Zandvoort Ordening en Advies, a Dutch environmental consulting firm.
Kenneth I. Hanf is Senior Lecturer in Public Administration at the Erasmus University of Rotterdam.
Alf-Inge Jansen is a political scientist, Professor at the Department of Administration and Organisation Theory, University of Bergen.
Corinne Larrue is Associate Professor at the Centre d'Etudes Supérieures d'Aménagement, University of Tours.
Rudolf Lewanski is Senior Lecturer in Public Administration at the University of Bologna.
Philip Lowe is Duke of Northumberland Professor of Rural Economy at the University of Newcastle upon Tyne.
Lennart J. Lundqvist is Professor of Environmental Policy and Administration at the Department of Political Science, Göteborg University.
Francesc Morata is Professor in Political Science at the Autonomous University of Barcelona.
Per Kristen Mydske is Professor of Political Science at the University of Oslo.
Oddgeir Osland is a sociologist, currently a Research Assistant at the Department of Administration and Organisation Theory, University of Bergen.
Heinrich Pehle is *Privatdozent* in the Department of Political Science, Universität Erlangen-Nürnberg.

Calliope Spanou is currently Assistant Professor in Public Administration at the Department of Political Science and Public Administration, University of Athens.

Søren Winter is Associate Professor in Political Science at the University of Aarhus.

Index

administration
  environmental policy 12
    Britain 17, 20–8, 34–5
    Denmark 42, 49–57
    EU 263, 266–9
    France 61–4, 66, 67–76
    Germany 84, 93–101
    Greece 115–22, 127
    Italy 138–47
    Netherlands 170–4
    Norway 190–4
    Spain 211–12, 214–19
    Sweden 232, 238–4
  professions in 51, 69, 100–1, 119–20, 201, 280, 302–3, 308
  staffing of, 51, 53, 68, 70–1, 100–1, 119–20, 139, 143, 171, 191–4, 200, 221, 241–2, 245
agriculture
  Britain 18, 26, 27, 34
  Denmark 40, 43, 46–8, 55–7
  EU 254, 261
  France 72, 77–9
  Germany 82, 101
  Greece 111
  Italy 140
  Netherlands 154–6, 178
  Norway 200
  Spain 208
  Sweden 230
air pollution
  Britain 20
  early legislation 279
  France 66, 72–4, 80
  Germany 83, 84–5, 88–9, 94, 98–9
  Greece 110–12, 114, 124
  Italy 133–4, 138, 149
  Norway 184
  Spain 211, 221
autonomous communities, Spain 217, 219–20

biologists' conceptions of the man–nature relationship 280, 319
Britain
  environmental policy 17–37, 278–9, 281–2, 286, 290, 297–300, 305
  EU directives 299–300
  market expansion policies 315–16
Brundtland Commission 185, 187, 189, 194
  see also World Commission on Environment and Development (WCED)
  report 64, 159, 196, 291–2, 295

carbon dioxide emissions 137, 298, 305–6, 319
  Britain 37
  Denmark 45–6, 56
  EU 253, 265, 266, 271
  France 60–1
  Germany 90–1, 97–8, 100, 106
  Italy 137
  Netherlands 175, 179
  Norway 187, 197–8
  Spain 210, 221–2, 227–8
  Sweden 231, 247
central government 3, 219 see also Ministries of Environment
  Britain 20, 31–7
  Denmark 42, 44, 49–50, 55
  France 61, 67–70, 72–3, 75, 77–9
  Germany 84, 93–8
  Greece 115–17, 119, 120–1, 124
  Italy 135–6, 138–41, 144–6
  Netherlands 158, 164–6, 170–2, 176
  Norway 183–5, 188–92, 199–201
  Spain 214–19, 222
  Sweden 232–4, 239–41, 249–50
citizenship 5–7, 299–300
civil rights 5–6, 15n2
climate change 226–8, 247–9, 269, 296
climate policy see also carbon dioxide emissions; Kyoto Conference

Britain 37, 305
Germany 90, 97–8
Norway 197–8, 200
Spain 226–8
Sweden 247–9
competition policy 314–17
sustainable development 195
consensus and consensus-making routines 8
  see also self-regulation; voluntary agreements
  Denmark 43, 52, 54
  France 74–6
  Netherlands 153, 158
  Sweden 238, 244, 248–9
cultural heritage preservation 278–9
  Britain 24, 27
  France 63, 71
  Germany 84
  Italy 149n2
  Norway 182

decentralisation see also federalism; local government; municipalities
  Britain 23–4
  Denmark 42, 43, 50–1, 52–4
  EU 268
  France 70–3, 78–9, 286
  Germany 102–4
  Greece 120–1
  Italy 138, 142–5
  Netherlands 170, 172–4
  Spain 217, 219–21, 225
  Sweden 237, 241–3
deforestation 133, 136, 156, 208
Denmark
  environmental policy 40–57, 279, 282, 296–7, 307–9
  Department of Environment (DoE), Britain 22–4, 32–3, 37, 281
deregulation 271, 314–17
desertification, Spain 208, 224, 227
developing countries 198, 270, 292

EC see European Community
ecological modernisation 14, 265, 277, 291–6, 300, 311–19, 320n5
  Sweden 248–50
ECU see European Currency Union
emission standards 66, 73, 76, 88–9, 297
emissions see also carbon dioxide emissions; sulphur dioxide emissions; pollution control
  pollutant
    Britain 25, 37, 305
    Denmark 56
    EU 253
    France 60–1, 73, 78
    Germany 83, 89
    Greece 111
  Netherlands 175, 178–9
  Norway 196–8
  Sweden 230–1, 247–8
energy policy
  Britain 37
  Denmark 40, 43, 56
  EU 261, 272
  France 60–1, 70, 285, 290
  Germany 82–3, 88, 91–2, 95, 278–83, 289–90, 297–8
  Italy 137–8
  Netherlands 171, 178
  Norway 184–5, 197–8, 283, 290
  Sweden 231, 247–9
environmental action programmes, EU 257–61
environmental information 300–2, 312 see also monitoring, environmental access 298, 300
  institutionalisation 300–2
environmentalism 280–92, 296
EU-fication 201
European Commission 254–5, 258–62, 266, 269–70, 291–2, 296
European Community (EC) 256–61 see also European Union (EU)
  directives 13, 66–7, 73
    Britain 26, 30–2, 34
    Germany 105
    Greece 126
    Italy 149, 150n14
    Spain 214, 225–6
European Court of Justice 255, 263
European Currency Union (ECU) 318
European Environment Agency 141, 268
European Parliament 262, 264, 268, 271
European Union (EU) 3, 12–13 see also European community (EC)
  Britain 30–2, 35
  Denmark 44–5, 54, 56
  directives 256, 258–9, 264, 266, 268, 271, 298–9
  Denmark 44–5, 54, 56
  environmental policy 1–3, 12–13, 253–74, 296–301, 306–7, 317–18
  France 66–7, 317
  Germany 89, 90, 92, 97, 105
  Greece 124, 127
  influence between the EU and member states 296–300
  institutions 254–5, 265
  international activities 269–70
  Netherlands 179
  Norway 185, 201, 283
  Spain 212–15, 218–19, 222–4, 226–7
  Sweden 235, 247–8, 250
  treaty base 256–7

federalism

Germany 84, 93, 102–4, 278–9
Italy 138, 150n10
financial markets, deregulation 314
financial resources
  Britain 18
  EU 270, 272
  Italy 136, 144–5
  Netherlands 173–4
  Norway 183, 195
  Spain 224, 226
  Sweden 241, 243, 249
flooding, Netherlands 154
France 7, 316–18
  environmental policy 60–80, 278–9, 282, 284–6, 290, 298, 304, 308
Friends of the Earth 28, 48, 212, 264, 269

genetic engineering 319
Germany 5, 7–8, 316
  environmental policy 82–107, 278–9, 282–4, 289–90, 297–8, 302, 305, 309–11
global warming 37, 90, 106, 226–8, 291, 319
globalisation 36, 313–19
governance 3–4
Greece, environmental policy 110–28, 286–8, 290, 297–8, 301–2, 304
Green parties
  Britain 28–9, 33, 36
  Denmark 49–50, 55
  European Parliament 262
  France 64–5
  Germany 87–90, 92, 105, 283–4, 289–90
  Greece 123
  Italy 135–7, 138, 148
  Netherlands 159
  Norway 186
  Spain 213
  Sweden 234–5, 248
  Western Europe 283–5
greenhouse effect 43, 45, 187
greenhouse gases 90–1, 175, 179, 270, 305
  see also carbon dioxide emissions
  Norway 187, 198
  Spain 227–8
Greenpeace 28, 158, 212, 235

hunting, Italy 139, 142
hydropower, Norway 182, 183, 184–6, 196

implementation see also administration
  environmental policies
    Britain 34
    Denmark 53–4
    France 72–80
    Germany 86, 102–4
    Greece 118, 125–6
    Italy 145–7
    Netherlands 175–7

  Norway 190–4, 198–201
  Spain 221–5
  Sweden 237–8, 243–5
  EU legislation 265, 266–9, 271–2, 274n11
industrial pollution 297
  Britain 19, 24, 30
  France 71–2, 76
  Greece 117
  Italy 131–3, 136, 140
  nineteenth century 278
  Spain 208, 216
industry
  Britain 18–19, 25–7
  Denmark 40, 44, 46–8, 55–7
  EU 260, 261, 265
  France 71–2, 76
  Germany 82, 98–9
  Greece 117
  Italy 131, 141
  Netherlands 154, 161–6, 168–9, 175–7
  Norway 188
  Spain 210, 218–19
  Sweden 230–1, 234, 236, 248
information, access 298, 300
institutions
  definition 4
  in relation to organisation 7–9
  institutional logics 4–8
  the changing of institutional orders 3, 13–19
interest balance principle, Sweden 232–4, 236
interrelationship, man–nature 280–1, 289
Italy, environmental policy 131–50, 286–7, 298–9, 303, 311

Kyoto Conference on Climate Change (1997) 37, 137, 179, 198, 227, 248, 270

land–use planning 20, 27, 82, 133, 139
legislation see also regulation
  Britain 20–2, 25, 286, 299
  Denmark 41–3, 47, 55
  EU 254, 261, 263, 265, 268–71
  European Commission 256, 258–60, 266, 270
  France 66, 70
  Germany 88, 93–4, 98–100, 102–3, 284
  Greece 111, 115–18, 126–7, 287–8
  inter-state differences 281–2
  Italy 133–8, 140, 142–3, 146, 148
  Netherlands 161–4, 170
  nineteenth century 278–9
  Norway 192, 194, 199, 203n1
  Spain 211, 214
  Sweden 237–8, 241–4, 246
local authorities see local government
local government 13, 140 see also municipalities
  Britain 20, 23, 31
  Denmark 41–2, 48, 49, 52–5

France 71–5, 78–9
Germany 102–4
Greece 121–2
Italy 136, 138, 141–6
Netherlands 170, 172–4
Norway 192–4
Spain 212, 219–22, 225
Sweden 238, 239, 241–5

Maastricht Treaty 35, 56, 222, 256–7, 260–1, 265, 270–2
market economies 4–5, 306, 312
  economic instruments 293–4, 306–8, 312–13
  institutional logic 4–5, 293–4, 313
  political imperatives 313–19
  social rights 6–7
ME *see* Ministries of Environment
media 46–7, 55, 89–90, 291
military dictatorship, Greece 111, 112–14
Ministries of Environment (ME) 295, 301–5
  Britain 22–4, 32–3, 37, 281
  Denmark 42, 44, 49–50, 55
  France 61–72, 74–5, 285–6, 298, 304
  Germany 94–9, 305
  Greece 116–20, 124–7, 287
  Italy 135–6, 139–41, 144–5, 287, 303
  Netherlands 158, 164–6, 169, 171–2, 174, 176, 288–9, 305
  Norway 183–5, 189–92, 198–203, 303
  Scandinavia 282–3
  Spain 215–17
  Sweden 240–1, 243, 245
monitoring, environmental 143, 200, 238 *see also* environmental information
movements *see also* environmentalism; nature conservation; non-governmental organisations
  environmentalist 157–61, 282–3
    Britain 28, 31–2
    Germany 86–7, 89, 107n6
    Italy 134–5
    Spain 212–13
  municipalities 3 *see also* local government
    Denmark 52–4, 55
    Greece 121–2
    Italy 143
    Netherlands 170, 173–4
    Norway 184, 190, 193–4, 200
    Spain 220–1
    Sweden 242–5

national parks 279
  Italy 136, 137
  Netherlands 156
  Norway 182
  Spain 209, 211
  Sweden 231

nature conservation 209, 278–9
  Britain 22, 24, 27, 279, 286
  Denmark 279
  France 62, 68–9, 71–2, 279
  Germany 84, 94–5, 98
  Greece 113, 123
  Italy 136–7, 142
  Netherlands 156–7, 171
  Norway 182–3, 192
  Spain 216
  Sweden 231, 279
Netherlands, environmental policy 152–80, 279, 282, 288–90, 298, 305, 308–10, 312
NGOs *see* non-governmental organisations
non-governmental organisations (NGOs) 298–9, 301 *see also* environmentalism; movement, environmental
  Britain 24
  Denmark 48–9
  EU 263–4
  France 61
  Greece 123, 124–5
  Netherlands 156, 157–9, 161, 289
  Norway 187–8, 204n8
  Spain 212–13
  Sweden 231–2, 235
Norway
  environmental policy 181–205, 279, 282–3, 303–4, 306, 312
  environmentalism 282–3
nuclear energy 283–5, 289–90
  France 60–1, 285, 290
  Germany 87, 90, 95, 102, 289–90
  Italy 134–5, 137
  Netherlands 161
  Spain 212
  Sweden 234–5, 249

OECD *see* Organisation for Economic Cooperation and Development
organisation
  basic assumptions 8
  in realtion to institutions 7–9
  institutional character 8
  mobilisation of bias 8
organisation environmental *see* administration
Organisation for Economic Cooperation and Development (OECD) 92, 184, 198, 201, 281, 291–5, 317
ozone layer, depletion 43, 45, 90, 291, 296

pesticides
  Denmark 57
  EU directives 263
  Germany 101
  Italy 140, 147, 149
  Netherlands 155, 161
  Spain 208
  Sweden 230

policy
   area 15
   environmental 1–4, 9–13
      Britain 17–37, 278–9, 286, 299–300
      Denmark 40–57
      emergence 278–80
      EU 253–74
      France 60–80, 278–9, 284–6
      Germany 278, 283–4, 289–90
      Greece 110–28, 286–8, 304
      Italy 131–50, 286–7
      Netherlands 152–80, 288–9
      Norway 181–205
      Spain 208–28, 286–7
      Sweden 230–50
   field 1, 3, 15
      instruments 11, 44–5, 77, 98–101, 162, 168–70, 195–201, 236–8, 292, 313, 320–9
   sector 15
   strategy 11
      ecological modernisation 291–313
      expanding the logic of the institutional order of the market 313–19
politics 1, 9–14, 281–4, 289–91, 302, 313–19
   Britain 21, 28–9, 32–7
   Denmark 43, 46–7, 49–50, 54–5
   EU 261–4, 271–3
   France 63, 64–5
   Germany 85, 86–92, 104–6
   Greece 112–14, 123, 125–6
   Italy 134–8, 141–2, 144–5, 147–9
   Netherlands 153, 157–60
   Norway 184–9
   Spain 212–14, 217, 224–5
   Sweden 232–6, 238–41, 248–50
'polluter pays' principle 42, 94, 183, 234, 236–7, 260
pollution
   definitions 281–2
   as industrial inefficiency 293
   transfrontier 40–1, 262
pollution control 11, 290, 319
   Britain 18–22, 25–7, 30–1, 34–5, 37, 279, 299
   Denmark 40–3, 46–8, 56–7
   France 61–2, 68–9, 71, 74
   Germany 99–101, 284
   Greece 117–18, 121
   Italy 132–4, 136, 142, 149
   Netherlands 155, 167, 175, 288
   Norway 183–5, 192–7, 199–200
   Spain 217–18
   Sweden 233–8, 240–5
pressure groups *see* movements
prevention policies 3, 163, 178, 236, 258–61
privatisation 19, 29, 300
public health 270–80
   Britain 18, 20
   Denmark 41
   France 62
   Germany 84
   Greece 112
   Italy 143
   Netherlands 156, 160–1, 164
   Norway 182
   Spain 211, 216
   Sweden 233, 240, 242
public opinion 291, 301, 313
   Britain 29
   Denmark 43, 46–7, 50, 55
   EU 264
   France 64
   Greece 122–4
   Italy 135
   Netherlands 157–9
   Spain 225

regulation 293, 311–12 *see also* legislation
   Britain 20, 23–32, 34, 286, 315
   Denmark 46, 56–7
   EU 257–60, 264, 270
   France 66, 70–2, 76–9
   Germany 98–101, 284
   Greece 112, 117–18
   Italy 140–1, 146–7
   Netherlands 153, 161–4, 168–70
   Norway 194–6
   Spain 217–18, 309–10
   Sweden 233–4, 236, 240–1
road traffic pollution *see also* transport policy
   Britain 19, 27, 33–4, 37
   EU 253
   Germany 83, 89, 98, 107n2
   Italy 141
   Sweden 235, 247
rural conservation, Britain 18, 19, 20

self–regulation *see also* voluntary agreements
   industry 153, 165–6, 168–70, 175–8, 294
   Britain 25–7, 31–2, 34–5
   Denmark 44
   Netherlands 35
social rights 6–7, 15n2
society-nature relations 280, 289, 291–2, 319
society-state relations 3, 4–8
soil erosion, Spain 208, 210, 227
Spain, environmental policy 208–28, 286–7, 290, 297–8, 302, 309–10
State
   institutional characteristics 5–8, 9, 281–91, 297–319
   institutional logics 7
   mode of governance 4
   state–society relations 2–3, 4–8, 313–19
sulphur dioxide emissions 56, 73, 78, 87–8, 99, 210, 230, 253, 285

sustainable development 2, 11, 277, 291–3, 295, 316, 320n6
  Britain 28, 31, 32–5
  definition 291
  EU 257, 260–1, 271, 273
  Netherlands 159, 167–9, 178
  Norway 185, 194–7, 202
  Sweden 237, 245–50
  weakness 195
Sweden, environmental policy 230–50, 281–2

taxes
  environmental 294, 290, 305–8, 320n9, 321n14, 321n19
    Britain 24, 27, 36–7, 298, 305, 312
    Denmark 44, 57, 307
    EU 265–6, 271
    France 64, 66
    Germany 91–2, 97, 305
    Italy 137, 140
    Norway 197, 201, 204n10, 306
    Sweden 237, 247, 307
tourism 110–11, 132, 209, 210, 261
toxic waste 61, 149, 210
transport policy *see also* road traffic pollution
  Britain 27, 33–4, 37
  EU 261
  Germany 98
  Italy 138
  Norway 200

United Nations (UN) 86, 184, 198, 291, 296
United States of America (USA) 25, 34, 280

voluntary agreements 308–11, 322n20 *see also* concensus and concensus-making routines
  Britain 25–7, 31–2, 34–5
  Denmark 44–5, 309
  Germany 100, 105
  Netherlands 168–9
  Norway 198
  Spain 218

waste management
  EU 253, 265
  France 70
  Germany 83, 94, 96, 98–9, 104
  Italy 137, 142
  Netherlands 155, 162, 164, 172–3
  Spain 208, 214, 217, 224
  Sweden 232
water management
  France 70–1, 75, 77
  Germany 94, 98, 101, 103
  Netherlands 154–5, 161, 170–4
  Norway 183, 185–6, 204n3–4
  Spain 208, 210, 217
  Sweden 246
water pollution
  France 61, 77–9
  Germany 83, 85–6, 101
  Greece 110–11
  Italy 134, 138, 140, 143, 287
  Netherlands 155
  Sweden 230, 232
water quality
  Denmark 40–1, 43, 48, 53, 56–7
  Italy 132, 139, 143
  Spain 208
  Sweden 232
welfare state 2–3, 6–7, 315–16
wetlands 126, 210
WHO *see* World Health Organization
wildlife protection
  Britain 18, 20–1, 24, 30, 286
  EC 259
  France 62
  Germany 84
  Italy 138–9, 142
  nineteenth century 278–9
  Norway 192–3
World Commission on Environment Development (WCED) 185 *see also* Brundtland Commission
World Health Organization (WHO) 114
World Wide Fund for Nature *see* WWF
WWF 134, 136, 158, 235, 264